SHAPING

RHETORIC OF THE HUMAN SCIENCES

Lying Down Together: Law, Metaphor, and Theology
Milner S. Ball

Shaping Written Knowledge: The Genre and Activity of the Experimental Article in Science
Charles Bazerman

Politics and Ambiguity
William E. Connolly

Machiavelli and the History of Prudence
Eugene Garver

The Rhetoric of Economics
Donald N. McCloskey

Therapeutic Discourse and Socratic Dialogue: A Cultural Critique
Tullio Maranhão

The Rhetoric of the Human Sciences: Language and Argument in Public Affairs
John S. Nelson, Allan Megill, and Donald N. McCloskey, eds.

The Politics of Representation: Writing Practices in Biography, Photography, and Policy Analysis
Michael J. Shapiro

The Legacy of Kenneth Burke
Herbert Simons and Trevor Melia, eds.

The Unspeakable: Discourse, Dialogue, and Rhetoric in the Postmodern World
Stephen A. Tyler

Heracles' Bow: Essays on Rhetoric and Poetics of the Law
James Boyd White

Shaping Written Knowledge

THE GENRE AND ACTIVITY OF THE

EXPERIMENTAL ARTICLE IN SCIENCE

CHARLES BAZERMAN

THE UNIVERSITY OF WISCONSIN PRESS

The University of Wisconsin Press
114 North Murray Street
Madison, Wisconsin 53715

The University of Wisconsin Press, Ltd.
1 Gower Street
London   WC1E 6HA, England

Copyright © 1988
The Board of Regents of the University of Wisconsin System
All rights reserved

5 4 3 2 1

Printed in the United States of America

Library of Congress Cataloging-in-Publication Data

Bazerman, Charles.
  Shaping written knowledge.
  (Rhetoric of the human sciences)
  Bibliography: pp. 333–346.
  Includes index.
  1. Technical writing. I. Title. II. Series.
  T11.B375   1988   808'.0666021   88-40187
  ISBN 0-299-11690-5
  ISBN 0-299-11694-8 (pbk.)

In memory of my parents
Solomon Bazerman (1916–1965)
Miriam Bazerman (1916–1975)

For Poesy alone can tell her dreams,
With the fine spell of words alone can save
Imagination from the sable charm
And dumb enchantment.
        John Keats, "The Fall of Hyperion"

Poets survive in fame.
But how can substance trade
The body for a name
Wherewith no soul's arrayed?

No form inspires the clay
Now breathless of what was
Save the imputed sway
Of some Pythagoras,

Some man so deftly mad
His metamorphosed shade,
Leaving the flesh it had,
Breathes on the words they made.
        J. V. Cunningham

CONTENTS

Acknowledgments xi

PART ONE WRITING MATTERS

Chapter 1 The Problem of Writing Knowledge 3

Chapter 2 What Written Knowledge Does: Three Examples of Academic Discourse 18

PART TWO THE EMERGENCE OF LITERARY AND SOCIAL FORMS IN EARLY MODERN SCIENCE

Chapter 3 Reporting the Experiment: The Changing Account of Scientific Doings in the *Philosophical Transactions of the Royal Society,* 1665–1800 59

Chapter 4 Between Books and Articles: Newton Faces Controversy 80

Chapter 5 Literate Acts and the Emergent Social Structure of Science 128

PART THREE TYPIFIED ACTIVITIES IN TWENTIETH-CENTURY PHYSICS

Chapter 6 Theoretical Integration in Experimental Reports in Twentieth-Century Physics: Spectroscopic Articles in *Physical Review,* 1893–1980 153

Chapter 7 Making Reference: Empirical Contexts, Choices, and Constraints in the Literary Creation of the Compton Effect 187

Chapter 8 Physicists Reading Physics: Schema-Laden Purposes and Purpose-Laden Schema 235

PART FOUR	THE REINTERPRETATION OF FORMS IN THE SOCIAL SCIENCES
Chapter 9	Codifying the Social Scientific Style: The APA *Publication Manual* as a Behaviorist Rhetoric 257
Chapter 10	Strains and Strategies in Writing a Science of Politics: The Unsettled Rhetoric of the *American Political Science Review,* 1979 278
PART FIVE	SCIENTIFIC WRITING AS A SOCIAL PRACTICE
Chapter 11	How Language Realizes the Work of Science: Science as a Naturally Situated, Social Semiotic System 291
Chapter 12	Writing Well, Scientifically and Rhetorically: Practical Consequences for Writers of Science and Their Teachers 318

References   333

Index   347

ACKNOWLEDGMENTS

My debts are greater than my memory. Among the many people who have offered support and wise criticism during the long gestation of this book are Fred Bauman, Susan Cozzens, Ed Davenport, Lester Faigley, Lim Teck Kah, David Mauzerall, Robert Merton, Carolyn Miller, Greg Myers, Charles Piltch, John Swales, Spencer Weart, and Harriet Zuckerman. For help on individual chapters I thank John Andreassi, David Butt, Fred Eidlin, P. K. Mohapatra, Roger Stuewer, and Alan Shapiro. I owe particular gratitude to the many librarians who have gone out of their way to aid my research, especially the librarians at the Center for the History of Physics, the National University of Singapore, and Baruch College. Much of this work has been made possible by the generous released time granted by the Dean of Liberal Arts and Sciences of Baruch College and the two fellowship leaves (1978-79 and 1986-87) also granted by the college.

Several of these chapters have appeared previously in different forms:

Chapter 2 in *Philosophy of the Social Science* 11 (1981): 361-88;
Chapter 5 in *Social Epistemology* 1 (1987): 295-310;
Chapter 6 in *Social Studies of Science* 14 (1984): 163-96, © Sage Publications;
Chapter 7 in *Pre/Text* 5 (1984): 39-66;
Chapter 8 in *Written Communication* 2 (1985): 3-24, © Sage Publications;
Chapter 9 in *The Rhetoric of the Human Sciences*, ed. Nelson, Megill, and McCloskey (Madison: University of Wisconsin Press, 1987).

For permission to reprint I thank the publishers.

I can not offer adequate thanks to my wife, Shirley Geok-lin Lim, who has had to pay all the costs while I crawled down into my basement study.

PART ONE

WRITING MATTERS

1 THE PROBLEM OF WRITING KNOWLEDGE

A simple practical problem within a single discipline began the line of inquiry that led to this book. As a university teacher of writing I was charged with preparing students to write academic essays for their courses in all disciplines. Since academic assignments bear a loose relationship to the writing done by mature members of the disciplines, a serious investigation of writing within disciplines promised to turn up information useful to teaching undergraduates. The investigation from the first was interdisciplinary by necessity, but only in a superficial sense, in that the writing examined came from a variety of academic disciplines. The concepts and analytical tools, however, did not extend beyond the typical repertoire of the English department.[1]

---

1. What constitutes the repertoire of the English department is no easy thing to categorize, nowhere codified, and nowhere discussed with methodological clarity. Rather, on the literary side it is embodied in the corpus of literary scholarship and criticism and in the seminar practices of textual discussion. Primarily it consists of close textual readings and historical contexting. The textual readings are all framed by recognition of traditional literary devices, and have been intensified by new critical insistence on the text in itself. However, other modes of criticism have suggested the application of interpretive frameworks from other disciplines, such as linguistics, psychology, sociology, anthropology, and philosophy. Such imported frameworks are justified in two ways: either they represent fundamental truths so that they cannot help but influence texts, or the writer on some level was aware of such ideas and constructed parts of the text upon them.

Historical contexting has served a variety of functions, from simply providing a decorative frame for a self-contained and independent text to offering a complete account for the creation and meaning of a historically bound text. On occasion text and context have been drawn more tightly together to view the text as a historical event within the unfolding context. Most often, contexting has served to make odd features of the text more accessible to the reader.

The recent concern for literary theory, while raising some fundamental questions, has done little to change the actual analytical tools of literary interpretation. Concepts such as self-referentiality, intertextuality, reader response, and binary oppositions simply put additional weight on existing analytical concepts and tools.

An extended repertoire of concepts and tools has also come out of the teaching of writing. The rhetorical approach to the teaching of writing has been particularly concerned with public argument; an approach loosely labelled composition has been con-

One: Writing Matters

Very soon into engaging this problem, I found that I could not understand what constituted an appropriate text in any discipline without considering the social and intellectual activity which the text was part of. Too much of the texts directly invoked and acted against these contexts to treat the features of texts simply as isolated conventions. Moreover, the rhetorical gist of entire texts evoked the larger framework of meanings within the active disciplines. That is, I couldn't see what a text was doing without looking at the worlds in which these texts served as significant activity. Sociology of science became an inevitable resource for understanding how communication was organized in academic communities and how texts fit in with the larger systems of disciplinary activity.[2] And philosophy of science became important, not for the ultimate questions of epistemology, but for more modest ones of how people conceived of disciplinary activity.[3] Understanding what people think they are doing gives insights into how they use words to accomplish those things.

History as well loomed large as I began to see that current writing practices (in conventional, interactional, and epistemological dimensions) build on a history of practice and speak to a historically condi-

---

cerned with the formal prescriptions of the school essay, but has in recent years also taken on a concern for the process of writing, as approached through a cognitive psychology model. Gary Tate, ed., *Teaching Composition: Twelve Bibliographic Essays*, offers the best and most current review of work in the field. I will discuss approaches to writing and the teaching of writing more fully in the final chapter of this book.

2. Robert Merton, in his personal generosity of spirit and his profound analytical clarity, has influenced my understanding of sociology deeply. As I will argue in chapter 5, his seminal thinking is consonant with much of more recent work, which has frequently attacked a straw man version of his work. Bazerman, "Scientific Writing as a Social Act," and Harry Collins, "The Sociology of Scientific Knowledge," provide reviews of sociological studies relevant to questions of text, language, and knowledge formation. I will refer to the literature of the sociology of science throughout this book, but see especially chapter 5.

3. Although my readings in the large and complex field of the philosophy of science have been limited, I have found myself most in sympathy with Thomas Kuhn's observation of communal interaction in the production of knowledge (*The Structure of Scientific Revolutions*), Karl Popper's concept of three worlds (*Objective Knowledge*), Imre Lakatos' relation of work to ongoing research programs (*The Methodology of Scientific Research Programs*), Stephen Toulmin's evolutionary view of the development of historically situated knowledge (*Human Understanding*), and Ian Hacking's emphasis on physical activity in science (*Representing and Intervening*). As will be evident throughout this book, I have been most profoundly influenced by Ludwik Fleck's *Genesis and Development of a Scientific Fact*. Further articles by and about Fleck appear in Cohen and Schnelle, *Cognition and Fact*. Explicit philosophic accounts of scientific texts include Joseph Agassi, *Faraday as a Natural Philosopher*; M. A. Finocchario, *Galileo and the Art of Reasoning*; and Edward Manier, "Darwin's Language and Logic."

The Problem of Writing Knowledge

tioned situation.[4] A political scientist or a medical researcher writes as part of an evolving discussion, with its own goals, issues, terms, arguments, and dialect. The history frames both the rhetorical moment and the rhetorical universe.

Psychology seemed also to have an important place. As a historically realized, social, epistemological activity, writing is carried on through people. People write. People read. What a text is must take into account how people create it and how people use it. The socially situated study of writing directly implies an interest in psychology, for in every situation, coming and going, writing vanishes into the black boxes of human nervous systems.[5]

All this contexting of writing as a multidimensional activity, finally, forced me to confront the traditional view of the word as a separable, textual fact. If the written word could only be understood within a historical, social moment, that would vex many of our habits of looking at language and texts as fixed structured systems of meaning. On the other hand, to conceive of meaning creation as fluid threatens to cast language loose on unchartable seas. Moreover, such an unmooring of language threatens to undermine the motivating impulse prompting this research. What does learning to write better mean if we cannot moor meaning to language? Thus I had to confront language theory.[6]

As the serious interdisciplinary base for the research broadened, fortunately the superficial interdisciplinary base narrowed a bit. Since context was becoming increasingly important to my understanding of knowledge texts, I sought some degree of uniformity of context by con-

---

4. Historical literature is cited throughout this book within the context of each study. Historical studies that specifically consider the role of text and language in the development of science include Peter Dear, "Totius in Verba"; B. Eastwood, "Descartes on Refraction"; Frederic Holmes, "Scientific Writing and Scientific Discovery"; Martin Rudwick, *The Great Devonian Controversy*; and Steven Shapin, "Pump and Circumstance."

This book can also be seen as part of the examination of the technology and consequences of literacy as historically developing processes. Landmark works in this area include Eric Havelock, *The Greek Concept of Justice*; Jack Goody and Ian Watt, *Literacy in Traditional Societies*; Jack Goody, *Domestication of the Savage Mind*; Elizabeth Eisenstein, *The Printing Press as an Agent of Change*; and Sylvia Scribner and Michael Cole, *The Psychological Consequences of Literacy*.

5. In social psychology I have been most influenced by the works of George Herbert Mead, Harry Stack Sullivan, and Lev Vygotsky. The latter has been of particular interest to me because of his analysis of symbolic behaviors as the concrete mechanism of social cognition. I will discuss some of his ideas in chapter 11.

6. Linguistic theory and its reflections in studies of scientific language are discussed in the beginnings of chapters 2, 6, and 7, and throughout chapter 11.

One: Writing Matters

sidering the sciences, with physics, and even more narrowly optics, becoming a central research site.

This decision was in part fostered by an early and continued contact with the sociology of science which offered many contextual maps to guide my way. Examining the writing in science seemed a particularly important challenge for several reasons. First, the statements made through scientific discourse have been socially and culturally important in ways I hardly need elaborate; we are constantly rebuilding our world upon the statements of science. Second, scientific methods of formulating knowledge have been highly successful in gaining almost universal assent to claims hardly accessible or persuasive to common sense. Third, as a result of science's great success, habits of scientific discourse have influenced almost all other areas of intellectual inquiry. By unpacking scientific language one can come to understand important influences in all disciplines. Finally, scientific language is a particularly hard case for rhetoric, for sciences have the reputation for eschewing rhetoric and simply reporting natural fact that transcends symbolic trappings. Scientific writing is often treated apart from other forms of writing, as a special code privileged through its reliance on mathematics (considered a purer symbolic system than natural language). If one can show the workings of formulating practices in sciences on the kinds of statements science produces, one can begin to mine important depths of rhetoric.[7]

Of course the sciences, or even one science, or a single specialty within science, is far from a single, unmixed discourse community. The more I looked at varieties of scientific texts, the more I saw, with Darwin, that variation is everywhere the rule. So I narrowed my view further, on a single mechanism generating similarity throughout the wide expanses of variation: Genre, and one genre in particular.[8] The emergence and

---

7. By rhetoric I mean most broadly the study of how people use language and other symbols to realize human goals and carry out human activities. Rhetoric is ultimately a practical study offering people greater control over their symbolic activity. Rhetoric has at times been associated with limited techniques appropriate to specific tasks of political and forensic persuasion within European legal institutions. Consequently, people concerned with other tasks have considered rhetoric to offer inappropriate analyses and techniques. These people have then tended to believe mistakenly that their rejection of political and forensic rhetoric has removed their own activity from the larger realm of situated, purposeful, strategic symbolic activity. I make no such narrowing and use rhetoric (for want of a more comprehensive term) to refer to the study of all areas of symbolic activity. I elaborate these views later in this chapter and in chapter 12.

8. In literary studies, attempts to understand and define genre have a long history, dating back to the first literary critic, Aristotle. In general these attempts have been either formal or essentialist, defining genre by a collection of recurrent features or by comprehensive typologies of literary types. Sometimes the two have been connected,

transformation of the single genre of the experimental report runs as a common thread throughout the natural sciences of the last three centuries and the social sciences of this century.

Clearly, many other genres of great significance have emerged in the sciences. Important stories remain to be told about theoretical articles, reviews of literature, speculative articles, handbooks and other reference works, proposals, and various pedagogic genres—their separate histories and interrelationships. Yet the experimental report has a ubiquity that seems to overshadow the others. The experimental report seems central to many conceptions of the sciences as empirical inquiry.[9] The experimental report has developed as a favored solution of the problem of how to present empirical experience as more than brute fact, as a mediated statement of inquiry and knowledge.

While features of the genre may emerge as individual solutions to

---

with the features seen as resulting from some more fundamental dynamic of the text, such as the structure of elegy derived from a psychology of grief and consolation (see, for example, Scaliger). Two recent volumes reviewing the debate over genre and adding many interesting observations about the workings of genre in literary contexts are Heather Dubrow, *Genre*, and Alastair Fowler, *Kinds of Literature*.

However, attempts to understand genre by the texts themselves are bound to fail, for they treat socially constructed categories as stable natural facts. Recently Ralph Cohen has argued against formalist and essentialist views and presented a more socially constructed view of literary genres as "historical assumptions constructed by authors, audiences and critics in order to serve communicative and aesthetic purposes"(210).

The most thoroughgoing analysis of genre as a social phenomenon, nonetheless, comes from rhetoric and not literary studies. Carolyn Miller in "Genre as Social Action" considers genres "as typified rhetorical actions based in recurrent situations" (159). The typification of rhetorical actions entails the emergence of recognizable text types marked by repeated formal elements. Recurrence of social situation is itself a socially constructed recognition. Thus the emergence of genre goes hand in hand with the emergence of generic situations, with the rhetorical action itself helping to define the situation. Miller, following Alfred Schutz, relates genre, as a social institution, hierarchically to other forms of social typification.

My analysis of genre follows Miller, both in the importance of social understanding of text and situation in the emergence of genre (see chapters 3 and 4) and in the interplay between typification of texts and typification of other social understandings (see chapter 5). A recent article by Paul DiMaggio develops important sociological consequences of a similar definition of genre. Unfortunately it came to my attention too late to be incorporated into my argument. In particular it has implications for the argument of chapter 5 here.

9. Theory testing through experimentation is a major premise of both positivist and Popperian philosophies of science and has roots going back to Isaac Newton's concept of crucial experiment (see chapter 4 below). Although all these have come under vigorous and valid criticisms, experimentation has had a robust and enduring role in science. Hacking's *Representing and Intervening* is a recent attempt to explain the central role of experiment in scientific practice.

various rhetorical problems, the regularities that appear in the genre come from the very historical presence of the emerging genre.[10] Writers find in existing models the solution to the recurring rhetorical problems of writing science. As these solutions become familiar, accepted, and molded through repeated use, they gain institutional force. Thus though genre emerges out of contexts, it becomes part of the context for future works. Thus the social fact of genre has given the study a peg to rest on. The emergence of the genre of experimental report is a social reality that helps shape discourse in a great range of disciplines. Now anyone with results to report must somehow address the context created by the social fact of this genre.

Yet we must be careful not to consider this genre as a unitary social fact. Formal definitions, expected features, institutional force, impact, and understandings of the genre vary through time, place, and situation. And that variation is an important part of the story. Each new text produced within a genre reinforces or remolds some aspect of the genre; each reading of a text reshapes the social understanding. The genre does not exist apart from its history, and that history continues with each new text invoking the genre. So the largest lesson that this study holds is not that there are simple genres that must be slavishly followed, that we must give students an appropriate set of cookie cutters for their anticipated careers, but rather that the student must understand and rethink the rhetorical choices embedded in each generic habit to master the genre. Although genre may help stabilize the multiform rhetorical situation of scientific writing and may simplify the many rhetorical choices to be made, the writer loses control of the writing when he or she does not understand the genre.

Since the genre I have chosen to study (like all genres) is no unitary thing, and since the canvas of scientific writing is vast and growing, this first inquiry is a spotty affair. I have investigated those spots which seemed to be crucial and about which I could gain some knowledge given my limited and happenstance resources. I did what I could. Major episodes of emergence and transformation are missing or only conjec-

---

10. A rhetorical problem is the set of constraints and goals recognized by a person framing a symbolic response within a rhetorical situation. A rhetorical situation consists of all the contextual factors shaping a moment in which a person feels called upon to make a symbolic statement. The identification and elaboration of rhetorical problem, situation, and moment are construed by the individual through that individual's perception, motivation, and imaginative construction, although the individual's desire to gain more information about the situation, problem, and moment can lead to more intimate understanding of these things (see Bitzer; Vatz; and Consigny). Jamieson makes an early (1974) connection between genre and regularization of rhetorical situation.

tured about; some parameters of variation are explored, others not; the range of variation is not mapped at all; some implications are explored, and others sidestepped. Further research may modify or reverse many of the claims made here. I see this work as a beginning, but a beginning that has afforded some insight into fundamental processes about writing in the sciences and about writing more generally. Using the tools and texts available to me, I have been seeing what kinds of things could be said.

Writing as an Interdisciplinary Concern

This account of increasing intellectual scope and sharpening research focus overlooks many of the thickets I found myself in along the way. Borrowing material and ideas from other disciplines comes at a price. The work in each discipline is framed around the problems and discussion internal to that field. In order to understand what I needed from the sociology of science or the philosophy of science or the history of science, I had to encounter them in the context of their own problematics. To steal random parts of different engines leaves one with a junkpile, even if one can create the appearance of a coordinated assembly.

Yet entertaining the discussion of a new discipline offers continual temptations of novel and important issues. The problematics of each discipline contain their own intrigue and motive. Keeping my own problematics clear while still taking seriously the problematics of others, translating from one conceptual system to another without distorting ideas beyond good conscience, is a struggle I cannot ever be certain of having won. Nonetheless, the struggle constantly poses the question, What is the fundamental goal of the study of writing? To that question I have been able to find no better answer than the practical goal of helping people (myself included) to write better. That goal suggests a facilitating question: How does writing work? The assumption linking the two is the naive one that writing improves through intelligent choice of the linguistic resources in any situation; the more we understand how writing works, the more intelligently we can control our choices.

Unfortunately for writing researchers, but fortunately for human beings, writing works socially, historically, philosophically, and psychologically. Writing occurs in writers and readers living in complex worlds. The page is no more than a score is to a Scarlatti sonata performed in a Santa Barbara living room or than a script to a production of *Oedipus Rex* in a Hyderabad auditorium—an archive mediating between

One: Writing Matters

an imagined event and a distant realization. To help people write more effectively we need to unpack the entire transaction and identify what the words are doing in the middle.

Nonetheless, as my findings started to take shape, I found they did start to reflect back on the problems of these other disciplines. Writing is a social action; texts help organize social activities and social structure; and reading is a form of social participation; thus, saying something about writing is saying something about sociology. In regard to philosophy, writing is the statement of what we know and reading is a way of learning; epistemological implications keep leaking out of the edges. Texts, as written and as read, are important historical events and the dynamics of the communication embody historical forces; in giving rhetorically sensitive accounts of historical events, we uncover new dimensions of history.[11] Any claims about how writing works are claims about how people handle words—a major issue in psychology and linguistics.

I found myself continually being drawn over the interdisciplinary cliff. I could not simply borrow without addressing. Particularly in the later chapters, as I draw the pieces of the puzzle together, the story becomes one that sits between disciplines, focused on an activity that is prior to the many branches of knowledge which are currently interested in it. The final conclusions I draw pertain to a praxis of writing, but a writing praxis so integrated with social, epistemological, psychological praxis and events-in-the-making that the problem of choosing which words to put on a page looks outward to the whole world rather than inward to a contained technology.

To anyone open to the gusts of intellectual zeitgeist, such an interdisciplinary location and import for the study of writing is hardly a breath of fresh air. Today, theory and research in many fields are claiming words to be the turtles upon which both the world and their disciplines rest. Wittgenstein, Derrida, Foucault, and other astral lights of the postmodernist pantheon remind us that we all talk in words, and words are just talk. Language is situated and ephemeral, a momentary realization of protean life forms. Rhetoric has again threatened, as in the scholastic middle ages, to become the queen of the sciences.

The academic atmosphere has been infused with linguistic structuring of textual organization, literary deconstructions of textual relations, sociological readings of social construction through language, historical reconstructions of rhetorical events, psychological restructuring of cognition, philosophical poststructuring of consciousness, and critical de-

---

11. See, for example, Hayden White, *Tropics of Discourse,* and D. La Capra, *History and Criticism.*

The Problem of Writing Knowledge

structions of entrenched discourse in all disciplines. The doings and undoings of language on all fronts have made this an exciting period in which to wonder about writing.

Yet it seems only in the last decade that such concerns have become general. Earlier in this century only a few philosophers, radical social scientists, and literary theorists seemed to hold these mysteries in their hands, despite the long preoccupation with rhetoric of the pre–twentieth-century world. When I began this inquiry, few people (except us drudges hired to teach composition) expressed any interest in nonliterary writing. Literary studies of nonfiction rarely ventured beyond belletristic biographies and autobiographies. Even linguistics had for a half-century abandoned written language as an unnatural phenomenon. Study of writing was considered necessary only for the grossly incompetent; the knowledge to be transmitted was of the kind already mastered by skilled junior high school students.

The renewed dignity for the written word, however, still maintains about it the aura of theory and philosophy. Rhetorical analysis has become the grounds for radical critique and epistemological ponderings. Concern for the role of the word in making our world has more often seemed a form of withdrawal or denial of the world, demonstrating that all these things we have once thought so solid were only the projections of evanescent symbols. The debunkers of illusions have exposed us all as charlatans of the word with only philosophic self-consciousness as a consolation. Proposals for the application of this new rhetorical self-consciousness to scholarly discourse recommend institutionalizing this critical disengagement in explicit required ironies and self-reflections, in encouraging fictionalizing freedoms and literary markers, in creating visible disjunctions and aporias.[12]

This apostasy from the world seems to me to miss the point of learning about language. For a writer the point of learning about language is engagement—doing it better. That words have great powers is hardly a secret to those who have wrestled with words to make worlds throughout history. Writers' self-consciousness about the power of words is what has allowed them to wield that power, to engage in the world

---

12. Two examples from sociology are Richard Brown, *A Poetic for Sociology,* and Michael Mulkay, *The Word and the World.* Some of the essays in *The Rhetoric of the Human Sciences,* ed. Nelson, Megill, and McCloskey, reflect similar views, but some present more balanced analysis and recommendations for rhetorical self-consciousness within the disciplines of the social sciences. Two of the contributors to that volume have published noteworthy books developing balanced views of language in the social sciences: Donald N. McCloskey, *The Rhetoric of Economics,* and James Boyd White, *Heracles' Bow.*

One: Writing Matters

through their words. Self-consciousness, reflexivity, to a writer is simply knowing what you are doing, not undermining what you do. This spirit of engagement in the world through language characterizes composition departments, and this is perhaps why they have not gained the status benefits of the new dignity of the word, despite a significant scholarly activity within composition. Put bluntly, composition research is too much committed to aiding language do the work of the world to mesh easily with critical exposé.

On the other hand, writers do have a dyspeptic, despairing, and cynical side. They know how recalcitrant a medium language is, how difficult audiences are, and how easily language can lead writer and reader down foolish paths. Words often fail. Messages go awry. Books remain unsold and unread. Finely hewed portraits of the conditions of this world gain no attention, while mindless hack work plays upon mass illusion. Skilled writers and readers know that language is a slippery affair. Whenever a text actually manages to accomplish anything admirable, it is a hard-won achievement. High hopes must constantly confront limited realities.

The world the writer wants to bring into being through words is often frustrated by the world that actually emerges. One way out of that frustration is the cynicism that finds the world a phantasm, that finds language manipulation a set of empty tricks. Another way out of the frustration is to limit ambitions; a hack is a respectable occupation that simply rehearses already available solutions to well-known writing problems. A hack reinforces the existent world, but does not extend it. But that frustration also can drive a writer back to do better, get it right, bring that more satisfying world into being. That motivation can be said to be the exact one that drives some scientists back to find the right formulation, find the compelling argument that will create a more satisfying world of living knowledge in the human community.

This attitude of engagement and positive concern for the use of language turns many of the issues of postmodernist criticism inside out, even while sharing a number of assumptions. Both the writer and the postmodernist critic consider language as a human activity shaping human consciousness with no necessary connection with objects beyond consciousness. But for the writer that is the opening situation and challenge rather than the final critique. Similarly, where both see language as socially conditioned, to the writer that is again a starting fact for a dialectical relationship between social givens and individual experiences, motives and inventiveness. While both see institutionalized social relations in received forms, the writer sees those institutions as prior achievements forming opportunities for new achievements.

While both see reading and textual interpretation as having as much to do with the readers as with the text, the writer sees responsibilities for both writers and readers to find in the text as much meeting ground as they can, rather than cutting each free to make of the text what they will. While the writer is impressed with the world of human consciousness created from nothing and thus feels responsible to participate in that creation of the human world, the postmodernist critic finds the human world made from no more than phantasms of nothing. In short, the writer is always looking with delight and surprise at what can be done with this fallen state.

Scientific Writing as an Accomplishment

The evaluative language of the last few paragraphs is no accident or methodological oversight. Writing is choice making, the evaluation of options. To view writing from the prospect of language users is to consider the benefit of some choices over others. Such an evaluative position would seem forbidden from both a social scientific objectivist position and a postmodernist relativist position—one would deny the propriety, the other the basis, for such judgments. Yet any praxis-oriented constructivist study cannot avoid evaluative assumptions built in somewhere. To mark human constructions as worthy of attention is to valorize accomplishments. To be curious as to how these things were accomplished implies a desire to imitate, incorporate, or outdo. To study choices is to notice what they accomplish and what they don't. To develop a praxis from such study is to encourage some lines of development for human society at the expense of other developments or nondevelopment. Finally, practical goals necessarily provide an evaluative framework for the entire scholarly endeavor.

A not-very-hidden assumption of this study is that the corpus of scientific writing is one of the more remarkable of human literary accomplishments. Innovation, complexity, intricacy, social influence, and simple extensiveness of the corpus make scientific writing interesting as an object of study and important as part of human society. The literary accomplishment is more narrow: the development of linguistic means for statements that move toward relatively stable meaning and assent among people sharing wide numbers of social variables (even while sharing participation in scientific activity). Moreover, these statements seem to give us increasingly immense control of the material world in which we reside. These symbolic representations have literally helped

One: Writing Matters

us move mountains and know when mountains might move on their own.

To someone who approaches scientific writing from the point of view of rhetoric, it is no surprise that people have different interests in communicating, that they disagree, that they will understand statements differently, that alternative descriptions are possible, that different contexts will lead to very different kinds of statements, statements so different as to seem to be contradictory. What else would one expect from human beings in contingent human society? What is remarkable is that statements emerge over time, that for all practical purposes these statements represent an overwhelming consensus as the best of currently available formulations, and that these formulations are sufficiently reliable to be near infallible for most practical purposes, such as operating microwave ovens.

The more I study scientific writing, the more I see how much work, thought, intelligent responsiveness to complex pressures, and fortunate concatenations of events went into creating this evolving and manifold linguistic system that could do these things. For the purposes of science, it is a remarkable achievement. Such a successful discourse system within its own domain, however, does not necessarily displace other linguistic systems in theirs. Poetry, law, and rhetorical analysis have developed their own discourse systems to meet their situations and goals. Recurring themes of this book are, in fact, the variety of discourse systems and their relation to evolving communities.

One peculiar aspect of the accomplishment of scientific discourse is that it appears to hide itself. We know that poetry, laws, and newspapers are the active products of word-hagglers. The only ploy to minimize human linguistic agency in these endeavors is to invoke divinity, muses, or the depths of the human psyche. Yet to write science is commonly thought not to write at all, just simply to record the natural facts. Even widely published scientists, responsible for the production of many texts over many years, often do not see themselves as accomplished writers, nor do they recognize any self-conscious control of their texts. The popular belief of this past century that scientific language is simply a transparent transmitter of natural facts is, of course, wrong; the evidence presented in this book only confirms this conclusion argued so forcefully and frequently in recent years. It is nonetheless fascinating that such a misconception could have thrived so well in the face of the massive linguistic work that has gone into scientific communication. This attests to the success of scientific language as an accomplished system. So much has already been done, and hides so far behind the

The Problem of Writing Knowledge

scenes of current practices, that using the language seems hardly an effort at all.

The apparent transparency of the system to the latecomers is something then imputed back to the firstcomers and makers of the system. This book, examining the many rhetorical choices evidenced over the last three centuries, should help dispel the view that scientists never have and never will write. Sometimes scientists' rhetorical choices are self-conscious responses to perceived rhetorical problems; sometimes they are unselfconscious impromptu inventions; sometimes they are slow and imperceptible shifts. In whatever way these writing choices are realized and become institutionalized, they shape the kind of thing we consider contributions to knowledge. To unpack what kind of thing a contribution to knowledge is, we need to see what these choices originally were and why they were made. We need to see what kinds of mechanisms are embodied in current unreflective practice. And by bringing unreflective practice to attention, we reassert conscious control over it.

The concern for actual practice leads to a smaller role for rhetorical theorists than is usual in rhetorical histories. The actual writers of scientific texts take center stage. Although a number of chapters here focus on scientific language in seventeenth-century England, Bacon appears only in his influence on practicing scientists as they interpret and attempt to realize his ambitions in their writing. Spratt and Wilkins are only minor background characters. Newton emerges in the forefront of actual innovation in rhetorical practice, and Oldenburg by rearranging the context of communication seems to wield great force in shaping communication.

No attempt is made to reread and reinterpret the classics of rhetorical thinking, except as they shed light on the rhetorical climate. Too often the history of rhetoric has meant the history of prescriptions and theories; the actual living practice has seemed less real than the prevailing theories. Certainly, prevailing theories bear important relationships to practice as social facts defining an intellectual climate of attitudes and understandings. But the history of rhetoric must be read more subtly and dialectically than has been the case.

This overreliance on theoretical statements read without concern for their impact on praxis has led to mistaking ambitions and goals for accomplished realities. This has been particularly the case with theories of scientific language. Bacon's desire to expunge the language of science from the four idols does not arise from the ease or even absolute possibility of doing so; quite the contrary, it arises from the contrariness of human language. Bacon's goal of finding better ways to describe that which is, rather than that which we imagine, helps create some interesting linguistic proposals, but it does not mean that epistemological magic has been performed. The attempt to realize these goals leads to particular kinds of rhetorical activity, even though the goals may be unreachable ontologically. Similarly, in epistemological terms Wilkins' attempt to create a philosophic dictionary of pure correspondence between words and things is a silly mistake, doomed to failure, but when we look at the project within the history of lexicography, we see his ambitions helping create the modern dictionary, which tries to establish the complete semantic range of a language, comprehensive of all words and meanings. Previously, only lists of difficult words had been compiled (Dolezal). What is important is the emerging practice; the contemporary theory is best understood as part of the historical dynamic—inspiring, encouraging, justifying, or hindering the practice.

Synopsis

In the attempt to understand what scientific language has become in practice, this book consists of a series of case studies. In chapter 2 the analysis of three texts will suggest how much differences in writing matter. The differences are not just on the page, but in how the page places itself with respect to social, psychological, textual, and natural worlds. By examining texts from three different disciplines, we see what very different textual objects they are and what different worlds they reside in. The remainder of the book will look more exclusively into scientific writing, concentrating on the genre of experimental report.

The second part of the book looks at the early emergence of the experimental article. One chapter examines the changing form of the article over the first hundred and thirty-five years of the *Philosophical Transactions of the Royal Society of London*, pointing to the shaping role of conflict. The next chapter examines Newton's struggles to find a textual form for his optical findings to contend with the controversial dynamics of journal publication. The last chapter of the section examines how the organization of scientific communication in journals had impact on the social structure of the scientific community.

The third part looks at more recent developments in the genre of experimental article within physics. A historical examination of spectroscopic articles in *Physical Review* suggests how the increasing role of theory has reshaped the experimental article. A study of the forces shaping an article by the early twentieth-century physicist Arthur Holly Compton considers how he used experiments as a resource and a constraint in arguing his views. An interview study of how contemporary physicists read research articles indicates how deeply their readings are embedded in their practice of science.

The fourth part examines the diffusion of the experimental report to the social sciences in this century. A historical survey of the development of writing in experimental psychology resulting in the American Psychological Association *Publication Manual* considers how the rhetoric of the experimental article is reshaped around the epistemology of the field adopting it. Finally, a look at some recent articles in political science reveals some tensions between the project of the discipline and the wholesale adoption of a transplanted form.

The closing chapters examine the implications of these studies for our understanding of language and our practice of writing.

These chapters are far from complete and I could just as well have written an anti-contents, of all the topics and issues not investigated. Yet the bits of the world I have tried to recreate here, I hope will begin a new world of rhetorical understanding of how we make statements about the world. It is for what comes after to give greater substance to that world or to let that world fade into the pale graveyard of failed visions.

2
WHAT WRITTEN KNOWLEDGE DOES THREE EXAMPLES OF ACADEMIC DISCOURSE

Knowledge produced by the academy is cast primarily in written language—now usually a national language augmented by mathematical and other specialized international symbols.[1] The written text, published in journal or book, serves as the definitive form of a claim or argument, following on earlier printed claims and leading to future claims. A traditional, although incomplete, form of history of knowledge has been simply to trace the record of printed claims. This book will argue that close attention to the textual form of written knowledge will tell us much about what kind of thing knowledge is, that the written form matters. The mode of argument here will be primarily close attention to the page, and persuasion (if it comes) will be through the force of what we find there.

But examination will not be of dormant symbols lying quietly on flat pages. The symbols will constantly lead us outward to the many worlds they interact with. Without use and activity there is no language. We will come to see how the word draws on and ties together writers, readers, prior texts, and experienced reality to constitute the domain symbolic knowledge.

---

1. Of all the contemporary national languages, English is by far the most commonly used in scientific and technical publication (Swales, "English as the International Language of Research"). However, in examining the technical literature on fisheries, Baldauf and Jernudd find "that despite the dominance of English as an international communicative medium, there was a strong national usage pattern . . . [which] cut across issues of international importance"(245).

Three Criticisms

The idea that writing matters, that different choices of what to put on a page result in different meanings, has been subject to three kinds of criticism that would diminish our estimation of the power and importance of written language. Each of these types of criticism has a long history and has been presented in many variants. Being neither a philosopher nor a historian of ideas, I cannot hope and do not desire to address the criticisms successfully in general terms, nor add any abstract arguments to centuries-old debate. I place these criticisms here to acknowledge the issues and suggest an orientation toward them consonant with the data to be presented later. Each of the criticisms point to a truth whose proper meaning, however, is not revealed until it is seen enmeshed with other truths in the living practice of language. After I have presented the specific studies that constitute the main argument of the book, I will in chapter 11 offer a more complete theoretical statement of my view of language.

The first criticism against finding much significance in written formulations argues that the meaning of texts lies somewhere outside of the symbols used to clothe them in the text. Some philosophers, theologians, artists, psychologists, and others have believed in direct apprehension of truths, ideas, or realities through direct nonsymbolic means. Symbols, they claim, only remind us of these meanings that we know from elsewhere. This argument, of course, is ancient, dating back to Plato and Moses, but it has gone through many transformations, finding primary meanings in such things as presymbolic imagination, biologic imperatives, and sensory apprehension of reality. Meaning is said to lie in these primary referents; once we grasp these referents, we can discard the clothing of public language that allows us to locate this presymbolic reality. From this perspective, the problem of language is only one of clarity and precision—to help us locate what we need and then to vanish.

From a modern, nontheological perspective, it is easy to scoff at a shadowy world of essences, of things in themselves, of authentic feelings, of positive reality—tantalizing our reach, but beyond our grasp. Yet people do use language as though they were referring to something other than their own linguistic practices. They do seem to have some loose grasp of a world they live in and premonitions of meanings that seem to reside within them. They in fact struggle with language to capture these external worlds and internal meanings, to get the words right. They are frustrated when their words fail to communicate their experience and vision. Mature writing can be said to begin with the realization of the need to struggle with words to make them do more fully what we wish them to do. The cases examined here indicate that this struggle is deeply conditioned in social and linguistic systems, and that the struggle only takes place at socially defined moments, around social activities, in social relations (no matter how displaced by internalization). Yet this struggle with meaning, a dialectic between the language system and the writer's knowledge, experience, ideas, and impressions of his reader, is a deeply creative force, constantly remaking our symbolic world.

The second criticism inverts the first. It claims the meaning of the text is enclosed entirely within a text, is purely a construct of the arbitrary signs brought together in a text. From this position, language becomes the entire contents of our minds and experience. Language, then unencumbered by any constraints other than the socially given linguistic system, can mean anything, which is as good as meaning nothing, for it is a web of illusions. Reference to objects, experiences, and ideas outside the sign system is only a deceiving appearance; the idea of reference is itself only a semiotic creation. With no grounding point of meaning outside the individual sign system, different sign systems create incomparably different worlds of consciousness. This vision of the world of human consciousness being constructed by human language-making goes back to the Sophists and to the Biblical description of Adam naming the animals, but finds its currently most influential form in the literary/linguistic theories of structuralism, poststructuralism, and deconstruction.

The power of language and other symbol systems to create our realities is certainly a cause for our fascination with language and an imperative for understanding. Otherwise language study would not extend beyond linguistics. The arts would be only an entertainment, literary studies would be trivial, and sociology, political science, anthropology, history, and philosophy would find no impulse to worry over our linguistic symbols. The symbol-makers of societies would neither be so adulated nor be so central in the operations of polity and culture. The cases examined in this volume indeed indicate how ways of perceiving and knowledge-making emerge out of sociolinguistic processes. Each community examined here finds its own way to formulate its knowledge and in so doing defines what it considers knowledge to be. As the community changes, so do the symbolic means.

Yet enough sharing of meaning occurs between communities of symbolic systems to make translation between Hindi and English a fruitful, although difficult and imperfect, endeavor. Certain common elements of life and the world allow occasional cooperation among people of different symbolic communities, although the meaning of such cooperative events may be interpreted variously by the participants. Enmeshed as we are in our own symbolic systems we can even gain shadowy glimpses into the worlds of others and expand our own symbolic repertoire by contact with different systems.[2] In surveying the symbolic options, we find some more apt to our experiences and needs, and others less. We choose among various possible meanings and rise above being unreflective automatons of our linguistic system. And we find that certain formulations, although not writ eternal, do have more staying power and wider cultural dispersion.

The cases of scholarly and scientific formulations examined here indicate that the symbolic developments within communities may depend on something more than arbitrary swings of cultural fashion. Symbolic systems react to experiences and situations, to contact with different communities and the formation of new communities, to struggles with old meanings deemed inadequate to account for emerging ideas and experiences, to the need to create shared understanding and agreement where none existed previously. The world of symbols and consciousness here is no blindfold, but a dynamic means of acting in the world. In the course of acting, there is even seeing, partial (yet focused and goal directed) as it may be in any instance.

The third argument against putting much stock in written texts is an extension of the second. Accepting language as a structured social creation, this position claims that the significant social and creative action occurs in the living moment of spoken language instead of on the dead written page. In some versions, informal personal communication, such as in letters, is granted some breath of life. Generally, however, this argument considers written language an epiphenomenon, a pale reduction of the living language of personal presence. Written texts appear contextless and socially meaningless in comparison with spoken language that arises out of the needs of a moment and has an observable effect on identifiable listeners. In the interactive dialogue of spoken conversation, community and communication seem to be born. This idea also has an ancient history going back to the early period of literacy. Biblical concepts of divine presence and Plato's preference for living dialectic over the death of wisdom that occurs in writing find their echoes in modern valorization of oral over written language in theology, linguistics, anthropology, and ethnomethodological sociology.

The important truth brought home by this criticism is that the power of language can only be understood in the context of social action in

---

2. Linguists have, of course, long observed that contact between people of different linguistic and dialect groups affects the language of both groups and is one of the main forces for linguistic change.

specific situations. But we should not be fooled by the distances travelled by written language, carrying messages across many miles and many years. Writing and reading may take place in privacy and composure, and they may carry out distant social actions, but they are still highly contextualized social actions, speaking very directly to social context and social goals. If written language didn't do anything, people would treat it only as an idle pleasure. The cases examined here uniformly demonstrate how much the writing and reading of texts are enmeshed in social activities. Moreover, the essential social purpose of the communities examined here is to produce statements of knowledge. That is, text production is the goal, and the activity cannot be understood without seeing the centrality of texts. In fact, as I will argue in chapter 5, the organization of textual activity can help generate many other features of social structure. The emergence of certain patterns of written communication give generic qualities not only to texts, but to the way the texts are used in situations, and even to the character of the situations themselves. Writing is social action. Regularized forms of writing are social institutions, interacting with other social institutions.

In communities organized around the production, reception, and use of texts, as in the cases examined here, much of the spoken interaction and even nonverbal behavior can be seen as in fact secondary to the written interaction. For example, chapters 3, 4, 7, and 9 suggest that emerging standards for the reporting of experiments create imperatives for experiments to be done in certain ways, so that an acceptable account may be given of them in an article. Similarly, chapter 7 suggests that specific debates in the literature create the impetus for new experiments. It is not a great stretch of the imagination to see talk occurring over the laboratory bench and even over morning coffee as bound together by the goal of producing written statements that would be found acceptable by the relevant audience (see, for example, Latour and Woolgar 151–86).

Although less formal oral and written linguistic events within "invisible colleges" (Crane) constitute significant moments along the way toward the public statement, the printed statement circulates beyond the inner circle, creating public knowledge out of esoteric knowledge. In the public forum the printed statement is what is held accountable and becomes the reference point for future discussion. Even within those fast moving and tightly structured scientific communities where preprints, letters, and chalk talks may be the primary forms of publication, with judgment of peers being passed long before the article reaches the archive of journal publication, the prejournal forms of publication must meet the essentials of public written argument to gain approval. The core of the argument must be inspected and approved by the relevant others.

Recent scholarship into the complex private and semiprivate activities of scientists has enriched our view of how knowledge is created, the impulses and processes that lead to public statements; these private moments indeed shed light on the public statement, and I shall often draw on such evidence (see, for example, Collins, *Changing Order*; Garfinkel et al.; Knorr-Cetina, *The Manufacture of Knowledge*; Latour and Woolgar; and Lynch). In helping show the construction of the public moment, insights into private activities do not deconstruct, devaluate, or invalidate the public moment. They would only be disillusioning if we held naive illusions that texts were to appear spontaneously and pristinely, and then were immediately to transsubstantiate, without being read, pondered, and acted on, into the pure world of truth. To recognize the rhetorical character of visually transmitted symbolic activity is only to recognize that we live and use our texts in a human world.

These three arguments against granting substantial importance to written texts are illuminating rather than damning. They help reveal the dialectical interconnectedness of written language with the worlds around it and point to the danger of seeing the printed page as an isolated, internally whole phenomenon. Written language can decay faster than the page it is printed on, although a powerful text can outlast multiple editions, translations, and reconstructions. The force of written language only maintains to the degree that contextual factors are properly aligned and the text is able to capitalize on these factors. That is why writing is hard. When we write with any success, the success is likely to be weak and transient. Only the rare statement has long-lasting social force.[3]

The regularization of writing genres and situations within specific communities can increase the likelihood of successful, forceful communication, as several of the case studies below will illustrate. If the communal wisdom of a discipline has stabilized the rhetorical situation, rhetorical goals, and rhetorical solutions for accomplishing those goals in those situations, the individual writer and reader no longer need make so many fundamental choices and perform virtuosities of communication. Writing up an experiment on visual perception may seem a more transparently easy activity to an experimental psychologist than framing an argument in aesthetics to a philosopher, but that has more to do

---

3. Our current eclectic hunger for texts from distant times and societies is a recent, sporadic, and incomplete phenomenon. Because a text exists in some archive does not mean it has living meaning for any readers.

One: Writing Matters

with the stabilization of the rhetorical world in one than the innate depth in the other.

This book examines the amount of difference writing makes in constituting what we consider knowledge. The different choices made in formulating knowledge under different conditions, the regularization of choices and contexts within communities, the modification of these regularities as they disperse through time and domain, and the implications of the rhetorical choices of individuals and communities, will I hope reveal how important it is that we attend to the rhetorical process in our understanding and production of knowledge texts. I doubt the fundamental philosophic questions surrounding language and knowledge will evaporate in confrontation with the evidence here; certainly since this book is not framed as an argument in philosophy, I would be surprised if it actually engaged recognizable philosophic questions. Yet I do hope that the concrete sense of the relations between language, social action, empirical experience, and knowledge will help us control our symbolic attempts at knowledge with increased skill.

Texts and Contexts

Here begins the examination of the ways in which writing matters. Three texts, from different sorts of knowledge creating communities, will be examined in relation to four contexts, as these contexts are referred to, invoked, or acted on in the texts: the object under study, the literature of the field, the anticipated audience, and the author's own self.[4] By examining how these four contexts are brought together in each text, we can see what is embodied in the language of the statement of knowledge. This method, although it gives no firm evidence about the actual intentions of the authors and the actual understanding of the readers, does nonetheless reveal the intentions and meanings available in the text.

This study also ranges beyond the scientific paper to examine knowledge-bearing texts in other disciplines in order to explore the possibilities of variation in what constitutes a statement of knowledge and to accentuate textual features through contrast. The differences in the examples reveal the resources of language to mediate the four contexts examined. The examples are not claimed to be typical of their disciplines, nor are the analyses to be taken as a simple model of the spectrum of knowledge.

4. This four-part analysis is a modification James Kinneavy's communication triangle. He sees language (or text) mediating among an encoder (or writer), a decoder (or audience) and reality; I have added the fourth item of the literature.

What Written Knowledge Does

How a text refers to, invokes, or responds to each context is explored here through specific features of language. First, the lexicon of an article is examined to find the types of information conveyed about the objects under discussion. The nature of the symbolization, the frameworks in which the objects are identified, the precision of identification, and the tightness of fit between name and object indicate the quality of tie between text and the world.

Second, explicit citation and implicit knowledge indicate an article's relationship to the previous literature on the subject.[5] About explicit references questions arise concerning the precision of meaning conveyed by the reference, the relation of the reference to the claim of the article, the use made of the reference, and the manner of discussion of the reference.[6] About implicitly used knowledge, questions arise concerning the extent of codification and the role the knowledge takes in the argument.[7]

Third, each article's attention to the anticipated audience can be seen in the knowledge and attitudes the text assumes that the readers will have, in the types of persuasion attempted, in the structuring of the argument, and in the charge given by the author to the readers (i.e., what the author would like the readers to do after being convinced by the article).[8]

Finally, the author is represented in several ways within the text. The human mind stands between the reality it perceives and the language it

---

5. Karl Popper in "Epistemology Without a Knowing Subject" in *Objective Knowledge* argues similarly that knowledge once created becomes largely autonomous, something separate from either reality or our subjective sense of it. Once created, knowledge can be treated as an object, upon which further intellectual operations may be made, much as a spider web once woven becomes an object in the world. In like manner, I consider the literature of the field as a fact in itself, a fact with which all new publications must contend, just as they must contend with the objects they presume to study. With respect to new publication the literature of the field has a status beyond simply the record of past subjective perception. The new publication, in criticizing, correcting, extending, and simply using the prior literature treats that literature as the "third world" Popper describes.

6. See G. Nigel Gilbert, "Referencing as Persuasion"; Henry G. Small, "Cited Documents as Concept Symbols"; and Susan Cozzens, "Comparing the Sciences" and "Life History of a Knowledge Claim."

7. Harriet Zuckerman and Robert Merton discuss codification in "Age, Aging, and Age Structure in Science," in Norman Storer, ed., *The Sociology of Science*, 510–19. Merton also discusses the implicit use of knowledge, or what he calls "obliteration by incorporation," in *Social Theory and Social Structure*, chap. 1, and in *Sociological Ambivalence and Other Essays*, 130.

8. Latour and Woolgar, Knorr-Cetina, "Producing and Reproducing Knowledge," and Knorr and Knorr, *From Scenes to Scripts*, seem most interested in the persuasive and other effects texts have on their audiences; the process of text creation is seen to have the primary goal of persuasion. In this they follow Joseph Gusfield, "The Literary Rhetoric of Science."

speaks in; statements reflect the thoughts, purposes, observations, and quirks of the individual. The individual can be seen in the breadth and originality of the article's claims, in the idiosyncrasies of cognitive framework, in reports of introspection, experience, and observation, and in value assumptions. These features add up to a persona, a public face, which makes the reader aware of the author as an individual statement-maker coming to terms with reality from a distinctive perspective.

Although the four contexts (and the features that indicate them) are separated here for analysis, they are mutually dependent in each text. An observation concerning one has implications for the others. The depth of the interdependence is evident if one considers that the perception and thought of both author and audience are shaped for the most part by the same literature, and that literature provides the accepted definition of the objects discussed. Similarly, shared interest in and observation of objects of study draw the literature, author, and audience together.

An author, in deciding which words to commit to paper, must weigh these four contexts and establish a workable balance among them. A text is, in a sense, a solution to the problem of how to make a statement that attends through the symbols of language to all essential contexts appropriately. More explicitly, an article is an answer to the question, Against the background of accumulated knowledge of the discipline, how can I present an original claim about a phenomenon to the appropriate audience convincingly so that thinking and behavior will be modified accordingly? A successful answer is rewarded by its becoming an accepted formulation.

Each of the contexts, when abstracted from the writer's task of embodying complex meaning in a specific text and when viewed singly as a theoretical problem in communication, can appear to raise overwhelming epistemological difficulties. The kinds of difficulties that arise from such monochrome analysis are suggested by a slight renaming of the four factors we have been considering: language and reality; language and tradition; language and society; and language and mind. Exclusive concern with the language-creating mind leads to a subjective view of knowledge which makes uncertain the reality perceived and which rejects the cognitive growth of cultures. Viewing in isolation the effect of tradition on statement-making may lead one to misjudge accumulated statements—whether called paradigms or authority—as juggernauts, flattening out observed anomalies and individual thought. Perceiving statements only within the process of social negotiation of a socially constructed reality ignores the individual's powers of observation and language's ability to adjust to observed reality. But the most common errors arise from language considered only in relation to reality: on one side the naive error of assuming that language is an unproblematic reflection of reality, and on the other side the sophistry that language is arbitrary, radically split from nature, with no perceiving cognitive selves and no trace of rational community to heal the split.

The three texts examined below represent three different solutions to the problem of writing knowledge: James Watson and Francis Crick, "A Structure for Deoxyribose Nucleic Acid"; Robert K. Merton, "The Ambivalence of Scientists"; and Geoffrey H. Hartman, "Blessing the Torrent: On Wordsworth's Later Style." The different balance of contexts established in each article derives in part from the differences in contexts—different types of objects studied, differently structured literatures, audiences of differing homogeneity, and different role expectations for the authors. The origin of the papers in separate fields (molecular biology, sociology, and literary criticism) representing the three traditional divisions of the academy (sciences, social sciences, and humanities) of course accentuates the differences on all fronts; however, these examples should not be overread as typical of large divisions of knowledge.

Suggesting a Molecular Structure

The article "A Structure for Deoxyribose Nucleic Acid" (see pp. 49–50) primarily describes a geometric model, elaborated in quantitative and qualitative terms, that is claimed to correspond to the structure of a substance found in nature. This act of geometric naming depends on the substance being discrete and robust and its structure being consistent through repeated observations, for otherwise the names will not convey a distinct and stable meaning to all observers.[9] Thus the primary context explicitly attended to by the language of the paper is the context of the objects of nature.

All other contexts are subordinated to this primary one so that the article may appear to speak univocally about nature. The previous literature on the subject is sorted out according to the criterion of closeness of fit between the observed phenomena and the claims made, and the accepted claims in the literature become assimilated into the language

9. Here I am not concerned with the reproducibility of individual experiments, but rather with the appearance of the phenomenon under a variety of circumstances. The more situations in which the phenomenon unmistakably appears, the more certain is the identification of its discrete existence.

used to describe the phenomena. The audience is assumed to share the same criteria of closeness of fit, discreteness, robustness, and reproducibility for acceptance of claims (or symbolic formulations) about phenomena; therefore, the audience can be relied on to have much the same assessment of the literature as the authors do, and persuasion may proceed by maintaining apparent focus on the object of study.[10] Further, because the audience has a well-established frame of reference in which to fit the new claim, they do not need to be given much guidance about the claim's implications. Finally, the authors' apparent presence is minimized by the common pursuit of authors, literature, and audience to establish a common, codified, symbolic analogue for nature. The authors seem only to be contributing a filler for a defined slot, and they are only in competition with a few other authors who are trying to fill the same slot. The persona, although proud among colleagues, is humbled before nature.

The opening sentence of Watson and Crick's article sets the task: "We wish to suggest a structure for the salt of deoxyribose nucleic acid." The task of identifying a structure assumes, first, that there is a distinct substance which can be isolated and inspected and which has qualities distinguishing it from other substances. By 1944 Avery, MacLeod, and McCarty had extracted a substance which they called "the transforming principle" and the method of extraction was standard by the time Watson and Crick began work.[11] Further, this substance is assumed to preexist the historical, human act of isolating and identifying the substance.

The ability to isolate the substance under repeatable conditions gives an ostensiveness to the name. Since the name only serves to point out or tag something distinctly and unmistakeably observable, the name need not convey any particular information. It can be arbitrary, whimsical, eponymic, or otherwise accidental; it need only be distinctive. The name, however, can do double service, conveying information as well as identifying. The name deoxyribose nucleic acid identifies elements of structure—e.g., the ribose configuration without an oxygen—as well as letting us know that the substance is to be found within cell nuclei. Thus the name is in this case overdetermined with respect to reality; we know more about the substance than we need to for purely identificatory purposes.

---

10. Latour and Woolgar, 75–76, suggest that scientific persuasion is successful when attention is drawn away from the circumstances of statement creation toward a "fact," which appears to be above the particularities of a specific circumstance. In the authors' terms, "the processes of literary inscription are forgotten."

11. Judson, 36. DNA was, in fact, first extracted by Johann Friedrich Miescher in 1869 (28). A more detailed account of the complex history of DNA can be found in Robert Olby, *The Road to the Double Helix*.

What Written Knowledge Does

At this point we can see how the accumulated knowledge of the field (represented by the literature) is incorporated into the language. The isolation of elements and the theory of chemical combination, as well as the idea that substances can be analyzed chemically, are all implicit in the name of the object. More than that, the name reveals the gradually emerging orientation of chemistry to describe most features and processes through structure. Even the linguistically oldest component of the name, *acid*, has been transformed through redefinition as chemical knowledge and orientation have changed. In Bacon's day the word *acid* meant only sour-tasting; then it came to mean a sour-tasting substance; then, a substance which reddens litmus; then, a compound that dissociates in aqueous solution to produce hydrogen ions; then, a compound or ion that can give protons to other substances; and most recently, a molecule or ion that can combine with another by forming a covalent bond with two electrons of the other (Oxford English Dictionary, 20; Webster's Collegiate Dictionary, 8; American Heritage Dictionary, 10). The tasting and taster vanish as the structure emerges.[12]

The task of assigning a structure relies on a further assumption, that nature arranges itself in geometrical ways; theories of forces account for this remarkable correspondence between the symbolic representation of geometric shapes and the repeating arrangement of matter in nature. Geometry as a study is the product of human consciousness, but geometric forms are claimed to preexist human invention. Thus the task of the molecular biologist is not to create a structure that approximates nature, but to discover and express in human terms the actual structure resulting from all the forces and accounting for the behavior and appearance of the molecule. The claim of representing an actual structure rather than creating an approximate model results in a strong requirement for correspondence between data and claim. This correspondence, as we shall see below, is the main criterion of persuasion offered to the audience.

The few words of text discussed so far convey much about the object and the knowledge developed through the history of chemistry and biology, yet such compact transmission of information reveals no literary genius on the part of the authors. The dense communication is inherent in the names of objects and tasks. That a mere naming of parts conveys such precise and full meaning indicates how much the historical genius of the discipline is embodied in the development of its language.

---

12. Notice also how the changing definitions of acid are tied to changing contextual knowledge as well as to changing procedures of identification of phenomena and interpretation of data.

One: Writing Matters

The analysis of the first sentence is not yet finished. The first five words, "we wish to suggest a . . . ," reveal much about the joint persona and contribution of the two authors. Despite the folk belief about the absence of the first person in scientific papers, the authors do assert their presence through the word *we*. That direct presence, however, is immediately subordinated to the object under consideration, the structure of DNA. Moreover, the authors are only *suggesting*, and the suggestion has only an indefinite article; whether *a* suggestion turns out to be *the* structure depends on nature. *Wish to suggest* is a form which implies humility before the facticity of the object, yet the phrase also has the boldness of the authors' presumption that their claim indeed will be confirmed by nature. Mild speech is possible because the suggestion will gain all the force it needs from the observation of reality; nature will stand up for scientists. The locution *wish to suggest,* appropriate here, might sound pompous in a branch of knowledge which does not find such immediate confirmation in nature.

Science will as well stand up for scientists, for the authors also subordinate themselves to scientific knowledge as currently constituted. By identifying their subject within the language of scientific disciplines, they are implicitly putting their original contribution within the framework of existing scientific knowledge. The placement and titling of the paper itself suggest how much the originality of the paper is subsumed within a highly structured framework of knowledge. The article is within a section entitled "Molecular Structure of Nucleic Acids" and is followed by another article of the same class, " Molecular Structure of Deoxypentose Nucleic Acid." The Watson-Crick article discusses only one particular substance in a larger class of substances, all being studied by colleagues to determine the same type of information.

The second sentence—"This structure has novel features which are of considerable biological interest"—places the chemical claim in the context of biological knowledge; this added context identifies the great importance of the paper. The knowledge of one field is not treated as the hermetic creation of that field, liable only to internal consistency within that field. Rather, other disciplines are subject to the discoveries about nature. Yet the specific implications of the discovery need not be discussed, for once the novel features of the structure are made known and referred to the codified knowledge of biology, any competent biologist would see a wide range of implications. Later in the article the authors comment, "It has not escaped our notice that the specific pairing we have postulated immediately suggests a copying mechanism for the genetic material." This brief comment invokes the knowledge of genetics and cellular mechanics and tells the biologist where to fit this structure into the open claims of the field. The single added piece of information will allow biology to move forward in directions determined by its own logic. It would be presumptuous, tedious, and unnecessary for Watson and Crick to lecture on the subject.

It is worth noting that although the subject of the paper is structural, the consequences and import are functional. From the shape of things, one can better understand how things happen.

It is also worth noting that all the uses of the first person are to indicate intellectual activities: statement making (opening words of paragraphs 1 and 4), making assumptions (later in paragraph 4), criticizing statements (paragraph 2), and placing knowledge claims within other intellectual frameworks (paragraphs 11 and 12). None of the first-person uses imply inconstancy in the object studied, but only changes or development of the authors' beliefs of what the appropriate claims about the object should be. The object is taken as given, independent of perception and knowing; all the human action is only in the process of coming to know the object—that is, in constructing, criticizing, and manipulating claims.

Once the claim about the object has been placed into its chemical slot, to define the inquiry, and its biological slot, to define the significant consequences, the competing claims that would fill the same slots must be eliminated. If the codified literatures of the relevant disciplines aim to represent the way nature is, a multiplicity of claims about the same phenomenon indicates an unresolved issue. Until a univocal formulation that describes the phenomenon in all its features is found, the phenomenon is not fully understood.

The grounds on which the two competing structures for DNA are rapidly dismissed in the second and third paragraphs reveal the central role of specific knowledge about the object of study. How any claim fits with what is or can be known about the object forms the chief constraint for originality, codification of the literature, and persuasion of the readers. The Pauling and Corey model, defined by a quick geometric description, is dismissed as impossible on two counts, both based on knowledge of features of such molecules well established in the literature: binding forces and van der Waals distances. Because Watson and Crick do not present their exact calculations, their criticisms must rely on the presumption that the features they invoke are commonly accepted and similarly understood well enough to allow reproducible calculations that will satisfy other researchers in the field. The codified knowledge about all aspects of the object presents clear constraints that must be met by any potential model. If a model does not match existing theory which is believed to accurately describe nature, then the model

One: Writing Matters

must be dismissed. If later the dismissed model is strongly supported by other evidence, the dismissing theory must be called into question.

The dismissal of the Fraser model on the grounds that it is "rather ill-defined" is even more interesting, for the ill-defined does not allow calculations of the kind invoked for the Pauling-Corey model. The Fraser model is not consequential enough. Since the model cannot then be discussed against the framework of codified knowledge or against measurable aspects of the object, there is no profit looking into it.

With the competition disposed of, Watson and Crick can proceed to the core of the paper, their suggested structure. The diagram to the left of the fourth paragraph gives the geometrical essence of the solution; the fourth through eighth paragraphs cast the geometry into words, add details, and clarify elements of the structure through reference to accepted causal statements, prior work, and other models. The five paragraphs are descriptive, recreating physical presence through the symbolic systems of words and numbers, but the symbols are more than approximate metaphors. The names point to discrete objects, and the geometry is of nature itself. Scientific language, as a symbolic system with a commitment to reform itself in accordance with replicable observation of nature, becomes more than an arbitrary symbolic system.[13]

After this long description of the model, only brief mention is made in paragraphs 9 and 11 of the evidence in hand that confirms the model and the evidence still needed to provide a rigorous test. Acceptance of the model depends on the confirming evidence; therefore, the sketchiness of the discussion of evidence might seem surprising. But once the model is described, the existing evidence needs only be referred to because it is generally available and can be interpreted by any competent molecular biologist. Similarly, the construction of new tests is within current technology. The other researchers must satisfy themselves that the model fits past evidence and new tests. It is up to nature to persuade the readers, not the authors.

Just as the ninth and eleventh paragraphs present only limited per-

---

13. Harriet Zuckerman, "Cognitive and Social Processes in Scientific Discovery: Recombination in Bacteria as a Prototypical Case," discusses the resistance to discovery created by misleading names and the processes by which definition is corrected through discovery. The inaccurate naming impedes, but does not prevent, discovery; ultimately, observation of the object leads to corrected knowledge. In the case Zuckerman studies, "bacteriologists believed that bacteria were asexual *by definition*" (emphasis hers) because bacteria were classified as schizomycetes, from the Greek meaning "fission fungi" (8). In 1946 Joshua Lederberg's discovery of sexual recombination in the bacteria *E. coli*, however, led to a revised definition of the classification schizomycetes, despite the literal meaning of the etymology.

suasion, the tenth paragraph presents only limited guidance to the readers about how the model might be applied. The comment that the model is probably not applicable to RNA may be primarily to eliminate RNA as a competitor for the biological slot of genetic carrier (as was then thought more likely than DNA).

After mentioning the genetic implications of the structure, the paper has finished its primary scientific business. The thirteenth paragraph promises greater detail in later publication. Later publications primarily were devoted to spelling out the genetic copying mechanisms (Watson and Crick, "Genetical Implications" and "Structures"; Crick and Watson, "Complementary Structure"). Nonetheless, it is this first short article that counts as the primary statement of knowledge and is the one usually cited.

The last paragraph pays its respects to some aspects of the social system of science: prepublication criticism, access to unpublished evidence and ideas, and funding. To those who know the history of this discovery, these few thanks and the earlier criticisms of competitive work recall a web of social intricacies and inchoate psychological reaching toward discovery.[14] These prepublication facts of life are recognized by working scientists as necessary preconditions of publishable work; nonetheless, these preconditions of discovery do not enter the actual argument of the publication. In the article, competition is dealt with only in cognitive terms, discovery is presented as a fait accompli, and the social system is appended only as a courtesy, a polite nod at the end.

Dependence on the community of the discipline is even more fundamental in the language used, the prior knowledge, and the accepted perception of the object of study, yet even this cognitive dependence on the scientific community is not given explicit recognition. The article cites only work immediately relevant to the assessment of claims made in the article. The six footnotes document only articles presenting competing claims that were criticized or offering supporting data.

In order to maximize the tightness of fit between nature and its symbolic representation, all the relations between language and other contexts—the literature, the audience, and the authors—are both harnessed to and driven by the relationship between language and nature. Society, self, and received knowledge are present in the research report, but they

---

14. The complex sociological, psychological, and historical specifics of the process of discovery in the case of DNA are extensively recounted in James Watson, *The Double Helix;* Anne Sayre, *Rosalind Franklin and DNA;* and Horace Judson, *The Eighth Day of Creation.*

One: Writing Matters

are subordinated to the representation of nature. The criterion of correspondence between statement and object governs all of the contexts.

Establishing the Ground beneath a Phenomenon

Robert K. Merton's essay in the sociology of science, "The Ambivalence of Scientists" (see pp. 51–53 for the beginning of this essay), presents a different kind of linguistic solution to a different kind of linguistic problem. In the DNA paper, except for the specific structure proposed, all aspects of the symbolic formulation are shared by author, audience, and literature. At the beginning of the ambivalence paper much less is shared; Merton must establish the ground on which his claim is to rest. The phenomenon which is the object of study is not universally recognized as a discrete phenomenon, and much of the language needed in the discussion does not have unmistakable ostensive reference. The literature of the field does not provide a generally recognized framework in which to place the current claim. The criteria the audience will apply are not clear-cut and universal, nor is it certain what intellectual framework they will bring to the reading. The author's perspective is, then, in many respects individual; nonetheless, through the medium of the paper he hopes to establish his claims as shared knowledge.

The particular subject of the article—the ambivalence of scientists (including social scientists) in observing and reporting certain aspects of behavior—adds an additional level of problem to be solved in the paper. The subject concerns the process of statement making and applies in a self-exemplifying fashion to the author's work in this essay, the statements in the literature, and the statements made by the readers. Thus, if the claims of the paper are correct, then the literature must be reinterpreted, the author must take into account his own ambivalence, and the readers must question their own statement making. Not only must Merton establish the grounds of the claim, he must carry the claim across shifting grounds.

In this article a wide range of linguistic choice is open to the author; little is predetermined by a knowledge of reality codified in language, literature, and criteria of judgment. Merton must develop at length original formulations to represent the phenomenon, to assemble and interpret the relevant literature, to establish his perspective, and to attend to the audience's perception.

The first specific difficulty faced by the essay is the identification of the topic and its placement in the discipline. Unlike the Watson-Crick topic, which is located at the intersection of two terms already within the lexicon of the discipline (i.e., "structure" and "DNA"), Merton's topic is doubly alien to his discipline. First, the topic depends on the recognition of a prior topic—multiples and priorities—not previously in the discipline;[15] then the topic inquires into why the prior topic has not obtained due recognition. Merton's solution to the importation of a topic which he claims to be indigenous, necessary just to set the stage for the true topic of the paper, is to rely on his own prior work on multiples and priorities and then to suggest that enough evidence already existed within documents familiar to the field such that the topic should have been raised earlier, except for the impeding mechanism of ambivalence.

The fact that the prior topic of multiples and priorities has a clear and substantial place in the author's own framework of knowledge, but does not yet have a fixed place in the codified literature of the discipline, leads to three consequences common in the social sciences. First, for clarification, readers are referred to the author's own works rather than the shared knowledge of the discipline. Second, the readers must be persuaded not only of the specific claims of the essay, but of the author's larger framework of thought in which the claims are placed. Finally, the author's new construction of the knowledge of the field requires a reconsideration of the validity of wide parts of the literature and not just of the specifically competing claims. Without a fixed, codified literature to place and constrain topics and claims, authors are both free and encouraged to frame their contributions in broad revolutionary terms, reordering large segments of knowledge. Paradoxically, the great power and broad implications of Watson and Crick's structure of DNA result from the claim's tight constraint within a highly elaborated framework of thought; the narrow claim reverberates through the whole system. A broader claim in a less tightly strung system may have a more damped effect.

In order to establish the phenomenon to be discussed, the opening paragraph of the ambivalence paper asks the scholarly reader to recall a wide range of evidentiary documents: "the diaries and letters, the notebooks, the scientific papers, and biographies of scientists" as well as the scholarly discussion of these documents. The reader of the Watson-

---

15. Brannigan (47) cites several precursors and sources for Merton's analysis of multiples, but these earlier discussions do not establish that multiples was a firmly entrenched topic in sociological discourse at the time of Merton's writing. Our concern here is with the rhetorical situation as perceived by the author.

Crick article must only make a highly directed scan of codified knowledge to locate and accept the topic. Here, however, the reader must review the literature from a critical perspective incorporating a new topic of priorities before he can place and accept the topic of ambivalence as worthy of study. Indeed, the large quantity of examples of the phenomenon cited throughout the essay are, in part, necessary to confirm to the reader that this topic does exist.

Since the topic of ambivalence involves a critique of the field, the writer has a special problem with respect to the scholarly audience, all of whom presumably are subject to the cognitive lapse which is under discussion. Merton must challenge the readers while still maintaining their good will and attentiveness. To overcome audience resistance and ease the shock of self-recognition, Merton creates a strong presence of his own viewpoint and an atmosphere of camaraderie that assumes temporarily that the audience is already with him. He begins with statements of great certitude and only later fills in the background of concepts that make the opening statement possible. This technique bears similarity to the way Hemingway opens *To Have and Have Not:* "You know how it is there early in the morning in Havana with the bums still asleep against the walls of the buildings; before even the ice wagons come by with ice for the bars" (1). The reader is drafted into a club, and only gradually is the reader filled in on the experience that reader presumably shared from the beginning. The reader is companionably drawn into the world populated by sleeping bums and bars and early morning adventures in Havana. In Merton's essay, the atmosphere of agreement takes the edge off the challenge and creates enough good will for the argument to unfold. Further, Merton withholds explicit discussion of sociologists' group involvement in the problem until the entire mechanism has been laid out, the giants of science implicated, a few confessions cited, and dispassion praised. Moreover, eminent psychologists and sociologists are identified as having the courage of self-examination on this matter before the readers are asked to consider their own cases.

After introducing the problem, in the second paragraph Merton identifies the mechanism of the ambivalence, thereby localizing the phenomenon in a theory of the operations of science. The metaphor of conflict of forces is drawn from physics, and Merton is careful to label it as metaphor by the phrase "can be conceived of." There is no claim here of measurable forces as there would be in physics. Metaphors are underconstrained in meaning; by their nature they are only suggestive and approximate. One resorts to metaphor only when the thing to be described is partially or imprecisely known, and one must look to correspondences with better known objects. Even in the best of meta-

What Written Knowledge Does

phors the correspondence between the thing being described and the metaphorical representation is only partial. In any specific case, however, the metaphor may be the best available description and, when combined with other underconstrained terms and contextural clues, may create a web of approximate meanings surrounding the actual thing, such that a meaning develops adequate to the situation. The second sentence provides a second underconstrained meaning to support the metaphor of resistance: "Such resistance is a sign of malintegration of the social institution of science which incorporates potentially incompatible values. . . ." Of all the sentences in the article, this sounds the most typically sociological, precisely because it attaches the topic to familiar sociological concepts. The terms of this sentence, however, are abstract, some of variable or disputed meaning, some metaphoric, and all in a complex syntactical relationship that makes the imprecision additive, if not geometrical. Further, resistance is only "a sign," not a particular sign or the only sign. Here the indefinite article is a true indefinite, unlike Watson and Crick's "a structure," where near at hand observations of nature can fix the structure as unique.

Such underdetermination of language provides further reason for requiring the good will of the audience. A sympathetic audience is more likely to expend the effort to reconstruct from partial indicators the meaning most congruent to the argument—a process that may be called reading in the intended spirit. The unsympathetic reader, however, can find in underconstrained meanings enough inconsistency, contradiction, and unacceptable thought to mount a serious attack. Even such ordinary appearing terms as "scientific accomplishment" or turns of phrases as "as happy as a scientist can be" rely on many loosely defined conceptual assumptions; they can easily disintegrate under a hostile reading.

In the third paragraph the author turns from an invisible social structure which is claimed to generate the ambivalence to the more visible "overt behavior that can be interpreted as expressions of such resistance." Even these overt manifestations of trivialization and distortion, nonetheless, are not directly measurable and discrete. Distortion, for example, is a conceptual term, requiring comparative judgments against a normative model, application of judgment criteria, imputation of thought, and similar interpretive procedures. The interpretation of the concrete evidence of contradictory statements by or about scientists on the matter of priorities requires the kind of analysis employed by psychologists and literary critics. Simple claims become indications of internal processes within the makers of the claims. Even the simple claims, that Halsted was overmodest about his work or Freud found questions of priority boring, are based on human judgment and the imputation of attitude.

The only direct evidentiary statements of the primary phenomenon of ambivalence are the confessions of the professionals of introspection, Freud and Moreno. On the less deeply embarrassing emotional conflicts discussed in the later part of the paper—fear of the joy of discovery being dashed and fear of unconscious plagiary—Merton is able to cite direct confessions of ambivalence by less trained observers of themselves. But even the evidence of introspection involves judgment, conceptual categories, and the naming of transitory and evanescent phenomena by the introspector. Claims of reproducibility of phenomena within the self require a kind of phenomenological sense memory, and claims of similarity between observers raises even greater difficulties of matching affect to language. On many levels we have only the introspectors' words to go by.

As the essay reaches its midpoint, the samples of irrational statement-making (analyzed as evidence of ambivalence) start coming from sociological sources: the literature of the discipline has become the evidentiary document. The practice of imputing psychological phenomena into the very record of the discipline is justifiable on the basis of social science's own discoveries, but it makes for great difficulties in establishing a codified body of knowledge from the literature. To draw the paradox more strongly, the desire to establish a professional literature that rises above the cognitive and perceptual limitations of individuals leads to self-examination, but that reflexivity only reveals the difficulty of codifying statements made by humans about human behavior.

Once Merton has indicated a similarity of structure in many examples and has moved the examples to the readers' discipline, he is ready to call on the readers for further analysis of this issue. Before the final peroration on the therapeutic value of the study of multiples, he has already steeled the courage and minds of those he wants to carry forth the investigation. He has also suggested the method: dispassionate observation of the self and others, aided upon occasion by collaboration. The final charge to the audience is quite directive: have courage to overcome your own ambivalence to begin a systematic study of priorities, for not only will this study add to knowledge, it will be therapeutic for all of science, including sociology. This kind of "follow my lead" is very different than the implicit charge to the reader offered by Watson and Crick: gather more evidence to see if we are right, then use the knowledge to advance science according to its own dictates.

The strength of Merton's directiveness at the end is typical of the entire essay, for he must establish a perception of reality and terms of discourse not universally shared in the discipline. He must persuade the readers not just of a specific claim, but an entire framework of knowledge. Language, rather than being highly determined by the discipline's shared perception of reality as it is in the Watson-Crick article, must be carefully shaped by the author to turn his own vision into the shared one of the discipline. Because of the originality of formulations, the author's presence is inevitably strong. If this were typical of the social sciences, one might see the consequences in authors being noted for a point of view or method of perception rather than a specific claim and in a greater tendency for schools to be formed around the most original authors. The differences in formulations among original authors may make reconciliation of viewpoints difficult, and many researchers may find the clearest direction by following in the footsteps of only a limited number of originators. There are, of course, many other economic, social, and cognitive reasons for the formation of schools in all disciplines.

Reading a Poem

Unlike the previous two articles, Geoffrey Hartman's "Blessing the Torrent: On Wordsworth's Later Style" (see pp. 54–55 for the beginning of the essay) unfixes our knowledge of its subject (a poem), to suggest an experience that goes beyond any claim we can make. Rather than taming its subject by creating a representation that will count as knowledge, the essay seeks to reinvigorate the poem by aiding the reader to experience the imaginative life embodied in it. Insofar as the poem can be reduced to easily understood, verifiable claims—"normalized," in Hartman's term—the poem is of little interest.

This concern with the aesthetic moment of the poem requires that an existential bond be created among poet, critic, and reader. In the process of conveying the poetic moment, the critic's sensibility plays the central role. The poem, the literature, and the audience's perception are all mediated through the critic's vision. The critic perceives new dimensions of the poem, uses the literature to allude to his own aesthetic experience, and asks the audience to accept a new way of reading the poem. The poetic text and its context, the accumulated experience of literary criticism and literary texts, and the audience's critical judgment and expectation of poetry do constrain what the critic can persuasively state, yet the critic has considerable power to transform all of them.

In one sense the object of investigation, a sonnet entitled "To the Torrent at the Devil's Bridge, North Wales, 1824," is a known and discrete phenomenon. It is printed in the collected works of William Wordsworth; apparently no scholar has questioned the attribution to Wordsworth, the dating, or the purity of the text. The poem is easily reproduced, as is done at the beginning of the essay. Moreover, some elementary literary techniques and a few well-known biographical facts seem to explain the apparent features of the poem, as Hartman demonstrates in the third through the sixth paragraphs. The topic of the essay, consequently, appears to be fixed in a framework even more complete than that which surrounds DNA, to the point where the topic appears trivial. Here, though, the essay sets the framework aside as not revealing the important knowledge of the poem.

That important knowledge is a complex state of mind beyond naming. Hartman can only try to reevoke it through description, contrast, analogy, and reconstruction of context. As Hartman states at the end of the second paragraph in what is the closest approximation of a thesis in the essay, "Uncertainty of reference gives way to a well-defined personal situation, that is easily described, though less easily understood." The outside of the situation, captured in the description, is distinguished from the inside of the moment, which counts as understanding. The poem, as verbal artifice, conveys something beyond the words.

The title of the essay indicates the true subject: "Blessing the Torrent" is an act accomplished through the poem. Six of the essay's seven sections are devoted to recreating the existential moment of blessing. The subtitle, "On Wordsworth's Later Style," indicates that the act of this poem is similar to the acts of others of Wordsworth's later poems, but this similarity is only discussed in the last section of the paper, and no other poem is examined in sufficient detail to establish that it is the vessel of a similar moment. This reading of one sonnet can only provide an analogy for the reading of others, making the other poems more accessible; any more specific claim of equivalence among poems would suggest a reductive normalization. Each poetic moment is itself and no other.

The essay is structured to make the poet's state of mind accessible in all its fullness to the reader, to widen gradually the reader's consciousness of the central issue of the poem. The essay opens with a consideration of the literal meaning of the opening question of the poem: "How art thou named?" Each of the following sections grows out of an issue raised in the previous one in order to open up the central, opening question. In a sense, each section progressively uncontains the flood.

The epigraphs of Hölderlin, Stevens, and Joyce prepare a first reading of the poem by setting the river in motion as one of a poetic family of floods, puzzling and uncontainable. The first section by raising issues of form—the untitled, unplaceable fragment versus the named, closed sonnet—localizes this particular flood, but raises the problem of understanding the localization. The second section takes up the theme of localization to examine biographical information that raises problems about what the poet could be meaning. At this point the critic brings in other samples of Wordsworth's writing to show the poet's way of thinking about these issues. The writings of other poets are examined to show what Wordsworth did not mean. By the end of the second section the formal solution to naming collapses as the critic points to the inadequacy of the poet's diction to fulfill the domesticating function of the sonnet.

The third section examines this dilemma through the text of the first half of the poem, where the poet explains the problem and proposes a first, inadequate solution. The fourth section discusses the acceptance of the inability of language to localize, as developed in the second half of the poem. Against this reading of the whole poem, Hartman reexamines a few phrases that appear to be clichés, but which now are seen to have unexpected depth, particularly in the context of Wordsworth's other writing. These phrases lead to a return to the problem of naming in the sixth section. Only after the full dynamics of the poem are revealed is the poem seen to represent a key part of Wordsworth's consciousness in his later career, deriving from the realizations of *The Prelude*.

The structure of Hartman's essay differs substantially from the structures of the two essays discussed earlier. In both of the earlier cases the arguments are built on claims to be placed, established, and applied—thereby achieving closure within a framework of knowledge. The two earlier essays differ primarily in the amount and directiveness of text required to define the framework and phenomenon, to establish the claim, and to indicate the applications of the claim. Hartman's essay, however, denies the reader the closure of a specific claim fixed within a coherent framework of knowledge. The essay only prepares the reader's sensibility to relive imaginatively the Wordsworthian sensibility. The essay ends with a method of reading and a promise of pleasure: "The later poems often require from us something close to a suppression of the image of creativity as 'burning bright' or full of glitter and communicated strife. Wordsworth's lucy-feric style, in its discretion and reserve, appears to be the opposite of luciferic. Can we say there is blessing in its gently breeze?"

The essay also denies closure in another way. The final test of Hartman's argument is whether it illuminates the poems. No hard evidence will determine whether he is right or wrong. Certain kinds of evidence are available to convince the reader of the plausibility of the argument, which evidence the critic violates only at his own risk. Hartman must show his reading is consistent with the wording and structure of the poems and harmonious with what we know of the poet and his period. Further, each interpretation has an implicit psychology and aesthetic which cannot, without extension rationale, violate readers' ideas of how people read and write poems; in his extensive writings on Wordsworth, Hartman has presented an intriguing and plausible phenomenological aesthetic, based on the Wordsworthian endeavor to feel a connectedness with nature through the poetic imagination (for example, *Wordsworth's Poetry, 1787–1814*). But all the argument is based on plausibility with no hard, provable answers. And even notions of plausibility can be changed if the essay succeeds in expanding the reader's poetic imagination.

As the object of investigation, the poem only gains importance in its subjective experience, so also with the literature, of which there are four relevant types. First is the critical literature, toward which Hartman's essay contributes. Yet the critical literature is used neither as a groundwork out of which the ideas of the essay grow nor as an orderly body of information into which the essay fits. The accumulated knowledge of the critical literature is implicitly dismissed in several ways, and the whole of Wordsworth criticism is treated as so inconsequential as not to require explicit discussion. In finding this one poem (and most of the other later poems as well) worth serious study, Hartman challenges the conventional wisdom which sees a collapse in Wordsworth's poetic powers after *The Prelude*. In addition, Hartman criticizes a normalized reading—i.e., conventional criticism—as inadequate to the poem. Finally, by locating the genesis of the later style in the perceptions of *The Prelude*, Hartman reverses the common view that the epic was the culmination of the early period and that Wordsworth almost immediately turned away from the great poem's realizations. In the text of the essay no explicit mention of Wordsworth criticism is made, and in the notes the only reference to any critics are to Longinus and Kenneth Burke, both of whom discussed concepts analogous to Hartman's. The references are brief, and serve only to illuminate Hartman's ideas. D. V. Erdman is also thanked for calling Hartman's "attention to a topographical tract published in London, 1796."

The second type of literature, used more extensively, provides contextual information, such as Wordsworth's activities at the time of the poem's composition and the typography of the poem's setting. These documents date primarily from Wordsworth's time. The argument does rely on this historical, nonliterary information, but only in service of Hartman's literary perception.

Third is the corpus of world poetry, quoted substantially throughout. The works of other poets are used to illuminate Wordsworth's work by analogy and contrast. Wordsworth's poetic moment is identified by setting it against other poetic moments. Even though a Hölderlin poem may shed light on a Wordsworth poem, however, they remain separate, with separate lives to be evoked and with no fixed relationship to each other. Hartman does not even attend to the historical task of tracing influence and literary tradition, which would establish at least some formal connections between poems.

The last type of literature is the testimony of Wordsworth and his intimates concerning his state of mind and poetic intentions. This category includes letters, journals, and Wordsworth's other poems when they are used in an evidentiary way. As with the previous types of literature, these documents are used only to illuminate Hartman's perception of the dynamics of the poem under study, and they are interpreted through that perception. Thus Hartman uses a letter in which Wordsworth copied the poem not as an honest reflection of the poet's state of mind, but to recall another time when Wordsworth criticized just such attitudes as expressed in the letter. This juxtaposition, not at all evident in Wordsworth's letter by itself, prepares Hartman's criticism of the absurdity of the conventional reading and introduces the existential paradox which becomes Hartman's theme. Thus all the references, from the most scholary historical geography to the most poetic evocations, serve only to recreate the consciousness Hartman perceives embodied in the poem.

The critical and poetic literatures have an additional important, but implicit, role: the language of the essay invokes and evokes concepts and aesthetic experiences from the entire history of poetry and poetic criticism. The literary vocabulary on one level appears to be purely technical, not unlike the technical vocabularies of molecular biology or sociology. Terms such as *topos, apostrophe, sonnet, turn, enjambment,* and *sublime* are the critic's basic conceptual equipment, learned as part of professional training. On another level, however, the literary terms are more than technical, for each reverberates with former uses and examples. One can know and understand *deoxyribose* on the basis of modern chemistry alone, but to understand the *sublime* one must not only have read Longinus and be familiar with the ensuing critical debate to modern times, one must have experienced a wide range of poems that embody the development and variation of that concept. Even terms that do not refer directly to experience—*sonnet,* for example—rely on wide literary experience. That a poem has fourteen lines, particular rhymes and meters, and a turn is of some outward interest, but of greater importance is that the poem stands in a tradition that began as a representation of love, became increasingly introspective and confessional, then took on religious and philosophic concerns, fell into disuse as uncongenial to the concerns of the eighteenth century, and was finally revived by the romantics. To understand the term *sonnet* is to be sensitive to the wide range of consciousness and experience it has served to realize. Moreover, to understand the term's use in a phrase such as "Though the sonnet as a form is a domesticating device . . ." one must remember the courtly lover torn by love yet graceful in his meters, Donne in religious turmoil tearing at the form, Herbert turning the sonnet in on itself, and Milton in grief, blindness, and civil war finding repose for the space of fourteen lines. In comparison, the sociological and psychological terms used by Merton—e.g., *ambivalence, denial,* and *integration*—do have histories in the literature, and familiarity with the original texts helps reveal how the terms are used, yet the history of the field and the experience of reading the entire corpus is not evoked in the use of the terms.

Because the experience embodied in the poetic literature and interpreted through the critical literature is implicit in the literary vocabulary, the terms take on an added subjective element. Not only does Hartman use the critical vocabulary to elucidate the subjective experience of the poem as he perceives it, his use embodies his own entire experience of literature—his experience of Longinus, Milton, and even Joyce. Moreover, in trying to communicate his perceptions he is relying on the subjective experiences each of his readers have of literature. Each reader has intimate familiarity with a different range of literature, and each reader gives each text a different reading. One's personal anthology personally interpreted comprises the individual's share of the corporate knowledge and is the basis of that individual's sensibility.

In the chain of consciousness from poet to critic to reader, the enterprise rests on the quality of the mediating critic's sensibility. Of course one can read a poem without benefit of a mediating critic, and some schools of thought suggest the best reading is the least tutored. If one turns to a critic, however, the reader must believe that the critic perceives things that would not be apparent to the reader. A critic's persuasiveness, therefore, depends in part on establishing a persona of perceptivity, if not brilliance. Reputation, which is prior to any given article, no doubt plays a significant role in fostering the persona. The content of the essay itself also provides a substantive basis for judging insight. But a persona of sensitivity and brilliance can also be fostered by stylistic habits. Hartman uses several techniques to increase the appearance of density of thought. First, like many critics, he prefers the elliptical argument to the fully delineated. Consider, for example, this sentence: "The word 'Viamala' has punctuated a pathfinding movement of thought and suggests a final station or resting point as it turns the sonnet toward the description of a single scene—though a scene that turns out to be a prospect rather than a terminus, with features that reach beyond time." The single sentence moves through many concepts cast in metaphorical terms, modifying and by the end even reversing the original imagery. A number of the key phrases, such as *pathfinding movement* and *features that reach beyond time,* are neither prepared for earlier in the paper nor spelled out later. No specifics are attached to any of the generalizations of the sentence; the reader is left to figure out how the complex point of the sentence applies to both the rest of the article and to the poem. The interpretation required of the reader is increased because the metaphor of the critical sentence turns the imagery of the poem around, suggesting that the poet, and not the river, is on a pathfinding journey. The sentence can suggest many thoughts to the reader, not all of which may be intended or supported by the argument. In contrast, although the Watson and Crick article does employ ellipsis, the items not spelled out, such as *van der Waals distances,* do have specific, univocal meanings with clear-cut application to the argument of the paper. The ellipsis runs through a single meaning rapidly rather than reverberating with many possible suggested meanings.

In the literary essay reverberative density is also achieved through allusive language, invoking concepts and experiences of other poets and implying connections between words. The *capable negativity* Hartman mentions at the beginning of section III is a Spoonerism for Keats's term "negative capability." The verbal play suggests a deep transformation of Keats's poetics, but the phrase seems actually to have only the simple meaning in the essay that the poem recognizes the impossibility of its task. The last sentence of the essay—"Can we say there is a blessing in its gentle breeze?"—refers to the opening line of *The Prelude* and the title of the essay as well as a contrast to the torrent. Puns run throughout the essay from the first epigraph (where the double meaning of the German *entsprungen* ties the river to a puzzle), through "the chasm that is like a chiasmus" in the fourth section, to the contrast of *luciferic* and *lucyferic* (referring to Wordsworth's Lucy poems) in the next to the last sentence of the essay. A plethora of connections attests to the fertile sensibility of the critic, and sensibility is essentially what the critic has to offer in the essay.

Final Comparisons

To recapitulate the major points of comparison among the three texts analyzed is to notice that the three statements of knowledge are three different things. In mediating reality, literature, audience, and self, each text seems to be making a different kind of move in a different kind of game. All three texts appear to show interest in phenomena which form the topics for the essays (as well as provide the titles). But the phenomena are not equally fixed prior to the essays. The substance DNA and the concept genetic carrier were well known (although not agreed to be synonymous) prior to Watson and Crick's essay. The Wordsworth poem was also well known, but Hartman claims what was known should not count as true knowledge, which can only come in the subjective recreation of the poetic moment. In the ambivalence essay Merton must first establish that the phenomenon exists and is consequential.

The chemical and biological literatures are codified and embedded in the language, problematics, and accepted modes of argumentation; consequently, the DNA essay does not need to discuss explicitly most of the relevant literature except for claims and evidence immediately bearing on the essay's claim. The sociological literature on scientific behavior is more diverse, unsettled, and open to interpretation; therefore, the essay must reconstruct the literature to establish a framework for discussion. The author attempts codification because codification is not a fact going into the essay. The literatures of poetry and its criticism tend to be particularistic and used in particularistic ways; the Wordsworth essay invokes both literatures idiosyncratically and only in support of the critic's vision of the particular poetic moment of consciousness being investigated. Codification, if it can be called that, is entirely personal.

The biological and biochemical audiences share an acceptance of much knowledge, evidence gathering techniques, and criteria of judgment against which to measure Watson and Crick's claims and to suggest how the claims might be applied; therefore, the authors do not urge, but rather leave the audience to judge and act according to the dictates of science. The sociological audience, sharing no uniform framework of thought or criteria of proof, must be urged, persuaded, and directed along the lines of the author's thoughts. The literary audience, concerned with private aesthetic experience, must find the critic's comments plausible, but more important must find the comments enriching the experience of reading; evocation of the richest experience is persuasion.

In their essay Watson and Crick take on a humble yet proud authorial presence: the humble servants of nature and their discipline, filling in only a small piece of a vast puzzle and subject to the hard evidence of nature and the cold judgment of their peers—yet the proud originators of claims that have the potential ring of natural truth and nearly universal professional acceptance. Merton stands more uncertainly before his discipline and nature, neither of which holds the promise of clear-cut judgment and unequivocal support, yet through the force of argument he hopes to establish some certainty. Curiously, the literary critic Hartman, who has the least responsibility to establish certainty, must take on the most demanding role: appearing to have insight greater than that of his readers. Since his contribution cannot be measured in terms of a claim to be judged right or wrong, the quality of his whole sensibility is up for judgment.

The diversity of the knowledge-producing activity embodied in these three texts suggests how important the form of knowledge is. Getting the words right is more than a fine tuning of grace and clarity; it is defining the entire enterprise. And getting the words right depends not just on an individual's choice. The words are shaped by the discipline—in its communally developed linguistic resources and expectations; in its stylized identification and structuring of realities to be discussed; in its literature; in its active procedures of reading, evaluating, and using texts; in its structured interactions between writer and reader. The words arise out of the activity, procedures, and relationships within the community.[16]

The solutions to the problem of how to write embodied in these articles are unique, even within their respective disciplines. Each article speaks to its own moment and own intellectual space; each actively realizes its own goals in that moment and space. Judgments about typicality and typologies of textual forms in different disciplines must be made cautiously, if at all. Certainly to declare any features of these three isolated articles as typical of their disciplines would be folly. Yet each does reveal something about its discipline, not so much in the specific writing choices as in the context in which each of those moves make sense; not in the moves, but in the hints about the gameboard revealed by the moves.

The gameboard of biochemistry as revealed by Watson and Crick's moves is far more defined and stable than the sociological gameboard Merton works on. Watson and Crick can count on many more regularities of the game than Merton can. And the most fundamental of those regularities have to do with the empirical basis of the game. Wat-

---

16. Other insights into disciplinary differences may be found in Becher.

son and Crick can rely on great agreement as to what empirical evidence is relevant to the claim and how that evidence is to be produced, represented, and applied in this situation. All the dimensions of Merton's game, however, are more fluid because of the lack of agreement over the relevant empirical experience, its production, its application, and its representation. And Hartman's game is open to even more idiosyncratic moves because the grounding evidence is displaced from the gameboard into the player; the fundamental reality to be experienced resides within the critic.

The contrasts among these three articles bring sharply into relief the accomplishment of the stable (though not static) rhetorical universe which makes possible Watson and Crick's precise, powerful, and highly successful formulation. The emergence of this rhetorical universe, its implications, and its variations will be the subject of the ensuing chapters. We will see the central role of text and genre not just in responding to the emerging regularities of rhetorical universe, but in helping indeed to create that rhetorical universe.

Appendix

No. 4356 April 25, 1953   NATURE

equipment, and to Dr. G. E. R. Deacon and the captain and officers of R.R.S. *Discovery II* for their part in making the observations.

[1] Young, F. B., Gerrard, H., and Jevons, W., *Phil. Mag.*, **40**, 149 (1920).
[2] Longuet-Higgins, M. S., *Mon. Not. Roy. Astro. Soc., Geophys. Supp.*, **5**, 285 (1949).
[3] Von Arx, W. S., Woods Hole Papers in Phys. Oceanog. Meteor., **11** (3) (1950).
[4] Ekman, V. W., *Arkiv. Mat. Astron. Fysik. (Stockholm)*, **2** (11) (1905).

MOLECULAR STRUCTURE OF NUCLEIC ACIDS

A Structure for Deoxyribose Nucleic Acid

WE wish to suggest a structure for the salt of deoxyribose nucleic acid (D.N.A.). This structure has novel features which are of considerable biological interest.

A structure for nucleic acid has already been proposed by Pauling and Corey[1]. They kindly made their manuscript available to us in advance of publication. Their model consists of three intertwined chains, with the phosphates near the fibre axis, and the bases on the outside. In our opinion, this structure is unsatisfactory for two reasons: (1) We believe that the material which gives the X-ray diagrams is the salt, not the free acid. Without the acidic hydrogen atoms it is not clear what forces would hold the structure together, especially as the negatively charged phosphates near the axis will repel each other. (2) Some of the van der Waals distances appear to be too small.

Another three-chain structure has also been suggested by Fraser (in the press). In his model the phosphates are on the outside and the bases on the inside, linked together by hydrogen bonds. This structure as described is rather ill-defined, and for this reason we shall not comment on it.

We wish to put forward a radically different structure for the salt of deoxyribose nucleic acid. This structure has two helical chains each coiled round the same axis (see diagram). We have made the usual chemical assumptions, namely, that each chain consists of phosphate diester groups joining β-D-deoxyribofuranose residues with 3',5' linkages. The two chains (but not their bases) are related by a dyad perpendicular to the fibre axis. Both chains follow right-handed helices, but owing to the dyad the sequences of the atoms in the two chains run in opposite directions. Each chain loosely resembles Furberg's[2] model No. 1; that is, the bases are on the inside of the helix and the phosphates on the outside. The configuration of the sugar and the atoms near it is close to Furberg's 'standard configuration', the sugar being roughly perpendicular to the attached base. There

This figure is purely diagrammatic. The two ribbons symbolize the two phosphate—sugar chains, and the horizontal rods the pairs of bases holding the chains together. The vertical line marks the fibre axis

is a residue on each chain every 3·4 A. in the z-direction. We have assumed an angle of 36° between adjacent residues in the same chain, so that the structure repeats after 10 residues on each chain, that is, after 34 A. The distance of a phosphorus atom from the fibre axis is 10 A. As the phosphates are on the outside, cations have easy access to them.

The structure is an open one, and its water content is rather high. At lower water contents we would expect the bases to tilt so that the structure could become more compact.

The novel feature of the structure is the manner in which the two chains are held together by the purine and pyrimidine bases. The planes of the bases are perpendicular to the fibre axis. They are joined together in pairs, a single base from one chain being hydrogen-bonded to a single base from the other chain, so that the two lie side by side with identical z-co-ordinates. One of the pair must be a purine and the other a pyrimidine for bonding to occur. The hydrogen bonds are made as follows : purine position 1 to pyrimidine position 1 ; purine position 6 to pyrimidine position 6.

If it is assumed that the bases only occur in the structure in the most plausible tautomeric forms (that is, with the keto rather than the enol configurations) it is found that only specific pairs of bases can bond together. These pairs are : adenine (purine) with thymine (pyrimidine), and guanine (purine) with cytosine (pyrimidine).

In other words, if an adenine forms one member of a pair, on either chain, then on these assumptions the other member must be thymine ; similarly for guanine and cytosine. The sequence of bases on a single chain does not appear to be restricted in any way. However, if only specific pairs of bases can be formed, it follows that if the sequence of bases on one chain is given, then the sequence on the other chain is automatically determined.

It has been found experimentally[3,4] that the ratio of the amounts of adenine to thymine, and the ratio of guanine to cytosine, are always very close to unity for deoxyribose nucleic acid.

It is probably impossible to build this structure with a ribose sugar in place of the deoxyribose, as the extra oxygen atom would make too close a van der Waals contact.

The previously published X-ray data[5,6] on deoxyribose nucleic acid are insufficient for a rigorous test of our structure. So far as we can tell, it is roughly compatible with the experimental data, but it must be regarded as unproved until it has been checked against more exact results. Some of these are given in the following communications. We were not aware of the details of the results presented there when we devised our structure, which rests mainly though not entirely on published experimental data and stereochemical arguments.

It has not escaped our notice that the specific pairing we have postulated immediately suggests a possible copying mechanism for the genetic material.

Full details of the structure, including the conditions assumed in building it, together with a set of co-ordinates for the atoms, will be published elsewhere.

We are much indebted to Dr. Jerry Donohue for constant advice and criticism, especially on interatomic distances. We have also been stimulated by a knowledge of the general nature of the unpublished experimental results and ideas of Dr. M. H. F. Wilkins, Dr. R. E. Franklin and their co-workers at

King's College, London. One of us (J. D. W.) has been aided by a fellowship from the National Foundation for Infantile Paralysis.

J. D. WATSON
F. H. C. CRICK

Medical Research Council Unit for the
Study of the Molecular Structure of
Biological Systems,
Cavendish Laboratory, Cambridge.
April 2.

[1] Pauling, L., and Corey, R. B., *Nature*, **171**, 346 (1953); *Proc. U.S. Nat. Acad. Sci.*, **39**, 84 (1953).
[2] Furberg, S., *Acta Chem. Scand.*, **6**, 634 (1952).
[3] Chargaff, E., for references see Zamenhof, S., Brawerman, G. and Chargaff, E., *Biochim. et Biophys. Acta*, **9**, 402 (1952).
[4] Wyatt, G. R., *J. Gen. Physiol.*, **36**, 201 (1952).
[5] Astbury, W. T., Symp. Soc. Exp. Biol. 1, Nucleic Acid, 66 (Camb. Univ. Press, 1947).
[6] Wilkins, M. H. F., and Randall, J. T., *Biochim. et Biophys. Acta*, **10**, 192 (1953).

Molecular Structure of Deoxypentose Nucleic Acids

WHILE the biological properties of deoxypentose nucleic acid suggest a molecular structure containing great complexity, X-ray diffraction studies described here (cf. Astbury[1]) show the basic molecular configuration has great simplicity. The purpose of this communication is to describe, in a preliminary way, some of the experimental evidence for the polynucleotide chain configuration being helical, and existing in this form when in the natural state. A fuller account of the work will be published shortly.

The structure of deoxypentose nucleic acid is the same in all species (although the nitrogen base ratios alter considerably) in nucleoprotein, extracted or in cells, and in purified nucleate. The same linear group of polynucleotide chains may pack together parallel in different ways to give crystalline[1-3], semi-crystalline or paracrystalline material. In all cases the X-ray diffraction photograph consists of two regions, one determined largely by the regular spacing of nucleotides along the chain, and the other by the longer spacings of the chain configuration. The sequence of different nitrogen bases along the chain is not made visible.

Oriented paracrystalline deoxypentose nucleic acid ('structure B' in the following communication by Franklin and Gosling) gives a fibre diagram as shown in Fig. 1 (cf. ref. 4). Astbury suggested that the strong 3·4-A. reflexion corresponded to the internucleotide repeat along the fibre axis. The ~ 34 A. layer lines, however, are not due to a repeat of a polynucleotide composition, but to the chain configuration repeat, which causes strong diffraction as the nucleotide chains have higher density than the interstitial water. The absence of reflexions on or near the meridian immediately suggests a helical structure with axis parallel to fibre length.

Diffraction by Helices

It may be shown[5] (also Stokes, unpublished) that the intensity distribution in the diffraction pattern of a series of points equally spaced along a helix is given by the squares of Bessel functions. A uniform continuous helix gives a series of layer lines of spacing corresponding to the helix pitch, the intensity distribution along the nth layer line being proportional to the square of J_n, the nth order Bessel function. A straight line may be drawn approximately through

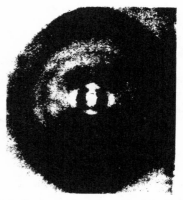

Fig. 1. Fibre diagram of deoxypentose nucleic acid from *B. coli*. Fibre axis vertical

the innermost maxima of each Bessel function and the origin. The angle this line makes with the equator is roughly equal to the angle between an element of the helix and the helix axis. If a unit repeats n times along the helix there will be a meridional reflexion (J_0^2) on the nth layer line. The helical configuration produces side-bands on this fundamental frequency, the effect[5] being to reproduce the intensity distribution about the origin around the new origin, on the nth layer line, corresponding to C in Fig. 2.

We will now briefly analyse in physical terms some of the effects of the shape and size of the repeat unit or nucleotide on the diffraction pattern. First, if the nucleotide consists of a unit having circular symmetry about an axis parallel to the helix axis, the whole diffraction pattern is modified by the form factor of the nucleotide. Second, if the nucleotide consists of a series of points on a radius at right-angles to the helix axis, the phases of radiation scattered by the helices of different diameter passing through each point are the same. Summation of the corresponding Bessel functions gives reinforcement for the inner-

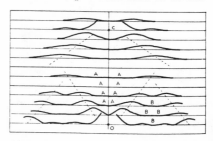

Fig. 2. Diffraction pattern of system of helices corresponding to structure of deoxypentose nucleic acid. The squares of Bessel functions are plotted about 0 on the equator and on the first, second, third and fifth layer lines for half of the nucleotide mass at 20 A. diameter and remainder distributed along a radius, the mass at a given radius being proportional to the radius. About C on the tenth layer line similar functions are plotted for an outer diameter of 12 A.

The Ambivalence of Scientists
1963

Many of the endlessly recurrent facts about multiples and priorities are readily accessible—in the diaries and letters, the note-books, scientific papers, and biographies of scientists. This only compounds the mystery of why so little systematic attention has been accorded the subject. The facts have been noted, for they are too conspicuous to remain unobserved, but then they have been quickly put aside, swept under the rug, and forgotten. We seem to have here something like motivated neglect of this aspect of the behavior of scientists and that is precisely the hypothesis I want to examine now.

This resistance to the study of multiples and priorities can be conceived as a resultant of intense forces pressing for public recognition of scientific accomplishments that are held in check by countervailing forces, inherent in the social role of scientists, which press for the modest acknowledgment of limitations, if not for downright humility. Such resistance is a sign of malintegration of the social institution of science which incorporates potentially incompatible values: among them, the value set upon originality, which leads scientists to want their priority to be recognized, and the value set upon due humility, which leads them to insist on how little they have in fact been able to accomplish. To blend these potential incompatibles into a single orientation and to reconcile them in practice is no easy matter. Rather, as we shall now see, the tension between these kindred values creates an inner conflict among men of science who have internalized both of them. Among other things, the tension generates a

First published as a part of "Resistance to the Systematic Study of Multiple Discoveries in Science." *European Journal of Sociology* 4 (1963): 250–82; reprinted with permission. A condensed version of part of this paper appears under this title in the *Bulletin* of the Johns Hopkins Hospital, 112 (February 1963): 77–97.

distinct resistance to the systematic study of multiples and often associated conflicts over priority.[1]

Various kinds of overt behavior can be interpreted as expressions of such resistance. For one thing, it is expressed in the recurrent pattern of trying to trivialize or to incidentalize the facts of multiples and priority in science. When these matters are discussed in print, they are typically treated as though they were either rare and aberrant (although they are extraordinarily frequent and typical) or as though they were inconsequential both for the lives of scientists and for the advancement of science (although they are demonstrably significant for both).

Understandably enough, many scientists themselves regard these matters as unfortunate interruptions to their getting on with the main job. Kelvin, for example, remarks that "questions of priority, however interesting they may be to the persons concerned, sink into insignificance" as one turns to the proper concern of advancing knowledge.[2] As indeed they do: but sentiments such as these also pervade the historical and sociological study of the behavior of scientists so that systematic inquiry into these matters also goes by default. Or again, it is felt that "the question of priority plays only an insignificant role in the scientific literature of our time"[3] so that, once again, this becomes regarded as a subject which can no longer provide a basis for clarifying the complex motivations and behavior of scientists (if indeed it ever was so regarded).

Now the practice of seeking to trivialize what can be shown to be significant is a well-known manifestation of resistance. Statements of this sort read almost as though they were a paraphrase of the old maxim that the law does not concern itself with exceedingly small matters; *de minimis non curat scientia* [*lex*]. Not that there has been a conspiracy of silence about these intensely human conflicts in the world of the intellect and especially in science. These have been far too conspicuous to be denied altogether. Rather, the repeated conflict behavior of great and small men of science has been incidentalized as not reflecting any conceivably significant aspects of their role as scientists.

Resistance is expressed also in various kinds of distortions: in motivated misperceptions or in an hiatus in recall and reporting. It often leads to those wish-fulfilling beliefs and false memories that we describe as illusions. And of such behavior the annals that treat of multiples and priorities are uncommonly full. So much so that I have arrived at a rule of thumb that

1. This paragraph draws upon a fuller account of the workings of these values in the social institution of science in "Priorities in Scientific Discovery," chapter 14 of this volume.
2. Silvanus P. Thompson, *The Life of William Thomson, Baron Kelvin of Largs* (London: Macmillan, 1910), 2:602.
3. Otto Blüh, "The Value of Inspiration: A Study of Julius Robert Mayer and Josef Popper-Lynkeus," *Isis* 43 (1952): 211–20, at 211.

seems to work out fairly well. The rule is this: whenever the biography or autobiography of a scientist announces that he had little or no concern with priority of discovery, there is a reasonably good chance that, not many pages later in the book, we shall find him deeply embroiled in one or another battle over priority. A few cases must stand here for many:

Of the great surgeon, W. S. Halsted (who together with Osler, Kelly, and Welch founded the Johns Hopkins Medical School), Harvey Cushing writes: he was "overmodest about his work, indifferent to matters of priority."[4] Our rule of thumb leads us to expect what we find: some twenty pages later in the book in which this is cited, we find a letter by Halsted about his work on cocaine as an anesthesia: "I anticipated all of Schleich's work by about six years (or five). . . . [In Vienna,] I showed Wölfler how to use cocaine. He had declared that it was useless in surgery. But before I left Vienna he published an enthusiastic article in one of the daily papers on the subject. It did not, however, occur to him to mention my name."[5]

Or again, the authoritative biography of that great psychiatrist of the Salpêtrière, Charcot, approvingly quotes the eulogy which says, among other things, that despite his many discoveries, Charcot "never thought for a moment to claim priority or reward." Alerted by our rule of thumb, we find some thirty pages later an account of Charcot insisting on his having been the first to recognize exophthalmic goiter and, a little later, emphatically affirming that he "would like to claim priority" for the idea of isolating patients who are suffering from hysteria.[6]

But perhaps the most apt case of such denial of an accessible reality is that of Ernest Jones, writing in his comprehensive biography that "although Freud was never interested in questions of priority, which he found merely boring"—surely this is a classic case of trivialization at work—"he was fond of exploring the source of what appeared to be original ideas, particularly his own."[7] This is an extraordinarily illuminating statement. For, of course, no one could have "known" better than Jones—"known" in the narrowly cognitive sense—how very often Freud turned to matters of priority: in his own work, in the work of his colleagues (both friends and enemies), and in the history of psychology altogether.

---

4. In his magisterial biography, *Harvey Cushing* (Springfield: Charles C. Thomas, 1946), pp. 119–20, John F. Fulton describes Cushing's biographical sketch of Halsted, from which this excerpt is quoted, as "an excellent description."

5. Ibid., p. 142.

6. Georges Gullain, *J.-M. Charcot: His Life, His Work*, ed. and trans. Pearce Bailey (New York: Paul B. Hoeber, 1959), pp. 61, 95–96, 142–43.

7. Ernest Jones, *Sigmund Freud: Life and Work*, 3 vols. (London: Hogarth Press, 1957), 3:105. Contrast David Riesman, who takes ample note of Freud's interest in priority, in *Individualism Reconsidered* (Glencoe: The Free Press, 1954), pp. 314–15, 378.

One: Writing Matters

GEOFFREY H. HARTMAN

Blessing the Torrent: On Wordsworth's Later Style

> Ein Räthsel ist Reinentsprungenes
>   Hölderlin
>
> The river is fateful,
> Like the last one. But there is no ferryman.
> He could not bend against its propelling force.
>   Wallace Stevens
>
> riverrun, past Eve and Adam's
>   James Joyce

I

> How art thou named? In search of what strange land,
> From what huge height, descending? Can such force
> Of waters issue from a British source,
> Or hath not Pindus fed thee, where the band
> Of Patriots scoop their freedom out, with hand
> Desperate as thine? Or come the incessant shocks
> From that young Stream, that smites the throbbing rocks,
> Of Viamala? There I seem to stand,
> As in life's morn; permitted to behold,
> From the dread chasm, woods climbing above woods,
> In pomp that fades not; everlasting snows;
> And skies that ne'er relinquish their repose;
> Such power possess the family of floods
> Over the minds of Poets, young or old!

IF THE TWO opening lines of this sonnet had been an untitled fragment, their referent would be uncertain. Whom is the poet talking to, what "thou" is addressed? Is the force natural or divine? And why should the act of naming be important?

But the lines are part of a sonnet titled specifically "To the Torrent at the Devil's Bridge, North Wales, 1824."[1] Moreover, as line 2 runs into line 3, the "force" is identified as a "force of waters," that is, a river or, more precisely, a waterfall. ("Force" was dialect in the North of England for "waterfall.") Describing the impact of a different sight, though it also involves naming or labeling, Wordsworth writes: "My mind turned round / As with the might of waters."[2] In the present poem the verse line itself turns round and naturalizes the poet's wonderment. Uncertainty of reference gives way to a well-defined personal situation that is easily described, though less easily understood.

II

In September 1824 Wordsworth traveled through North Wales on one of the many sentimental journeys he was fond of taking. They were sentimental in the sense of covering old ground in order to reflect on the changes time had wrought in him or the scene; and "Tintern Abbey" was the earliest and most remarkable issue of such memorial visits. On this particular trip Wordsworth saw a friend of his youth, Robert Jones, who had shared with him two determining moments in his life: the ascent of Snowdon in 1791 and the tour of 1790 through revolutionary France and the Alps, with its complex seeding in his mind of experiences in the Simplon/Viamala region. Both journeys were now over thirty years old, and had already been described: the Snowdon climb in Book XIII of the unpublished *Prelude*, and the Continental tour in Book VI, as well as in *Descriptive Sketches* (1793). In 1820, moreover, Wordsworth retraced his journey through the Alps with his sister, Dorothy, and his wife, Mary, both of whom kept journals of the visit.

On a portion of this new trip to Wales the poet was accompanied by Robert Jones; and it was with him (as well as with Mary and Dora Wordsworth) that he viewed the waterfall described in the sonnet. No wonder, then, that as he stands at the torrent's edge, he feels he is back "in life's morn," and what he sees with the eyes of an aging man (he is fifty-four years old) is not a local river but "the young stream that smites the throbbing rocks, / Of Viamala,"

What Written Knowledge Does

which had giddied him when his own mind was young and in turmoil.

We can normalize this sonnet then; and the fact that it is a sonnet, one of so many written during the poet's later career, tempts us to give it a nod of esteem and pass on. There is little on first reading to hold the attention. Formal features of a conventional sort abound: opening and closing apostrophes; a first half comprising a cascade of questions that receive their resolution or coda in the second half, which is introduced by an efficient turn in the eighth line; enjambments that reflect the passion or perplexity of the utterance; and the abbreviated effect of sublimity created by a broken series of descriptive phrases characterizing his memory of the Viamala region (ll. 10-12).

In line with this we can also normalize the initial "How art thou named?" as a rhetorical or animating movement that is a residue of sublime style and so risks bathos. The poet must have known the name; he is obtruding the question to express a momentary ecstacy or disorientation. Still, this trace of sublime diction makes us uneasy; and the discomfort spreads if we read the letter Wordsworth wrote to his noble painter friend, Sir George Beaumont. We learn that "It rained heavily in the night, and we saw the waterfalls in perfection. While Dora was attempting to make a sketch from the chasm in the rain, I composed by her side the following address to the torrent."[3] There is a calming or distancing effect in the phrase "waterfalls in perfection" that reminds us of Wordsworth's own earlier critique of the picturesque artist's superficial mastery of landscape; there is also the subdued paradox of making "a sketch from the chasm" and "composing" an "address to the torrent."

Even if "compose" is used here without the overtone of "repose," two further sonnets written during the visit to Wales stress that "expression of repose" with which nature or time endows wild places.[4] And there is, I would suggest, something faintly absurd about an "address to the torrent." How does one address a *torrent*? To do so, one hears Alice or some Wonderland Creature saying—to do so one must have its name and know where it lives. And, indeed, Wordsworth is not asking for an actual name. His opening question is in search of something existential rather than informational. If Lucy lives among untrodden ways near the Springs of Dove, where do I live? Where now, in 1824? Near what springs or feeding-sources? Like the torrent itself, he seems uncertain of origin or direction, and the questioning mood of the next lines confirms that.

Yet his opening cry is not "What art thou?" nor as in a moving poem of Hölderlin's "Where art thou?" ("Wo bist Du? Trunken dämmert die Seele mir . . ."). It is "How art thou named?" What force, then, lies in the naming of a force? One of the other sonnets written in Wales describes a stream that mingles with the Dee and flows along the "Vale of Meditation," or "Glyn Myrvr"—a "sanctifying name," comments Wordsworth. As in his early "Poems on the Naming of Places" (1800), he then invents a name in Welsh for the place he wishes to single out. Yet the sonnet before us bestows no name, even though "Devil's Bridge" and "Viamala" might have encouraged a man called Wordsworth.

To "address the torrent" means, clearly enough, to domesticate the sublime: to contain it in the form of picturesque sketch or reflective sonnet; and the opening exclamation, at once perplexed and marveling, is expressive of Wordsworth's problem. The sublime, moreover, is not a quality of place alone but also of time: a bewildering memory seems to decompose the name of the torrent or any that might be given. Though the sonnet as a form is a domesticating device and though Wordsworth emulates Milton's "soul-animating strains" when he first chooses the sonnet as a verse instrument, his diction falters or condenses under the strain. But the significance of this cannot be discussed without attending carefully to the strangeness of Wordsworth's later verse, indeed to the verbal style of the sonnet in its entirety, from title to final exclamation. The title already suggests the problems of (1) naming and (2) localization. It anticipates the question of how a "force" can be localized in place, time, or language.

III

It is when we realize what naming implies that this poem betrays its significant failure, its capable negativity: it cannot name the stream. Acts

PART TWO

THE EMERGENCE OF LITERARY AND SOCIAL FORMS IN EARLY MODERN SCIENCE

3

REPORTING THE EXPERIMENT

THE CHANGING ACCOUNT OF

SCIENTIFIC DOINGS IN THE

PHILOSOPHICAL TRANSACTIONS OF

THE ROYAL SOCIETY, 1665–1800

Experimental reports tell a special kind of story, of an event created so that it might be told. The story creates pictures of the immediate laboratory world in which the experiment takes place, of the happenings of the experiment, and of the larger, structured world of which the experimental events are exemplary. The story must wend its way through the existing knowledge and critical attitude of its readers in order to say something new and persuasive, yet can excite imaginations to see new possibilities in the smaller world of the laboratory and the larger world of nature. And these stories are avidly sought by every research scientist who must constantly keep up with the literature.

If each individual writer does not think originally and creatively about how to master recalcitrant language in order to create such powerful stories, it is only because the genre already embodies the linguistic achievement of the three hundred years since the invention of the scientific journal necessitated the invention of the scientific article.[1] The experimental report, as any other literary genre, was invented in response to a literary situation and evolved through the needs, conceptions, and creativity of the many authors who took it up. The corpus of the genre is not only immense, it is rich and varied, synchronically and diachron-

---

1. For a comprehensive view of the rise of scientific journals see David A. Kronick, *A History of Scientific and Technical Periodicals*. A. J. Meadows, *Development of Science Publishing in Europe* suggests some of the historical variety of scientific publication. A. J. Meadows, *Communication in Science*, and William Garvey, *Communication: The Essence of Science*, describe some features of the current system of journal communication.

Two: Literary and Social Forms in Early Modern Science

ically. Despite familiar pedagogical prescriptions, the experimental report is no single narrow form.

Fiction, Nonfiction, and Accountabilities

The extent of literary construction is not diminished for the genre's being nonfiction. Nonfiction—a concept defined only negatively, for its not being the regular meat of literary investigation—presents serious literary questions of the representations of worlds in words. Given modern critical understanding and modern epistemology, the traditional distinction between that which is made up (and therefore of literary interest) and that which reflects the world (and therefore trivial linquistically), obscures rather than illuminates. Few today would contend that signs are unmistakable and predetermined reflections of things.

Some contemporary theorists would in fact reduce all texts to fiction, claiming reference itself a fiction.[2] While much may be said for this position, nonfiction creation incorporates procedures tying texts to various realities. An introspective phenomenology of religious experience or a political speech or an annual report is no less nonfiction than an account of doings in a room at a physics laboratory. Differences of nonfictions hang on differences of accountabilities (of both degree and kind) that connect texts to the various worlds they represent and act on.

The concept of accountabilities will run throughout this book, as we look at how the various writers and readers, situated in certain communities, following the habits and procedures of observation and representation, are restricted in what they say, do and think by empirical experience. Many mechanisms (of training, argument, criticism, normative behavior, application, sanction, and reward) realize and elaborate this fundamental commitment of the discourse. The chapters that follow will examine numerous accounts of empirical experience that play crucial roles in scientific communication, the emergence of standards and procedures for those accounts, means for reconciling accounts and developing more generalized accounts consistent with more specific accounts, situations where discrepancies or uncertainties within or between accounts call for further accounts. The scientific enterprise is built on accounts of nature, and the development of scientific discourse can be seen as the development of ways of presenting accounts.

2. Core documents for this position are Jacques Derrida, *Of Grammatology*, and Michel Foucault, *The Order of Things*.

Reporting the Experiment

Other types of communities may have other fundamental accountabilities and means of enforcing and elaborating these accountabilities. Sacred texts, for example, provide the constant ground, pattern, and reference point for communication in some religious communities; all discourse is held accountable to the sacred text by means of discourse style, conceptual assumptions, overt quotation and paraphrase, psychological rewards of certainty, social rewards for piety, and ostracism for blasphemy. Legal discourse is held accountable on one hand to a hierarchically arranged series of court decisions, laws, and constitutions, and on the other to evidence gathered through procedures defined by the system and represented in a manner established by tradition and explicit rule. In certain types of literary critical discourse, as exemplified by one text examined in the last chapter, the fundamental reference point is a subjective experience of the text; Hartman's article mobilizes many mechanisms to identify that experience and transfer it to the reader. The whole enterprise rests on that experience and is elaborated through the socially recognized means of developing such accounts.

As developed here, the concept of accountabilities is closely related to Ludwik Fleck's definition of a fact as a "stylized signal of resistance in thinking" within a thought collective (98). That is, following the thought style (including styles of perception, cognition, and representation) of a group of people engaged in intellectual interchange, certain statements limit what can be appropriately said and thought within the collective. Certain of the constraints are what Fleck calls active elements, actively produced by the thought style; others are passive, where the discourse system so to speak bumps into objects outside itself, which by the thought style must be respected by the thought collective. Facts are perceived and represented through the actively constructed thought style, but reflect the passive constraint imposed by external conditions. (A more complete discussion of Fleck's analysis of facts is presented in chapter 11.)

These facts accepted by the community form the basis for the accountability, as I use the term. These facts, outside the immediate active elements of discourse, must be brought into the discourse and accounted for. The process of holding the text accountable to these facts serves to shape the discourse. The mechanisms of accountability permeate the creation, reception, and textual form of statements in the collectives holding themselves accountable in this way.

Fleck goes on to characterize the thought style of contemporary science as actively seeking to include a maximum of passive elements despite their tendency to disrupt other accepted active elements. Put more simply, the fundamental commitment is to empirical experience. Scientific discourse, therefore, is built on accountability to empirical fact (as of course characterized within the thought style of science) over all other possible accountabilities (such as to ancient texts, theory, social networks, grant-giving agencies), and must subordinate other forms of accountability (that is, those other forms of accountability which do form part of the scientific thought style) to the empirical accountability.

The Experimental Report as a Historical Creation

Although many kinds of communication pass within scientific communities, experimental reports are close to the heart of the accountability process, for experimental reports present primary accounts of empirical experience. Experimental reports attach themselves to the nature that surrounds the text through the representation of the doings, or experiment. How does the world of events get reduced to the virtual world of words? How did the conventions and procedures for this reduction develop? What are the motives and assumptions implicit in the rhetoric and procedure? And what are the accountabilities that limit statements, ensuring the influence of the evidence of the world on human conception? These are equally questions of literary theory and rhetoric as of philosophy of science, for what appears to philosophy of science as the problem of empiricism, appears to rhetoric as the problem of persuasive evidence, and to literary theory as the problem of representation.

One place to turn for answers to these questions is the early history of the experimental report, for the formation of a genre reveals the forces to which textual features respond. A genre consists of something beyond simple similarity of formal characteristics among a number of texts. A genre is a socially recognized, repeated strategy for achieving similar goals in situations socially perceived as being similar (Miller). A genre provides a writer with a way of formulating responses in certain circumstances and a reader a way of recognizing the kind of message being transmitted. A genre is a social construct that regularizes communication, interaction, and relations. Thus the formal features that are shared by the corpus of texts in a genre and by which we usually recognize a text's inclusion in a genre, are the linguistic/symbolic solution to a problem in social interaction.

That a well-established, successful genre is usually realized in relatively static formal features should not hide the social meaning and dynamics of a genre, no more than the active reality of a performed Beethoven quartet should be obscured by the sheet music. By examining the emergence of a genre we can identify the kinds of problems the genre was attempting to solve and how it went about solving them. The history of the experimental report shows how a certain kind of detailed picture of a laboratory event became the standard and how particular information became essential to a successful telling. We can also see forming, as the genre takes shape, a particular literary community with certain critical expectations.

The *Philosophic Transactions of the Royal Society of London*, the first scientific journal in English, carries the main line of the development of scientific journal writing in English through the nineteenth century. Here I follow the development of the genre of experimental report in the pages of the *Transactions* from its founding in 1665 until 1800, when a number of familiar features of the experimental report were firmly in place.

This chapter focuses entirely on the internal development of the genre. Although the genre of experimental article has origins in essay, epistolary, and journalistic writing of the seventeenth century (Frank; Houghton; Kronick; Paradis; Sutherland), the internal dynamics of scientific communication within a journal forum reshape the initial sources to create a new communicative form, powerful enough to influence other forms of communication and the social structure of the community which uses it. Chapter 4 will begin to explore the relations between the existing book publication of scientific arguments and the newly emerging journal article. Chapter 5 will consider the kind of social structure out of which journal publication arose and the power of journal communication to transform the social structure of science.

Method

This study is based on an examination of all articles (about 1000 altogether over 7000 pages) in volumes 1, 5, 10, 15, 20, 25, 30, 35, 40, 50, 60, 70, 80, and 90 of the *Transactions*. From these volumes all articles using the word experiment either in the title or running text were then selected for closer examination. Then those articles providing only secondary accounts of experiments were eliminated, leaving only articles written by the experimenter reporting on new experiments. This procedure left a remainder of about 100 articles to be analyzed.

Because of the changing character of the writing in the articles and because of the individual character of each separate article no quantitative comparisons appeared useful, so I resorted to the traditional method of literary criticism, descriptive analysis of each of the separate articles. This method did allow me to explore the varying features of writing as they presented themselves. However, such individual description makes generalization difficult. In order to facilitate the comparison and continuity among the many cases, I narrowed my descriptive analysis to a set of specific questions.

1. To what kind of event does the term "experiment" refer?
2. How fully and in what manner are experimental events described?
3. How fully are apparatus and methodology described? How fully and in what way are methodological concerns discussed?
4. How precisely and completely are results presented? What criteria of selectivity are used? How much and what kind of discussion and interpretation are present?
5. Is the experiment presented as a single event or as part of a series of experiments? In a series, what is the principle of continuity?
6. How is the account of the experiment organized? How are series of experiments organized? Where does the account of experiment or experiments fit within the organization of the entire article?
7. What is the rhetorical function of the experiment within the article?

To facilitate the organization of the material of these separate analyses, particularly with the intention of clarifying historical trends, I then synthesized the analyses from each volume examined, thus forming generalizations about the character of the experimental reporting in each time period. From this collection of chronologically arranged syntheses I extracted the major themes and trends as presented below. The story I will be presenting thus has been filtered several times through my own personal interpretive, selective, and synthetic judgments. I will present detailed evidence from the texts to illustrate and support the story I present, but I will not be presenting all the trees in the forest. If I were to tell more I would risk the reader losing sight of the shape of the forest I believe I have found. On the other hand, I have no more impersonal way of either reconnoitering the shape of the forest or communicating and demonstrating that shape. This is always the dilemma of attempting to make sense of historical and literary material which incorporates the complex actions of many individuals. In terms of persuasion, this essay must rest in the short term only on the impression it gives of a plausible story and in the long term only on whether others crossing the same terrain find the shapes presented here recognizable and useful.

Another consequence of working from individual accounts of the products of many individuals, each reacting to specifics of individual situations, is that the overall trends are likely to wash out many individual variations as well as to appear more uniform than they in fact are. When looking at all the trees in the forest, I find a somewhat more ragged shape than will emerge here, although I will attempt to indicate where the raggednesses are.

The Changing Experiment

Despite our current belief in experiment as one of the foundations of science, only a small part of the volumes examined up to 1800 were devoted to reporting on experiments. Both in terms of the percentage of total articles and percentage of pages, experimental articles accounted for only 5 to 20 percent of each volume through volume 80. Only in volume 90, opening the nineteenth century, did the percentages rise substantially to 39 percent of the articles and 38 percent of the pages.

Until 1800, however, it is clear that experiments were only one of many types of information to be transmitted among those interested in science. The most articles and pages were devoted to observations and reports of natural events, ranging from remarkable fetuses and earthquakes, through astronomical sightings, anatomical dissections, and microscopical observations. Human accomplishments received attention with accounts of technological and medical advances, travelogues of journeys to China and Japan, and an interview with the prodigy Mozart. The reportable business of natural philosophers was hardly restricted to experimenting or even theorizing, which received even less space than experiments.

The relative paucity of experimental accounts should remind us how much the importance we attach to experiments is a function of the rise of the experimental article as a favored way of formulating and discussing science. Although experiments may have their ancient precursors and early books may have experimental accounts embedded within them, the creation of the experimental article has helped create our modern concept of experiment.

Those reported events identified as experiments change in character over the period 1665–1800. The definition of experiment moves from any made or done thing, to an intentional investigation, to a test of a theory,

Two: Literary and Social Forms in Early Modern Science

to finally a proof of, or evidence for, a claim. The early definitions seem to include any disturbance or manipulation of nature, not necessarily focused on demonstration of any stated preexisting belief, nor even with the intention of discovery. With time, experiments are represented as more clearly investigative, corroborative, and argumentative.

In the first volume of the *Transactions*, a number of experiments reported are simply cookbook recipes for creating marvellous effects or effects of practical use, such as the directions for coloring marble internally "of use to artisans" (1:125).[3] Elsewhere experiments are a method of investigating nature, treated on a par with observations, as in the formula often appearing in the pages of the journals: "experiments and observations." Observations were made upon undisturbed or unmanipulated nature while experiments involved human intervention. That intervention need not imply intention of investigation; for example, one series of experiments grew out of a cook's pickling of mackerels. Only after several days, when the cook noticed that the broth had turned luminescent, did the master of the house identify the phenomenon as something worth observing (1:226–28).

By volumes 5 and 10 the definition of experiment had narrowed in most cases to a conscious investigation of phenomena involving some doings or manipulations, even though cookbook novelties appeared as late as volume 20 (20:42–44, 87–90, 363–65). These experiments, however, are presented as simply allowing the conditions for brute nature to reveal itself. The meaning of the experiment is simply what is observed upon its occurrence. For example, in volume 10 Christian Huygens and Denis Papin report on a series of experiments to "know, whether the Vacuum would be of use to the Preservation of Bodies" so they placed various flowers, fruits, and other comestibles in vacuums for various periods of time and observed (10:492–95). Retrospectively, such experiments seem part of a broader investigation of the atmosphere, but nowhere do the reports of these or similar vacuum experiments suggest that questions, theories, problems, or hypotheses were being explicitly explored or tested.

Only in cases of overt controversy are assumptions or hypotheses explicitly set out to be tested, for then the experiment becomes a means of adjudicating between two or more proposed views. Again in volume 10, as part of a report of another series of vacuum experiments, Huygens and Papin, prompted by comments by another investigator,

---

3. This and similar articles are part of a regular program of reporting on the trades, according to Kathleen H. Ochs, "The Failed Revolution in Applied Science," and Marie Boas Hall, "Oldenburg, the *Philosophical Transactions*, and Technology."

present their alternative view of the reasons for collapse of lungs and then describe a specific experiment that led them to their conclusion.

By volume 20 several experiments have clear hypothesis-testing or debate-solving functions. Experiments are being recognized as created events designed with specific claims about nature in mind. In volume 25, for example, Francis Hauksbee[4] comments, with some pleasure, at the use of experiment as a way to test hypotheses: ". . . the greatest Satisfaction and Demonstration that can be given for the Credit of any Hypothesis, is, That the Experiments made to prove the same, agree with it in all Respects, without force" (25:2415–17).

In volume 30 five articles place their experiments in the context of extensive discussions of debates which the experiments are set to resolve, such as whether a vacuum is truly empty. In this particular case, the experimenter Jean T. Desaguliers spends a full page reporting on a previous vacuum experiment he had made and the particular objections a group of plenists had to his procedure, against which he sets his current work (30:717–18). In most cases the experiments provide rather direct observations concerning the issue at hand, as in the preceding example where pairs of different objects were dropped in an evacuated column to see whether the time of fall were the same for each. But at least in one case the experiments were at some remove from the issue of contention, indicating that experiments were now accepted within the context of a complex of accepted knowledge rather than simply as brute demonstrations. Desaguliers, in order to dispute Leibniz's explanation of barometric measurements during rain, enters into a theoretical discussion of the weights of bodies falling through a medium, which discussion he then supports through a series of ingenious experiments employing neither barometers nor atmosphere. The experiment stands on established background knowledge for its construction and interpretation (30:570–79).

At this point the experiment's role of adjudicating disputes as to the brute truth of nature starts to shift toward establishing the truth of general propositions that are not necessarily disputed by anyone. Experiments stop being a clear window to a self-revealing nature, but become a way of tying down uncertain claims about an opaque and uncertain

---

4. Francis Hauksbee and John Desaguliers, as frequent contributors to the *Transactions* throughout the early part of the eighteenth century, seem to have had significant influence on the development of the experimental article. Studies of their development as writers of experimental science would help fill out the story of the evolution of the experimental report. Similarly, a study of the innovations and influence of William Herschel as a scientific writer ought to reveal significant trends in the latter half of the eighteenth century.

nature. The meaning of an experiment is no longer the simple observation of what happens. An experiment is to be understood only in terms of the ideas that motivate it, for nature is no longer considered to be so easy to find. Through volumes 50 and 60 the experiments become increasingly couched in terms of problems—things that despite our familiarity with phenomena we do not understand. In volume 60, Joseph Priestley describes his puzzlement concerning the nature of the electrical phenomenon he calls "lateral explosion" which did not behave the expected way in a series of simple exploratory experiments: "I do not remember that I was ever more puzzled with any appearance in nature than I was with this; and, in the night following these experiments, endless were the schemes that occurred to me, of accounting for them, and the methods with which I proposed to diversify them the next morning, in order to find out the cause of this strange phaenomenon" (60:195). A series of experiments follow, logically solving the puzzle, part by part. Experiments are now clearly represented as part of a process of coming to conclusions. Priestley in another article on electricity comments on the personal intellectual consequences of an experiment: "With respect to the main object of my inquiry, I presently satisfied myself, that the conducting power of charcoal . . ." (60:214).

By volume 80, experiments are subordinated to the conclusions the authors have come to; that is, the experiments are ways of proving or supporting general claims. Hypotheses are presented up front and the series of experiments follow. Priestley, for example, now adopts language of proof rather than of discovery: "That my former supposition . . . is true, will appear, I presume, from the experiments which I shall presently recite" (80:107).

By volume 90 authors talk about the necessity of establishing general knowledge and the role of experiment in testing our beliefs as well as filling out knowledge. William Henry comments at the beginning of his report of experiments analyzing muriatic acid, "The theory of the formation of acids . . . must be regarded as incomplete, and liable to subversion, till the individual acids now alluded to have been resolved into their constituent principles" (90:188). Experiments test and justify the general claim, which is part of a larger system of general claims. The language of general proof holds sway: "The above facts prove, that the combination of oxygen and muriatic acid . . ." (90:194).

Methodological Concern

As experiments gain an argumentative function, the reports explain more fully how the experiment was done and why the particular methods were chosen. How nature is prodded is recognized as affecting nature's response. Debates over differing results focus attention on differences in experimental methods and conditions. Methodological care enables experiments to be used as investigative tools and then as proofs. Investigators, in order to satisfy their own problems, make subtle methodological distinctions among different experiments within the same series. Then toward the end of the period, when experimenters start arguing general propositions, the meaning and validity of the experiment depends on proper methodology.

In the early volumes how an experiment was performed was generally mentioned in passing, simply to let the reader know what kind of experiment was done. In volume 1, issue 3, for example, the editor, Henry Oldenburg, describes a series of observations and experiments made by Thomas Henshaw on the putrefaction of May-Dew. In each of the series, the procedure is described in an introductory clause or modifying phrase only, such as "Dew newly gathered and filtered through a clean Linnen cloth, though it be not very clear, is of yellowish color . . . (1:34). The procedure only serves to identify the dew. Only when directions are for practical use (and not, I emphasize, replication) are more detailed instructions given, though these are still vague by modern cookbook standards. Robert Boyle, for example, in volume 1, issue 15, appearing in mid-July, explains, for the benefit of sweltering Londoners, his new method for producing cold, useful for chilling drinks: "Take one pound of Sal Armoniack and about three Pints (or pounds) of Water, put the Salt into the liquors, and stir altogether, if your design be to produce an intense, though but a short coldness; or at two, three, or four several times, if you desire, that the produced coldness should rather last somewhat longer . . ." (1:256–57).

Even as early as the fifth volume, challenged by disagreements, authors demonstrate their experimental care and account for differences in results by describing in greater detail their experimental procedures and the conditions. Disagreements over experimental results on sap flow in sycamores lead Willoughby to consider both the date and weather conditions when the trees were bled. In volume 10, fear of challenge leads Robert Boyle to report that he took great care that a copper mixture was not shaken in the course of the experiment, although he himself does not believe that disturbance of the mixture to be of any consequence (10:468). The most explicit presentation of technique results from Francis Line's challenge of Isaac Newton's results. Newton in response lays out in much greater detail the method of his earlier experiment and the conditions under which the experiment occurred. He further suggests additional experiments and challenges Line to replicate them all. In presenting the method in such great detail, Newton insinuates that Line in

Two: Literary and Social Forms in Early Modern Science

doing his first set of experiments got things wrong. Since the debate is over whether such things as reported in Newton's account happen, the method and conditions to make them happen are crucial to the argument. (This incident and the surrounding story are examined more fully in chapter 4.)

By volume 30 authors claim they design experiments to meet specific objections of opponents. Desaguliers, for example, attempts to answer objections of the plenists that earlier experiments concerning bodies falling in vacuo were done over too short a distance. Desaguliers reports: "To obviate this I contriv'd a machine to this purpose, which consisted of a strong wooden frame . . ." (30:718). Variations in apparatus that might cause variations in results are also noted, to indicate that the author is not misled or confused by such variations (for example, 30:1078). Delicate parts of the procedures are noted, so as to distinguish the author's proper procedures from his opponent's less careful ones and to indicate specific points where the opposition may have erred. Desaguliers, for example, defending Newtonian optics at length against an extensive attack by John Rizzetti, points out many places where Rizzetti's procedures were misguided and where his judgment may have failed. In one instance Desaguliers comments, "This must have been Signior Rizzetti's mistake . . . for several of the Persons present at my Experiments made the same Mistake at first before they could perform the Experiment in manner above-mentioned; which they at last did. . . . This mistaking a Reflection for a Refraction has been the Occasion of several more Errors, and Difficulties to be met with in Signior Rizzetti's Book" (35:610).

Articles not engaged in overt contention continue to discuss method only sketchily. However, as experiments become incorporated into stories of discovery, the distinctions between trials become important as events in consciousness, so at least the crucial differences between trials become defined. Richard Watson, for example, introduces the sixth experiment of his series on the solution of salts with the comment: "Thinking that the difference in the bulks of the water before and after solution might be owing to the separation and escape of some volatile principle; I took care to balance as accurately as I could, water and sal gemmae, water, and the salt of tartar, water and vitriolated tartar &c . . ." (60:335). Since persuasion comes through the audience's willingness to accept the experimenter's experience of discovery, detailed accounts of method indicate both the experimenter's care and that he was convinced of his discoveries for good reasons.

Finally, in volumes 80 and 90, as articles present proofs of general hypotheses, the details of the experiments demonstrate care, exactness

Reporting the Experiment

of results, relevance to thesis, and the elimination of alternatives. Henry, for example, gives a complex rationale for a particular method on the bases of precision and clarity of results:

> I employed the electric fluid, as an agent much preferable to artificial heat. This mode of operating enables us to confine accurately the gases submitted to experiment; the phaenomena that occur during the process may be distinctly observed; and the comparison of the products with the original gases, may be instituted with great exactness. The action of the electric fluid itself, as a decomponent, is extremely powerful; for it is capable of separating from each other, the constituent parts of water, of the nitric and sulfuric acids, of the volatile alkali, of nitrous gas, and of other several bodies, whose components are strongly united. (90:189)

William Herschel, in his experiments on the distinction between the visible and radiant spectrum, takes his measurements in several different configurations to prove that his results are caused by the principle he is trying to prove. Not only that, he rotates the position of the thermometers to ensure the results are not artifacts of faulty measuring devices. In such duplication and varying of measurements to ensure validity of results and to eliminate all other possible variables, Herschel presents his work in a way that approaches the modern concept of controls (90:255-326).

Indeed, throughout the period, the increasingly expressed awareness of possible variables seems to reach toward an unexpressed concept of controls. In recognizing differences of conditions or execution of the experiment that might affect results, the reports started comparing results from different situations. As more experiments report multiple trials with only slight variation of experiment, crucial factors are isolated. Then, as we have seen, multiple trials are explicitly designed to establish distinctions between two sets of conditions. The practice of experimental controls—running an experiment twice, identically except for an isolated crucial variable—is only the next step in argumentative clarity through the representation of method.

The changes in illustrations through the period also express the growing importance of methods. The early issues of the journal frequently illustrate the phenomena being reported on or new technological marvels, but rarely is the apparatus used for an experiment considered worth a picture. However, as experiments become more ingenious, elaborate, or just simply careful, illustrations follow. The first apparatus illustrations I found were in volume 25, showing the brushes and vacuum devices used by Hauksbee to generate static electricity in vacuum. Although not all experiments are illustrated with apparatus diagrams, they do become a prominent feature, as in Desaguliers's answer to Rizzetti, allowing the reader to visualize the experimental procedures and the results (35:575opp.). Herschel's articles in volume 90, punctuated by a number of quite realistic apparatus illustrations, give a concrete feel of what was done. The realism of illustration becomes particularly important as the account or story of the experiment becomes the reader's vicarious surrogate for the actual experiment, as will be discussed below.

Precision and Completeness of Results

As with method, results of the experiments are reported with increasing detail, care, and quantitativeness as the experiment bears more and more weight of argument, persuasion, and then proof. Early results are described vaguely and qualitatively, as though the phenomena of nature were robust, uniform, and self-evident. As disputes arise over reported results, writers become more careful about reporting what they see, and measurement takes a greater role. With the proliferation of quantitatively comparable results, experimenters begin puzzling over subtle variations in results; detailed results become a means of figuring out exactly what is going on. Finally, detailed quantitative experimental results, fitting quantitative theoretical results, form the empirical proof of general hypotheses.

In the early volumes, those experiments that provide directions for achieving certain wondrous effects have no explicit results at all, for it is simply assumed that following the recipe will lead to the desired effect. Where results are given they are in the form of general qualitative observations, such as in the example of luminous mackerel broth: "As soon as the Cooks hand was thrust into the water, it began to have a glimmering. . . . they who look'd on it at some distance, from the further end of another room, thought verily, it was the shining of the Moon through a Window upon a Vessel of Milk; and by brisker Circulation it seem'd to flame" (1:227). Even where quantification of results seems a rather simple matter, as in two experiments in volume 5 concerning expansion of a freezing solution and the timing of respiration, the results were given in purely qualitative terms.

Again, debate and conflict push results to greater detail and precision in exactly the same articles with more detailed accounts of method. Newton, for example, in answering Line, spends a lengthy paragraph describing the three different images cast by a prism, distinguishing the character of these different images, so that anyone repeating the experiment can find the oblong image which was Newton's particular concern and which Line disputed (10:503).

By volumes 15 and 20 quantitative results in measurements of the speed of sound, barometric air pressure, and specific gravity enable comparisons. With the increase of multiple trials, distinctions among results of various trials become a practical expository tool. In volume 50, for example, William Lewis, in investigating mixtures of platinum and gold, creates nine different mixtures of different proportions from 1:1 to 1:95 in order to compare both qualitative and quantitative properties (50:148–55).

By volume 60 the results sought and reported have specific relation to the hypotheses being investigated and tested. James Johnstone reports a series of experiments designed to test hypotheses concerning the function of nerve ganglions; not only are the results found consistent with the hypotheses, but he adds results from experiments reported by other authors. These additional results also support the hypotheses, even though the experimenters did not have the same questions in mind; Johnstone was already aware of the potential for bias in experimental design (60:30–35).

Near the end of the eighteenth century, as arguments move toward proof, the precise reporting of results enable them to be compared to quantitative predictions of hypotheses and thus to serve as direct evidence. Herschel, in volume 90, in order to prove that the radiant heat is distinct from the visible spectrum, provides extensive quantitative results, to the point of inductive tedium (90:255–322).

The Ocular Proof and Communal Validation

The increasing precision and detail of method and result accompanies a major change in how accessible the experimental demonstrations are for the readers of the journal. As the actual experiment becomes more of a private affair for the investigator and close associates, verisimilitude of the report reassures the readers that the events happened, and happened in the way reported.

In the early years many of the experiments reported in the *Transactions* were demonstrated before the assembled body of the Royal Society at regular meetings. The demonstration is its own meaning, for all to witness and agree it did take place. The report of the experiment is little more than a news report that such an event took place and was witnessed by the assembled body. The validity of events rests on the communal witness and not the story told.

The communal witness remains important validation of the experimental events for much of the earlier part of the period, but as experiments gain subtlety and face conflicting results, experimenters need to control the particular conditions of demonstration. Experiments stay in the laboratory, remote from the lecture hall. Designated competent witnesses travel to the experiment to represent the general membership and a prestigious list of witnesses becomes an important feature of the report. Thus in volume 30 Desaguliers's witnesses include the king and queen as well as the chief members of the Royal Society. Witnesses, however, no matter how prestigious, can always be opposed by equally impressive witnesses who attest to conflicting results, so a precise account of methodology with detailed results, allowing critique, comparison, and replication become part of the argument.

The next change occurs when the problem shifts from the simple existence of phenomena to the meaning of baffling, troublesome phenomena. The experiment, no longer an end in itself, certainly no longer performed in public, becomes a private affair, an event in the individual intellectual journey of the investigator. In the volumes 40 onward there is almost no direct conflict over results, but rather only over theories, and even the theories are presented more as the results of individual research programs rather than highly combative claims and counterclaims. Conflicts and comparisons of results are more likely to occur within the series of experiments of a single scientist trying to work out the subtleties of a complex phenomenon. The series of experiments are not presented as being likely to be replicated. For example, Tiberius Cavallo in volume 70 reports his earlier experiments as part of his puzzling through a problem in electricity, not giving adequate instructions for replication, but at the end he does give detailed replication instructions for one final, contrived experiment so others can convince themselves of his conclusions and can explore the phenomenon further (70:15–29).

Nonetheless, specificity, detail, and plausibility of the experiments are important as part of the story of the intellectual journey of the investigator. Since neither the reader nor any surrogates or representatives, except for the author himself, has witnessed the series of experiments, the account must stand in place of the witness. The reader in order to understand the experimental argument must vicariously witness the experiment through the account. In order to earn the trust of the reader, the story of the experiments must be told plausibly if not persuasively, and the events reported on must provide sufficiently good cause for the investigator to come to the conclusions he reports.

Reporting the Experiment

When finally the structure of the series of experiments turns from a representative personal journey to a retrospective guided tour of conclusions and experimental evidence, the account of the experiment has come, at least for the time being, to stand as the proof. In the long run, the experiment or series of experiments may be replicated, but in the persuasive experience of the reading of the argument, the story of the experiment must serve as a surrogate for the actual experiment.[5] By this time papers were read to the Royal Society, but experiments were conducted in private, simply to be reported on.

Organization of the Articles

The organization of the experimental articles serves as an outward manifestation of all the trends discussed to this point. Articles tend to grow longer throughout the period as the argument surrounding the experiment grows and individual investigations rely more and more on series of logically connected experiments rather than single events.

In the first issues most of the information passes through the voice of the editor who simply reports on things he has found out about from a variety of sources. Typically, Oldenburg announces that, "The Ingenious Mr. Hook, did, some months since, intimate to a friend of his, that he had . . ." (1:3). By the end of the first volume authored articles appear, with much the format that would maintain through volume 25. The article opens with a short statement of what was done, followed by a narrative of results. Often articles end there, although some discussion of

---

5. Steven Shapin, "Pump and Circumstance: Robert Boyle's Literary Technology," reports that a number of the features of virtual representation that did not appear regularly in journals until the second half of the eighteenth century were mobilized in books by Robert Boyle almost a century earlier. This uncoordinated development of book and journal publication raises two questions. First, are the dynamics, constraints, form, and literary situation of book publication significantly different than that of journal publication, so as to encourage the emergence of different textual features at any particular time or to cause any one common feature to emerge at different times? Second, what are the formal interplay and mutual influence of journals and books? Chapter 4 begins an investigation of such questions.

Chapter 2 of Shapin and Schaffer, *The Leviathan and the Air-Pump*, offers a more complete view of the proper form of communal knowledge and the importance of empirical experience, direct and virtual, for successful public debate and evaluation of knowledge claims. Much of Boyle's attitude toward empiricism and public debate seems to have been carried out in the history of the rhetoric of the *Transactions* analyzed in this chapter.

cause or meaning may follow. Articles tend to be short, often only a page or two, and usually discuss only a single experiment or trial.

In those articles reporting on a series of experiments, however, continuity does increase over this early period. At first experiments in a series are only loosely connected, being concerned with the same general phenomenon, as in the many series of experiments putting different objects in vacuum reported on in volumes 5 and 10 or as called for in William Petty's "Miscellanious Catalogue of Mean, vulgar, cheap and simple Experiments," all loosely related to weights and specific gravities (15:849–53). Occasionally a rationale explains specific trials, as when Boyle provides reasons for choosing particular animals to deprive of air: the duck, which breathes in air and dives; the viper which has lungs but is coldblooded; a newborn kitten, recently in the womb without access to the atmosphere, etc. (5:2011–31, 2035–56). A result in an early experiment in the series may lead to new questions to be explored in later trials, such as when a Captain Hall notes that a rattlesnake is less lethal on successive bites; unfortunately, the experimental program was cut short when neighbors began complaining about missing dogs (35:309–15).

As experiments begin to respond to conflicts, their reports focus on the issue in contention. Typically, the report starts with a statement of the phenomenon in dispute and then a discussion of the opponent's work or position. The author's own position with consequent experimental method and supporting results follow, with perhaps some general conclusions, as in Desaguliers's article arguing with Leibniz' explanation of barometric fall in wet weather (30:570–79). By volume 40, the hypothesis or meaning of an experiment often precedes the account of the experiment, even where no particular issue is in contention. Desaguliers, for example, not only begins a paper on statics with a general proposition, but he promises to provide elsewhere a more general theory (40:62–69).

As phenomena are treated as more problematic, articles take on a different organization, opening with an introduction to the problematic phenomenon, often substantiated with the story of an experiment that did not go as expected. With the problem established, the article chronologically describes a series of experiments aimed at getting to the bottom of the mystery. Transitions between pairs of experiments draw conclusions from the previous experiment and point to the rationale or need for the consequent one. In the process of continuous reasoning, the experimenter gradually comes to an adequate understanding of the phenomenon, which is pulled together in a concluding synthesis or explanation of the phenomenon, as in William Hewson's investigations into the nature of blood (60:368–83).

At the end of the period, articles using experiments to prove general claims often begin with philosophic statements about general knowledge. Then a problem is presented, either through a surprising experimental result or through the exposition of a gap in current knowledge. Then a series of claims resolve the problem, followed by supporting experiments. Although a subseries of experiments may be presented chronologically, the larger structure of the article is based on the logical order of the claims to be proved. The conclusion may discuss consequences of these claims, but no synthetic set of conclusions is needed because the claims have already been presented at the beginning. Henry's investigation of muriatic acid (90:188–203) and Herschel's investigations of radiant heat (90:239–326) conform to this general pattern.

Forging Persuasive Forms and a Collective Literature

One early article (15:856–59) that in many respects resembles articles from a century later reveals the rhetorical function of the features of the experimental article that emerge by 1800. The anonymous article starts with a general proposition concerning the ease with which larger wheels may be drawn over an obstacle. The experiment, clearly designed as a demonstration of that proposition, is presented in great quantitative detail, as are the results. Moreover, many different trials are set out, isolating variables and allowing exact comparisons proving the general proposition. However, the theoretical point was already well established in the literature (it is attributed to three authors, both ancient and contemporary) and this experiment and the article reporting it are only to convince practical people—wagonmakers and wagon purchasers—of the advantages of what was already known theoretically. The point is not to prove the truth of the statement, but to persuade recalcitrant craftsmen to use a well-established truth.

In those early years, argumentative persuasion could be used for the ignorant artisan, but for those actively pursuing nature, nature was portrayed as speaking for herself. The scientific report was simply a matter of news. Just as an earthquake or passage of a meteor needed to be reported, so did experiments. Not until nature was treated as a matter of contention and then a puzzle could the experiment become part of an

Two: Literary and Social Forms in Early Modern Science

argument and could theory or claims hierarchically and intellectually dominate experiments.

With the journal serving as a forum, contention grows. This contention pushes the individual author into recognizing that he is not simply reporting the self-evident truth of events, but rather is telling a story that can be questioned and that has a meaning which itself can be mooted. The most significant task becomes to present that meaning and persuade others of it. Persuasion of claims then lies in a story of personal discovery, supported by good reasons and careful work. Since all people, however, have good reasons, the persuasive story must shift to more universal grounds: the proof of a claim transcending the particulars of an investigation.

To draw the historical lines even more sharply, we observe four loose and overlapping stages in the development of the experimental report. In the first stage, most evident through volume 20 (c. 1665–1700), articles consisted of uncontested reports of events. In the second stage, most evident from volumes 20 through 50 (c. 1700–1760), experimental articles tended to argue over results. Beginning in about volume 50 through volume 70 (c. 1760–1780), articles explored the meaning of unusual events through discovery accounts. Finally, in volumes 80 and 90 (1790 and 1800), experimental articles offered claims and experimental proofs.

In this process we find the beginnings of something like Karl Popper's third world of claims, separate from both nature and the individuals who perceive it. The earliest reports—accounts of what happened, as witnessed by many—recognize only the first world of nature. Contention draws attention to the second world of human perception and consciousness, throwing the authors back on to their own experience and thought (although hedged with the proper respect for nature and empirical methods) as the essence of their reports. Finally the claim or conclusion—Popper's third world—becomes the central item to be constructed within the article, to be supported by empirical evidence from the first world and proper method and reasoning from the second world.

Yet to the end of the period, experimenters present their claims as purely products of their individual interactions with nature, not explicitly recognizing the communal project of constructing a world of claims. In most of the articles the literature is still not treated in any explicitly codified way, as we have become familiar with in the twentieth century. The experiment still appears solely the result of the individual's invention and understanding. Although the individual scientist has an interest in convincing readers of a particular set of claims, he does not yet explicitly acknowledge the exact placement of the claims in a larger framework of claims representing the shared knowledge of the discipline. Herschel does not relate his theories and findings to a large body of knowledge other than his own, except in the most general way. He presents himself as the only explorer of his terrain and the experiments are thus the confirmation of the general truths he has discovered in his particular travels. The only consistent use of other literature occurs in debates where discussion of the literature serves to draw lines and marshal forces rather than construct an edifice beyond the immediate claims.

Although the collective intelligence of the scientific community before 1800 is not regularly displayed in explicit codifications of the literature, the collective intelligence is embodied in the way the members of the community have chosen to communicate with one another. Whether the emergence of an argumentative community necessitated a conventional genre in which to carry on that argument or whether the clarification of forms of argument allowed a coherent community to coalesce in discussion is an unanswerable dialectical conundrum. A more exact formulation might be that a community constitutes itself in developing its modes of regular discourse.

In this particular case, the kind of argument the community engaged in, over the regular appearances of natural phenomena, seemed best pursued by increasing descriptive detail and precision, re-creating events increasingly designed to display particular features of that nature. But regularity and particularity proved at odds, creating new problems in symbolizing nature. Particular and general formulations did not always fit together easily, so new modes of discourse were needed to expose the regularities hidden in the anomalous particulars and to demonstrate that general formulations offered precise representations of particulars. The emerging form of experimental report offered a way to harness stories of the smaller world of the laboratory to general claims about the regularities of the larger world of nature. In the attempt to satisfy the objections and desires of the growing scientific community, the experimental report kept changing in form, as it continues to do today—for objections and desires grow with the ability to formulate them. And what is a science without objections and desires?

4 BETWEEN BOOKS AND ARTICLES

NEWTON FACES CONTROVERSY

The appearance of the scientific journal in 1665 did not immediately displace books as the primary means of communicating scientific findings. Books remained the more substantial source for scientific information for many years, interacting with the emerging journals. Currently we have only an impressionistic overview of this transformation, as expressed by A. J. Meadows: "Major research continued to be written up in monograph form throughout the eighteenth century, but the habit began to die out in the nineteenth century, at least among the physical sciences" (*Communication in Science* 67). This broadstroke characterization carries some broad-stroke truth, but a few pieces of information suggest a much more complex picture that needs investigation.

Even during the late seventeenth century some major findings first appeared in the *Philosophical Transactions of the Royal Society* rather than in books, such as Anton Leeuwenhoek's microscopical investigations and some of Boyle's vacuum experiments. Indeed Leeuwenhoek published exclusively through correspondence printed in journals, primarily in the *Transactions* beginning in the 1670s. His books were only collections of his letters (*DSB* 8:126–30). Other lesser seventeenth- and eighteenth-century scientists, such as Desaguliers (*DSB* 4:43–46) and Hauksbee (*DSB* 6:169–75) published primarily in journals. Certainly, as discussed in the previous chapter, the genre of experimental report developed fairly rapidly toward the presentation of primary research, with the generic features being shaped by the dynamics of controversy that would only attend primary publication for a professional audience. As we shall see in a later chapter, the journal article appears in fact to have from early on played an important role in organizing the scientific research community. Further, there seems to have been a great proliferation of journals during the eighteenth century. According to Kronick, the number of active, substantive scientific journals in Europe increased from 7 in 1710 to 27 in 1750 and 118 in 1790 (89).

On the other hand, at the end of the nineteenth century, some journals, including *Physical Review*, still carried book reviews, treating the books under review as major research contributions. Even well into the new physics of the twentieth century, books like Arnold Sommerfeld's *Atembau und Spektralinien* (going through six German editions) and Linus Pauling's *The Nature of the Chemical Bond* presented major theoretical advances as well as primary reports of research.

The more closely one looks at the shift from book to article science, the more the story seems a complex one, with different findings and different kinds of work going to different venues. Nor will Kuhn's association of mature science with journal science fully sort out the complex historical facts. The first two hundred pages of the first volume of the *Dictionary of Scientific Biography*, for example, reveal many exceptions to the expected overall pattern. For example, the eighteenth-century naturalist Michel Adanson, mathematician-physicists André Ampère and Franz Aepinus, and chemist Franz Karl Achard had mixed patterns of articles and books cited for their primary findings. The same mixed pattern pertains in the cases of twentieth-century astronomer Eugen Antoniadi, chemist Richard Anschuetz, paleobiologist Othenio Abel, and radio physicist Edward Appleton. The twentieth-century astronomer Robert Aitken made his most important contribution in book form, and the eighteenth-century polymath José Antonio Alzate y Ramírez contributed through journals. The data for nineteenth-century contributors are even more unpredictable, by date or by specialty.

Moreover, the forms of books and articles are not always distinct and insulated from each other. Although journal articles started off as generally quite short, some became rather long, such as Robert Boyle's "New Pneumatical Experiments about Respiration," which during 1670 filled most of issues 62 and 63 in the fifth volume of the *Transactions*. Such long articles resembled pamphlets of the period in form. By the eighteenth century the long article became common, with volume 90, for example, comprised of only 18 articles, averaging over twenty-five pages in length each. Moreover, Kronick reports some eighteenth-century journals that bear close resemblance to books, with each issue devoted to a single topic, and perhaps written by a single author (92). Similarly, books early show the influence of article styles of experimental presentation and adopt new functions to coordinate with journal publication, as might be observed in Joseph Priestley's *History and Present State of Electricity* (1775).

Thus there seem to be many kinds of books and many kinds of articles with complex relationships to each other. Much historical and textual work remains to be done before a clear picture can emerge.

The following is one attempt to look at an early moment in the book-article dialectic, shedding light on the dynamics and form of both book

Two: Literary and Social Forms in Early Modern Science

and article publication at the time.[1] We will consider how Isaac Newton—an intelligent, rhetorically sensitive, creative, and highly motivated individual—understood the two forms and made linguistic choices on the basis of his understanding. Moreover, we will see how he reconsidered his rhetorical problem and strategy, on the basis of readers' responses expressed within a structured communications forum. His reconsiderations influenced both book and article forms. Thus the story is of active reshaping of the form of communication with long-range impact on generic resources and expectations.

Newton's Optical Publications

From a biographical perspective, Newton seems to have dallied only once with journal publication, got burned badly, and never returned.[2] That is, he first published his optical findings in a 1672 *Transactions* article, entitled "A New Theory of Light and Colours," which sparked a controversy with much of the correspondence printed in later issues of the *Transactions*; afterward Newton refused to publish in journals and withheld further publication of his optical findings for thirty years until the *Opticks* appeared in 1704.

But from the perspective of the history of the journal, the "New Theory" article is the earliest significant finding published in the *Transactions*, and is treated as an exemplary piece of scientific writing.[3] Thus Newton's biography suggests that article publication was a failure for Newton, who found the book a more congenial medium, while the history of science judges the article a success. However, a closer examination of Newton's papers reveals the biographical and historical judgments as consistent and related. Newton, perceiving journal publication as a platform, created a forceful statement, but the bitter experience of controversy taught him that journal publication meant entry into an agonistic forum. To address this newly perceived situation, he developed new rhetorical resources to answer criticisms in following issues of the *Transactions*. These rhetorical innovations provided a mode of argument that shaped his book presentation and provided a model for fu-

---

1. James Paradis, "Montaigne, Boyle, and the Essay of Experience," examines another closely related moment in the early history of the relationship between longer book forms and the shorter article form. He finds the roots of the article in Montaigne's invention of the essay, which for many reasons appealed to the empirical skepticism of the Royal Society.
2. See, for example, Westfall, *Never at Rest*, chapter 7.
3. See, for example, both Cohen's and Kuhn's introductions to Cohen's edition of *Isaac Newton's Papers*.

ture scientific publication by others. The form of compelling argument he developed relied on creating a closed system of experience, perception, thought, and representation that reduced opposing arguments to error. The closed system Newton developed was his own, framed by the worlds represented in his powerful books. Only later was the scientific community to develop the means to construct communally developed closed systems; nonetheless, the Newtonian model of argument provided a powerful way of arguing for general truths from empirical experience. Newton shaped the science that came after him on many levels.

More specifically, this chapter will examine the different forms Newton used to describe his prismatic experiments and related findings about the spectral colors and the composition of white light. This work, forming the matter of book 1 of the *Opticks*, is the most deeply documented of Newton's optical investigations and has appeared in the most forms, including the forms occasioned by controversy. The material which composes books 2 and 3 of the *Opticks* has a shorter and less documented experimental history, has not undergone so many literary transformations by Newton, nor has it faced such extensive public controversy, requiring Newton's defence.[4] Moreover, in the *Opticks*, book 1 is presented confidently and compellingly, whereas books 2 and 3 are presented with greater hesitancy, noncompelling speculation, and open-endedness—indicating Newton's inability to harness the latter material to his newly minted conception of compelling scientific argumentation, realized in book 1. The judgment of history seems to have born out Newton's rhetorical judgment, for the argument of book 1 still stands, whereas in the last two centuries only the observations and not the theoretical arguments of the latter books are given scientific credence.

We currently have, depending on how you count, at least seven significantly different versions of the material of book 1 by Newton's hand:

1. entries in his private notebook, *Questiones quaedam Philosophicae*, circa 1664 (Add. 3996);[5]
2. a private manuscript, "Of Colours," circa 1666 (Add 3975);[6]
3. university lectures, first version, circa 1670–71;
4. university lectures, second version, prepared with intent to publish in book form, circa 1671–72;[7]

---

4. For a discussion of the material leading to the second book of the *Opticks* see Westfall, "Isaac Newton's Coloured Circles twixt two Contiguous Glasses," 13–14.

5. McGuire and Tamny have edited these notebooks under the title *Certain Philosophical Questions: Newton's Trinity Notebook*. I have used this edition throughout.

6. Also in McGuire and Tamny, 466–89.

7. Both versions of the university lectures, in Latin, are published with English trans-

Two: Literary and Social Forms in Early Modern Science

5. a letter to Oldenburg, dated February 1672, (*Correspondence* 1: 92–102) published in slightly edited form in the *Transactions* of 19 February 1672, under the title "A New Theory of Light and Colours";
6. consequent exchanges of correspondence, via Oldenburg, much of it published in the *Transactions*, 1672–76. The details of these exchanges will be provided later;
7. *Opticks*, Book 1, written circa 1690, including multiple extant drafts in English and a partial draft in Latin; published during Newton's life in 1704, 1717, and 1721.

Newton also reported that an additional book-length manuscript on the subject, presumably written after the 1672–76 controversy, was destroyed by fire before work began on the circa 1690 draft.[8]

The story of Newton as a self-conscious and flexible writer revealed in these documents as well as other Newton papers is a rich one, which I hope in future publications to be able to lay out with all the detail and leisure it deserves. Here I discuss only those events and textual transformations that shed light on the dynamic interaction between book and article publication as experienced by Newton.

The basic claims that Newton presents in these various forms were set by the first university lectures, even though later controversy and developments of the argument would cause some drawing back, some further elaboration, and some further precision. The simple substance is the now familiar observation that light of different colors is refracted to different degrees when passed through a prism. Thus light composed of a combination of colors, such as white light, upon passing through a prism will be broken into its various component colors, displayed as a spectrum. The modern understanding is that color is only our perception of light waves of different wavelengths. Thus we can easily conceive of the difference between color produced by light of a single wavelength and color produced by light of a number of wavelengths. At the time of Newton, color was seen as a unitary phenomenon. Newton's association of color with differing refractive indices (or as he called it refrangibilities) and consequent need to distinguish between simple and compound colors created conceptual difficulties for his contemporaries. Much of the controversy and Newton's rhetorical innovation hinges, in fact, on this problem.

---

lation in a modern edition as *The Optical Papers of Isaac Newton*, vol. 1, edited by Alan Shapiro. The introduction, pages 16–20, discusses the dating of the two versions.

8. In *Never At Rest*, Westfall dates work on this manuscript to 1677–78, with the fire in 1678 (276–78).

Student Explorations in Optics

Prior to the "New Theory" article, Newton's formulations of his prismatic investigations were free of the exigencies of open public debate. The first mention of a prismatic experiment comes in the middle of a private notebook kept by Newton while a student at Cambridge, circa 1664–65. The earliest part of the notebook consists only of summary notes in Latin and Greek of the required reading in his scholastic curriculum, but in the middle he turns to independent contemporary reading. Not only are these later notes in English, but they tend to represent Newton's own thinking and experiences set in motion by the reading.

Among the notes on many subjects, Newton speculates on the nature of light. These speculations are set in motion by his reading of Boyle and Descartes on the subject, and perhaps by his attention at Isaac Barrow's lectures (McGuire 241–44). In his notes Newton develops a mechanical, corpuscular description of light and he includes a diagram of a light particle moving through ether (384–85), paralleling an earlier diagram he had made of a body moving through water (366–69). He follows these speculations with several observations from his experiences and some queries (386–89). It is in the context of this speculative, theoretical, private musing about commonly experienced phenomena, as inspired by his reading, that we must interpret his accounts of prismatic experiments some pages later in the notebook.

His first prismatic experiment is presented only as a proposal, in the imperative mode: "Try if two prismas, ye one casting blue upon ye other's red, doe not produce a white" (430). He continues with a diagram and more than a dozen additional similar combinations (432–33). His comments thereafter are highly speculative and theoretical, giving an interpretation based on the speed of moving light globuli affecting both the amount of refraction and the impact on the optic nerve. A chain of reasoning follows, in which is embedded an experiment he clearly represents himself as having done: viewing through a prism a thread—half its length colored red, the other half blue. One half appears higher than the other. After three more pages of theoretical speculation, this set of notes trails off into a set of diverse observations about colors exhibited under varying situations (432–45). Another more extensive list of observations of colors in various situations appears later in the notebook (452–65).

In 1666 Newton reorganized and expanded these notes into a more coherent private document entitled, "Of Colours." The twenty-two folio sheets, divided into sixty-four numbered experiments and comments,

Two: Literary and Social Forms in Early Modern Science

contain fifty prismatic and related observations (number 6–55). The organizing principle here, rather than being the associations of explanatory theorizing, is the apparent similarity of observed phenomena. This is solely an account of actual observations and experiments, until the end when dissection of an eye leads to speculations about the operation of the visual faculty.

Although the mode is now empirical rather than speculative and the theoretical literature inciting the investigation has now dropped from sight, the ordering of observations is still exploratory, as one experiment suggests another of similar format or pursuing a related idea. Topics of recurrent interest keep reemerging, but in no obviously planned manner nor with any clear argumentative order. Descriptions remain largely brief and qualitative. Newton has not yet sorted out what he has into an ordering theory.

Professorial Expositions

After these first student explorations, the next record of Newton's prismatic investigations consists of his lectures delivered at Cambridge University under the terms of the Lucasian Chair of Mathematics and Natural Philosophy, which he took up in late 1669 (replacing Isaac Barrow, who had stepped down in his favor). Manuscripts of these lectures were deposited, according to the terms of the chair, at the university library some time later—the first version perhaps in 1672 and a revised version perhaps in 1674 (*Optical Papers* 1:19). It is unclear how intensively Newton carried out prismatic investigations between 1666 and 1669, but his responsibilities as newly appointed chair occasioned a new formulation of what he had learned to that point.

This formulation was shaped by the situation and goals of the university lecture. The authoritative voice of the professor, introducing students into a coherent and comprehensive understanding of a subject leaves little room for serious challenge. The usual authority relations of the classroom that acknowledge the lecturer as the unquestioned source of knowledge, were further supported by both the dispirited intellectual atmosphere at Cambridge at the time and Newton's already established campus reputation for brilliance (Westfall, *Never at Rest*, 185–95). Newton's lectures, consequently, were expository in organization and tone, rather than persuasive or argumentative.

By the time of the lectures, Newton was no longer uncertain about the meaning of his experiments: different colors are differently refrangible—that is, they suffer different amounts of refraction when passing from one medium to another. This meaning beomes the center of his expository organization of both versions. In the first lecture, immediately after a motivating introduction, Newton presents this basic principle through a schematic diagram not attached to any specific experiment (48; 282). The rest of the text follows as an explanation and elaboration of that opening principle. Topics are presented sequentially, generally moving from the simple to the complex, divided into separate lectures and further divided by section headlines. About half of the exposition is mathematical, offering geometric demonstrations, derivations, and calculations. Proofs serve as elaborations rather than arguments. The other half is experimental, using the experiments to demonstrate features and consequences of the basic principle.

Because both mathematics and experiments are presented as elaborations of a consistent and coherent explanation and because these elaborations are so extensive (the first version comprised of eighteen lectures, and the second comprised of thirty-one), Newton can rely on the massiveness of the overall vision as a device both of persuasion and pleasure. Typically, the lecturer comments at one point, "I now repeat the experiment, however, so that I may pursue its various features that are no less pleasant for the experimenter than they are informative for our purpose" (63). Alternative theories are dismissed rapidly, in passing, steamrollered by the weight of the exposition and the lecturer's authority.

Newton's Perception of Journal Publication

Through his private journals and then his lectures, Newton had produced confident formulations, coherently connecting many experimental details and mathematical elaborations around a central principle. Yet the rhetorical situations of journal and lecture had not necessitated that Newton prepare a public argument persuasive to other experienced and confident natural philosophers holding contrary beliefs. When a student talks to his notebook and a monopoly professor talks to his class, the speaker in satisfying himself, satisfies all relevant critics.

Although not prepared for the contentiousness he was to meet, Newton nonetheless perceived journal publication as presenting a new kind of rhetorical situation, for he chose an entirely different form of presentation, as we will examine below. But before we examine the rhetorical understanding realized in the "New Theory" article, we should examine evidence indicating Newton's perception of publication in the *Transactions*.

Two: Literary and Social Forms in Early Modern Science

First we have some reading notes made by Newton on the first twenty-four issues of the *Transactions* (Add 3958). These notes seem to date from a single period, probably 1668–69. The notes consisting of thirteen pages of close handwriting summarize all the articles in the issues. The summaries range from only a five-word general description to a three-hundred-word discussion. In general Newton gives fairly detailed attention to concrete observations, findings, and inventions, no matter what the subject, even if rather far removed from his apparent interests, such as whales found in Bermuda or ores found in Germany. But he is especially attentive to all claims about lenses, telescopes, and astronomical observations. On the other hand, he is generally rather brief on theoretical or speculative articles. Thus he seems to treat the *Transactions* as a repository of concrete reports. In only a few cases does he comment on these reports—they are simply taken as reported facts. In answer to Boyle's article on hydrostatics in issue 10, Newton comments "Descartes answer to this unsatisfactory," without giving reasons for his judgment. More notably he adds a twenty-four line parenthesis to his summary of Wallis' account of diurnal and annual motion in issue 16, giving his own opposed account: "Saith Dr. Wallis (But I observe . . .)." He offers no arguments, just his contrary account. His comments in neither case suggest that he felt that his opposition needed support through close argumentation.

If he read the *Transactions* as a collection of concrete facts, he may well have seen publication in it as an opportunity to present his own findings in preview of the book version of his lectures he was preparing. Oldenburg first wrote to Newton on 2 January 1672 requesting additional information about his reflecting telescope, a version of which had been brought down to London at the end of 1671 by Barrow and demonstrated before the Royal Society in late December (*Correspondence* 1:29).[9] There had been no prior contact between Newton and the Royal Society as far as we know except for Newton's reading of the *Transactions*.

Newton provided the requested details about his telescope in a letter of 6 January (79–81). On 18 January he sent a follow-up letter, adding further details about the telescope, but also including a promise of "an accompt of a Philosophicall discovery wch induced mee to the making of the said Telescope, & wch I doubt not but will prove much more gratefull than the communication of that instrument, being in my judgment the oddest if not most considerable detection wch hath hitherto beene

---

9. Several other letters published in Newton's *Correspondence* indicate the wide fame of his reflecting telescope in this period before publication of its details (1:4, 5, 72, 78, 88, 89).

made in the operations of Nature" (82–83). In a return letter of 20 January, Oldenburg responded to all the particulars of Newton's letter except this last one (83). Yet in a 29 January letter, Newton renewed his promise: "I hope I shall get some spare howers to send you also suddenly that accompt wch I promised in my last letter" (84). He fulfilled the promise in a letter of 6 February, which with a few editorial changes became the "New Theory" article (92–102).

Newton's persistence in pressing unrequested material on Oldenburg suggests that Newton saw in the Royal Society interest in his telescope, expressed in Oldenburg's letters, an opportunity to publicize what he considered a more significant finding. Having been offered an open door, he was prepared to make most of it. Moreover, before any of this interest in him had been expressed, he had already indicated his intention of publishing his nearly completed revised lectures.[10] Thus we must consider the "New Theory" article not as a preliminary finding of a work in progress but as a summary announcement of a much larger, essentially completed work.

Newton saw this completed work as true, consistent, massive, and important, but even more he saw it as concrete fact. The confidence and coherence of the lectures, presenting original work as what we would now call textbook knowledge—chosen by him as his initial topic in the only lectures on mathematics and natural philosophy being given at Cambridge University—combined with his hardly modest characterization of his findings in the letter to Oldenburg quoted above, suggest the depth of his conviction. Moreover, in the controversies to follow he was repeatedly to insist his claims were not hypotheses, but fact. A look at the character of his prismatic work can offer some insight into his sense of concrete conviction. His theory of colors—that the white light entering the prism is composed of all the colors that separate in the prism because of different degrees of refraction—is clearly a second order abstraction from simpler observations, such as that white light entering a prism emerges multicolored. Yet having once postulated that theory, Newton not only could explain a wide range of results, he could construct endless other experiments that always work out correctly. He could prismatically analyze and recombine light in a dazzling array of ways. And he did so, as he reported in his notebook, lectures, and later *Opticks*. This plethora of evidence and manipulation of the phenomenon can plausibly leave one, as it apparently did Newton, with a concrete

---

10. Alan Shapiro, in the introduction to *Optical Papers* 1:18, gives the evidence for Newton's intentions to publish the lectures in 1672.

sense that one knows exactly what is going on, that one's hands literally hold the phenomenon.

The Discovery Account as Rhetorical Strategy: The Opening of the "New Theory" Article

Newton's perception of the *Transactions* as a vehicle for concrete findings and his sense of the facticity of his own findings frame his solution of how to represent his claims in a letter to Oldenburg, which as he well understood would likely appear in print.[11] His overall rhetorical problem is to give an account of his findings so that they appear as concrete fact, as real as an earthquake or ore found in Germany, even though the events that made these facts visible to Newton occurred in a private laboratory as the result of speculative ponderings and active experimental manipulations. Moreover, the conclusions that he wishes to present as facts are based on complex interrelated statements, forming a detailed, elaborated picture with implications for many related phenomena, as he spelled out in his lectures.

Newton attempts to make his findings appear as concrete facts by establishing in a discovery narrative his own authority as a proper observer of concrete facts. This narrative presents him stumbling across a natural fact, as one would stumble across a rock. Then the narrative presents him as pursuing the oddity of this fact in a systematic way until he completes a proper description of the concrete fact. The article begins:

> Sir,
> To perform my late promise to you, I shall without further ceremony acquaint you, that in the beginning of the Year 1666 (at which time I applyed myself to the grinding of Optick glasses of other figures than Spherical,) I procured me a Triangular glass-Prisme, to try therewith the celebrated Phaenomena of Colours. And in order thereto having darkened my chamber, and made a small hole in my window-shuts, to let in a convenient quantity of the Suns light, I placed my Prisme at his entrance, that it might be thereby refracted to the opposite wall. It was at first a very pleasing divertisement, to view the vivid and intense colours

---

11. In the correspondence over the account of the reflecting telescope, Oldenburg has already requested Newton's permission to publish (20 January; 83) and Newton had replied that he was "willing to submit my private considerations in any thing that may be thought of publick concernment" (29 January; 84).

> produced thereby; but after a while applying my self to consider them more circumspectly, I became surprised to see them in an oblong form; which according to the received laws of refraction, I expected should have been circular.
>
> . . .
>
> Comparing the length of this coloured spectrum with its breadth, I found it about five times greater; a disproportion so extravagent, that it excited me to a more then ordinary curiousity of examining, from whence it might proceed. (92)

The narrative continues his pursuit of the cause of this elongation for three pages until he reaches one experiment (which he calls "the experimentum crucis") that gets to the bottom of the matter.

The personal account of stumbling across an unusual fact was a common one used in the early *Transactions,* such as in the accounts in the first volume of the luminescent pickled mackerel and the putrefaction of maydew, as described in the previous chapter. Since Newton had taken notes on and summarized a number of such articles, imitating that model need not have been a highly reflective act.

This earlier part of the article relies heavily on the language of personal thought and agency as it unfolds the attempts of a baffled investigator to come to terms with a robustly visible phenomenon. The first person followed by an active verb forms the armature of most sentences: "I suspected," "I thought," "I took another Prisme," "I then proceeded to examine more critically," "Having made these observations, I first computed from them." At key moments he offers quantitative descriptions of his experiments, switching to third person existential statements: "Its distance from the hole or Prisme was 22 feet; its utmost length $13^{1}/_{4}$ inches. . . ." But even experimental quantities are framed by his limited agency: "The refractions . . . were as near, as I could make them, equal and consequently about 54 deg. 4' " (93)

The orderliness with which he pursues and isolates the phenomenon gives rhetorical warrant to the degree of facticity of language Newton allows himself in this section. That is, the credibility of the investigation helps establish the credibility of the fact and the credibility of the investigator. The procedure Newton presents himself as following, moreover, is exactly that of exclusions, as prescribed by Bacon: "What the sciences stand in need of is a form of induction which shall analyse experience and take it to pieces, and by a due process of exclusion and rejection lead to an inevitable conclusion" (Great Instauration B, 1, 137).[12] Newton, in

---

12. Sabra in *Theories of Light from Descartes to Newton* gives an exemplary explanation of Bacon's method of exclusions which Newton presents himself as following (175-84).

an orderly narrative, presents himself as analyzing the possible causes of the elongation of the prismatic image, and rejecting them one by one, until he settles on the final, inevitable cause as revealed by the experimentum crucis. He examines and excludes causation from the varying thickness of the parts of the prism, unevenness of the glass, and the breadth of the sun's image before he finally examines the differing refrangibility of the several colors. By presenting himself as acting as any proper Baconian should, Newton establishes an authority which he will rely on in the latter part of the article.

Most interestingly, Newton's persuasive structure here seems in many respects a close precursor of the kind of articles appearing a hundred years later in volumes 60 and 70 of the *Transactions*, as I have discussed in the previous chapter. Then the rhetorical problem had seemed rather similar to that perceived by Newton: presenting work done out of sight of peers, that gave novel accounts of newly found anomalous phenomena. In those cases, the narrative of the scientist operating under procedures, as any proper scientist might and ought to have done, is the main rhetorical resource to establish the credibility of the events and conclusions. Strikingly, Newton also offers a demonstration experiment at the end, although truncated, just as some of the later writers do. Whether this congruence is a matter of Newton serving as a model or similarity of rhetorical situation suggesting similar rhetorical strategy remains unclear.

What is clear is that much of Newton's account of his investigation in the "New Theory" article differs from details of his earlier accounts. In viewing these differences we need keep in mind that Newton was writing a number of years after the event when memory of dates and sequence may have faded and more significantly after his memory may have been restructured around later meanings. Yet parts of this autobiographical rewriting may reflect a conscious rhetorical strategy adopted for the current account.

Also we need distinguish between accounts of individual experimental events and accounts of the contexts—intellectual, emotional, autobiographical, sequential—in which the experimental report might be placed. The differences we are about to look at all develop contexts concerning the order, motivation, and interpretation of experiments—but not the actual results. As we shall see, the ensuing controversy leads Newton to focus increasingly detailed attention on the experimental events and on the superstructure of claims that can be constructed on those events, rather than on the kinds of contexts in which the events occurred. Thus the kinds of issues in which we see distortion here, fade from importance in later versions.

The first difference is over the dating of the first prismatic experiments. The article dates the purchase of prisms and experiments to 1666, but McGuire and Tamny date the notebook account of Newton's early prismatic experiments in late 1664 (13).[13] The 1666 date shortens the discovery period, emphasizing the lucky-find interpretation and obscuring the longer term interest.

More significantly, Newton represents his motives and attitudes in beginning this work as different from those evident in the notebooks. In the notebooks a theoretical investigation of the motion of particles and light as a form of corpuscular motion very clearly motivates the early experiments. Here, however, Newton presents himself as being moved by the phenomenon of colors itself and as having an attitude of naive wonder at the spectacle of nature: "a very pleasing divertisement to view the vivid and intense colours produced thereby" (92). He presents his observations as incidental to an interest in grinding nonspherical lenses. He does not present himself as trying to find out anything in particular until he stumbles across the surprise of an oblong projection, rather than the expected circular projection. Thus he presents himself as the Baconian collector, free of prior theoretical impulse, being only led into inquiry by the observed facts themselves.

Discrepancies also appear about the sequencing of experiments. According to the article, his first experiment was projecting a narrowed beam of sunlight through the prism against a wall, almost immediately leading to the discovery of the oblong projection. The notebooks describe no such experiment involving projection. The two experiments presented in the notebooks involved looking through the prism at bicolored objects (432–35). The later 1666 paper, "Of Colours," does record a pair of projection experiments, producing the oblong image (#7 and #8), but again only after a pair of experiments (#6) looking through the prism at a bicolored line and a bicolored thread. In the lectures, the projection experiment is presented first of the actual experiments, but not with the claim that it was chronologically first. Newton instead gives the pedagogical rationale that it was the experiment that enabled him to figure out what was happening and would therefore be most helpful to others' understanding (50–53; 284–85). The implication is, of course, that this experiment was preceded by others less easily intelligible.

The article then represents a series of experiments as following al-

---

13. Westfall also doubts the 1666 purchase date and offers evidence for earlier dating of the interest in lens-grinding (*Never at Rest*, 156n) and speculates on the possibility of both earlier and later dating for the purchase of prisms (157–58). In any event Newton would have had to possess some prisms before 1666 to have carried out the experiments reported in the notebook.

Two: Literary and Social Forms in Early Modern Science

most immediately to eliminate possible causes of the effect—in order: thickness, termination with shadow or darkness, unevenness of glass, the curving of the light corpuscles. No such follow-up projection experiments are reported in the notebook, and only a pair in the 1666 paper (#15 and #16), involving a square water-filled prism rather than a triangular glass one. They also appear only after an intervening sequence of experiments, unrelated to the sequence discussed in the "New Theory" article. Further on, another experiment (#24) bears some similarity in method to one of the experiments reported in the article, but the context is entirely different, with Newton looking for color rather than shape. Nor is the finding reported in the later article even recorded in the earlier paper. In fact, the "New Theory" article very carefully separates the issue of differing refraction from that of colors (as does the later version of the optical lectures), but no such separation appears in the private notebooks or the manuscript "Of Colours." These earlier documents are more directly concerned with colors, with differing refrangibility appearing only in explanation of color phenomena.

Further, Newton in the article presents himself as withholding interpretation and belief concerning differing refraction until after the experimentum crucis, while in "Of Colours" after reporting only two very similar projection experiments (#7 and #8), Newton quickly announces his conclusion of differential refraction: "And therefore if theire sines of incidence (out of glass into air) be ye same, theire sines of refraction will generally bee in ye proportion of 285 to 286, & for ye most extreamely red & blew rays, they will be as 130 to 131+" (468). Having achieved closure, Newton moves immediately on to a different sequence of look-through experiments.

In the early paper, a series of experiments resembling what Newton later labelled the experimentum crucis, is presented much after the conclusion of differential refraction (#44–#46). These experiments are, morever, treated as a separate series with no explicit connection to the oblong observation. In the lectures, Newton also describes two experimental arrangements similar, but not exactly the same, to that of the experimentum crucis. Moreover, these variants appear in subordinate positions in the exposition, elaborating different propositions than in the "New Theory" article (96–97; 134–35; 448–51; 496–97).[14]

---

14. Lohne also discusses the ephemeral appearance of the experimentum crucis as a persuasive device in the "New Theory" article ("Experimentum Crucis"). Lohne also points out, that although Newton nowhere else in his optical writings uses the crucial experiment as a form of argument, the experiment so designated in the "New Theory" article becomes emblematic for his optical findings. As an emblem, the illustration of

These many discrepancies strongly suggest that Newton's discovery account was deliberately shaped for this occasion, to create the appearance of the discovery of a naturally found object, described by proper Baconian procedures. This does not mean Newton is lying about what he has found or is fabricating results. It is the sequence of thoughts and experiments that were fabricated. In manipulating the context in which he places his results, Newton revealed awareness that not only must he be convinced of the factual truth of what he has to say, but he must make it appear so to others. As we shall discuss below, the strategy he first chose to create that appearance did not forestall the kinds of criticism journal publication made possible. But in showing awareness of the rhetorical necessity of persuasion, Newton was setting himself on the path that would lead to a more compelling form of argument.

From Discovery to Theory: The Latter Part of the Article

Two sections of the article following the discovery account are easily identified, although unmarked by formal divisions or headings. An account of the invention of the reflecting telescope and a general exposition of the doctrine of colors solidify and extend the conclusions in the narrative.

The presentation of the invention of reflecting telescope as a direct consequence of his discovery of differential refraction helps reinforce the sense of concrete reality of the finding. First, it makes it clear that Newton was so sure of his discovery that he gave up his attempt to solve chromatic aberation through nonspherical lenses and set out on a whole new line of invention. Second, because the reflecting telescope not only worked but was a current sensation, it added certain persuasive force to the refraction findings, as though such a wonder could not be invented without that theory. (This persuasive connection is not only not necessitated by logic, Cassegrain had already independently discovered the reflecting telescope, without needing the push of a theory of differential refraction.)[15] Finally, bringing in the reflecting telescope in a subordi-

---

the experiment remains in increasingly schematic (and imprecise) versions in later optical publications, such as the 1722 Paris edition of the *Opticks* ("The Increasing Corruption of Newton's Diagrams").

15. Moreover, Newton's analysis of the incorrigibility of dispersion in lenses was in itself faulted, as shown by John Dolland a century later. The story of Newton's construction of his faulted argument is described in Shapiro, "Newton's Achromatic Dispersion Law," and Bechler, "Newton's Search" and "A Less Agreeable Matter."

Two: Literary and Social Forms in Early Modern Science

nate way graphically emphasizes Newton's judgments about the greater importance of the refraction findings: if the telescope is considered of consequence how much more important is this doctrine.

The last section shifts from a discovery/invention narrative into a list of abstract propositions, supported by only limited concrete material. Newton calls these propositions stated authoritatively a "doctrine" and makes little attempt to persuade. For his and the propositions' authority, he relies on his credentials for proper method established in the earlier discovery narrative. Whereas the first two sections bear some resemblance to other articles in the *Transactions*, this last section seems in direct contrast with the stated principles and general practice of the journal.

To see how this shift is accomplished and the nature and consequences of this shift, we need first look at the turning point between the second and third sections. The second, just-completed section on the telescope is presented as a continuation of the chronological narrative of the opening, with Newton seemingly just turning his attention from the fundamental discovery to the technological consequence. This narrative continues through observations with the telescope, the current presentation of the telescope in London, and future plans for a reflecting microscope.

At this point Newton switches organization (from narrative to expository list), vantage point (from first person active to third object existential), level of discussion (from discovery and invention process to general claims) and specific topic (from differential refraction to colors) while seeming to be simply continuing his prior discussion. He does this by labelling the telescope narrative a digression and using the concluding sentence of the prior discovery narrative as an assumption for a generalized exposition. That narrative ended with a general statement, which as we have discussed is made to appear a natural experimental fact. The experimental particularity is now, however, left behind, as Newton treats the claim as a general principle which sets the terms for another general statement to be elaborated:

> But to return from this digression, I told you, that light is not similar, or homogeneal, but consists of difform rays, some of which are more refrangible than others. . . .
> I shall now proceed to acquaint you with another more notable difformity in rays, wherein the Origin of Colours is infolded. . . .
> The Doctrine you will find comprehended and illustrated in the following propositions. . . . (96–97)

The Doctrine is then elaborated in thirteen numbered general propositions with only passing reference to experiments or other empirical evidence. The vocabulary is general and statements are lawlike:

> 1. As the Rays of light differ in degrees of Refrangibility, so they also differ in their disposition to exhibit this or that particular colour. . . .
> 2. To the same degree of Refrangibility ever belongs the same colour, and to the same colour ever belongs the same degree of refrangibility. . . .
> 3. The species of colour, and the degree of Refrangibility proper to any particular sort of rays, is not mutable by Refraction, nor by Reflection from natural bodies. . . . (97)

There is a logical and expository sequencing among these statements as Newton elaborates the difference between pure prismatic colors and mixed colors, leading to an explanation of white as a compound, the functioning of the prism, the appearance of the rainbow, and several other related phenomena. After the end of the numbered list appears a statement of even greater theoretical character and generality about the nature of light itself, following on the proposition that light is a quality and not a modification. This generalization revives his earliest speculations on the corpuscular character of light as raised in his notebooks:

> These things being so, it can be no longer disputed, whether there be colours in the dark, nor whether they be qualities of the object we see, no nor perhaps, whether Light be a Body. For, since Colours are the qualities of Light, having its Rays for their intire and immediate subject, how can we think those Rays qualities also, unless one quality may be the subject of and sustain another; which in effect is to call it a Substance. (100)

Newton seems careful to have excluded this deduction from his list of propositions of Doctrine, but neither does he label it a theory, speculation, or hypothesis. Rather he treats it as indisputable fact, a necessary consequence. Except for one qualifying "perhaps" (which will be discussed below) he has been rather careful to avoid any language admitting of uncertainty. In the next paragraph he in fact breaks off the discussion when he feels himself on less firm ground: "And I shall not mingle conjectures with certainties." Even the descriptive title appearing in the *Transactions* "A New Theory of Light and Colours" is Oldenburg's editorial addition.

Newton never uses the word theory or an equivalent. The term doctrine, which Newton does use to describe his generalizations, avoids any possibility of questioning or uncertainty. In his original letter to

Two: Literary and Social Forms in Early Modern Science

Oldenburg, Newton is even more explicit about the facticity of his generalizations, but Oldenburg in his one major substantive editorial change[16] deleted this passage, appearing near the beginning of the third section:

> A naturalist would scearce expect to see ye science of those [Origin of Colours] become mathematicall, & yet I dare affirm there is as much certainty in it as any other part of Opticks. For what I shall tell concerning them is not an Hypothesis but most rigid consequence, not conjectured by barely inferring 'tis thus because not otherwise or because it satisfies all phaenomena (The Philosophers universall Topick,) but evinced by ye mediation of experiments concluding directly & without any suspicion of doubt. To continue the historicall narration of these experiments would make a discourse too tedious & confused, & therefore. (96–97)

Then, as in the published text, he continues "I shall lay down the Doctrine First, and then, for its examination, give you an instance or two of the Experiments, as a specimen for the rest" (97).

This excised passage not only asserts Newton's certainty of his claims, but also characterizes the claims, gives his grounds for belief and explicitly discusses his strategy of presentation. Both the character of the claims—that they are mathematical—and the grounds for his belief—direct experimental proof—are nowhere in evidence in the article and can only be considered plausible in light of his manuscript of the optical lectures, which he anticipated publishing. Those lectures include extensive mathematical derivations, proofs, and calculations as well as pages of experimental demonstrations. Although they are not arranged argumentatively as a definitive proof, they carry the enormous weight of a coherent and empirically responsible system, as discussed earlier. In this article, however, we have little more than Newton's word to go on, relying primarily on the credibility he has established in the first two parts. Even the experimental evidence he calls into play is only sketched in a passing phrase, again throwing us back on his credibility for accuracy, method, and interpretation. Only a single demonstration experiment is described in any detail. A demonstration experiment is of course very different in character than a proof by experiment. The demonstration experiment simply puts the phenomenon on display; it does not resolve any question nor directly argue for any proposition.

His reasons for adopting this strategy are apparent and admitted.

---

16. The only other changes were Oldenburg's deletion of the words "& others" and the signature (*Correspondence* 1:102).

First, he would need a book (his lectures) to lay out his full system and evidence—that I take to be part of the meaning of his phrase "too tedious." But the other part of his reasons is suggested by the continuation "& too confused." That is, although he has been able to recount some of the experiments in the first part of the article to suggest an orderly discovery procedure of differing refractions, he cannot create as neat and pointed a story out of this other half of his claims. His lectures, because aiming at a complete exposition, do have a structure, but not an argumentative one—they are tedious and argumentatively confused in the accepted, pedagogically useful, academic sense. Here he has neither the space nor the appropriate relation with his audience to be the tedious professor.

He seems rather to have a collegial estimation of his readers, relying on them to fill in the necessary details. He reveals his assumption about readers being able to grasp the consequences and implications of his claims just before the demonstration experiment, when he states "I see the discourse it self will lead to divers experiments sufficient for its examination. And therefore I shall not trouble you further than to describe one of those, which I have already insinuated" (100). Reasonable readers should be able to follow his lead properly on their own. In a letter written to Oldenburg four days later (on 10 February), Newton confirms this perception of his readers and his relation to them: "I designed [the letter] onely to those that know how to improve upon the hints of things, & therefore to shun tediousnesse omitted many such remarques & experiments as might be collected by considering the assigned laws of refractions" (109).

Thus the persuasiveness of the whole seems to rely on a confidence Newton's voice maintains about the facticity of the specific events and general claims made. The first part of the article narrates the discovery of a general claim as a natural fact stumbled across and described through proper method. The last part presents an entire system of claims based only on the authority Newton has established earlier and the anticipation that, having read this article, readers will go out and see exactly what Newton saw. The article ends with an invitation that others indeed do that. Although Newton raises the possibility of admitting error, the article ends with a self-assured final clause: "If anything seem to be defective, or to thwart this relation, I may have an opportunity of giving further direction about it, or of acknowledging my errors, if I have made any" (102).

Two: Literary and Social Forms in Early Modern Science

First Responses and Newton's Answers

The rhetorical strategy of establishing personal authority to underpin broad claims (technically known in rhetoric as the argument from ethos) seemed to have worked when Newton's letter was read aloud at the Royal Society meeting of 8 February, for it was met with general approbation. Oldenburg reports in a letter later that day to Newton, that the account was "mett both with a singular attention and an uncommon applause, insomuch that after they had order'd me to returne to you very solmne and ample thankes in their name" (107).

As soon as the account met more careful inspection, however, it came under question from many quarters, most immediately by Robert Hooke who took a copy home and within a few hours had written a long critique which he read at the next meeting of the society on 15 February. The controversy over the "New Theory" article, initiated by Hooke, lasted four years, into 1676, and seems to fall into three periods. In response to each set of criticisms, Newton develops a related set of rhetorical strategies, such that by the close of the period, the main features of the presentation of the *Opticks*, Book 1, are set.

The first set of criticisms, as outlined in the table below, are immediate responses to the reading of the text, in either manuscript or printed version. These were all initiated within two months of the article's publication. Newton had access to them all before writing a response to any of them, and Newton's answers were published in the *Transactions* before the end of the year.

Critic	Date of criticism	Date Newton answered	Date criticism published	Date answer published
Robert Hooke	Feb. 15	June 11	—[17]	Nov. 18
Robert Moray	?	April 13	May 20	May 20
Ignace Pardies	April 9	April 13	June 17	June 17
	May 11	June 10	July 15	July 15
	June 30	—	July 15	—
Generalized response to all three				
	—	July 6	—	July 15

Of this first round of controversy, Hooke's criticism was most significant, done first, yet answered in print last. Newton's attempt to formu-

---

17. Although at the time Hooke's critique was only read aloud and then circulated in manuscript, it has since been published in Birch (10–15), Newton *Papers* (110–15), and Newton *Correspondence* 1 (110–15).

Between Books and Articles

late a proper answer to him influences all the other responses Newton makes during this period. That is, Newton received Hooke's critique (20 February) two weeks to the day after dispatching the original article. He immediately promised a rapid reply. (116) According to a letter of 19 March from Newton to Oldenburg, Newton appears to have drafted some comments, which he did not find sufficient (122).[18] He did not send Oldenburg his completed comments until June. In the meantime, while several times renewing his promise to answer Hooke forthwith and otherwise making reference to a task obviously very much on his mind (see *Correspondence* 1:137, 155, 159), he received and answered two other sets of correspondence on the same subject. When he finally replied to Hooke he relied on all the rhetorical tactics he had developed in the interim correspondence. He then reduced these rhetorical lessons to a single strategy embodied in a list of queries proposed in a letter to Oldenburg, which Oldenburg printed long before the reply to Hooke. In order to analyze the development of rhetorical tactics, we will examine the correspondences in the order of Newton's answers.

Newton first answered Sir Robert Moray, the first president of the Royal Society (1660–62) and a continued active member. Moray had proposed a series of four experiments to be carried out by Newton. The purpose of these variations of Newton's reported experiments is not spelled out, but they seem aimed at establishing whether Newton's results may have arisen from other causes or may have been contaminated in some fashion. Newton handled these proposals by spelling out in

---

18. Newton was probably referring to the manuscript on folios 445 to 447 in Add 3970. This manuscript reflects several of the features of Newton's eventual response, such as the appeal to the common ground of plain inquiry, the calculation of the relative errors of refracting and reflecting telescopes, the attempt to distance himself from the corpuscular hypothesis, the exploration of analogies, and an attempt to distinguish between compounded and uncompounded light. In this early draft, however, Newton's attempt to disown the mention of corpuscularity is awkward and involuted, his distinction between compounded and uncompounded light is not as crisply drawn, and his use of analogy is not contained by his later-developed argument that arguing by analogies is futile. Thus there is no attempt to switch the discussion from theoretical grounds to empirical ones, although he does complain that Hooke seems to be more concerned with asserting his hypothesis than inquiring after the truth. By recognizing his rhetorical problem in trying to put a wedge between Hooke's commitment to his hypothesis and the evaluation of Newton's own claims, Newton is only a step away from finding the rhetorical solution of discrediting hypotheses. Nevertheless, in this early attempt to answer Hooke, Newton tries to meet Hooke more on Hooke's own grounds. In the final version, discussed later in this chapter, Newton's newly developed strategies of disowning hypothetical discussion and reducing issues to empirical questions allows him to distance himself from the complaints Hooke makes and to mount more elaborate and forceful arguments against Hooke.

greater detail the methods and results of relevant experiments briefly mentioned in the earlier article and adding to them other relevant experiments he had already done but had not mentioned in the article. In this manner he demonstrated that he had already taken Moray's concerns into consideration (*Correspondence* 1:136-39; *Transactions* 7:4059-62).

In the first letter, dated 9 April, Ignace Gaston Pardies (professor at College Clermont in Paris and a committed Cartesian) objected to Newton's theory on the grounds that the experimental results were explicable by existing laws of optics and that certain other common experiments contradicted Newton's conclusions, which Pardies labelled a hypothesis. The largest part of the letter offers a geometrical derivation showing that according to received principles the expected shape of the image projected through a prism in Newton's experiment should be an oblong; therefore, Newton's results are unsurprising and do not require any new theory. Newton responded to this quickly (on 13 April) and simply, by adding an experimental detail he neglected to put in the article (but was in the account of the experiment in the lectures) and by redoing Pardies' geometric derivation (again paralleling an expansion in the lectures). To Pardies' other criticisms, he gives further detailed explanation and interpretation of experiments he had done and the common experiments mentioned by Pardies (*Correspondence* 1:140-44; *Transactions* 7:4091-93).

Here, as in the response to Moray, Newton is discovering the limitations of the elliptical style he had adopted for the article, and is returning to the fuller exposition of the lectures. As students may need full details as part of their education so that they can comprehend fully, so do one's peers, for although they are likely to fill in the details on their own, they are likely to do it in their own way, according to their own lights. Newton is discovering he cannot rely on shared visions and shared experience. Although he still insists that he is not here hypothesizing, he does willingly label his claims a "theory." By categorizing the phenomena he presents as "certain properties of light, which, now discovered, I think not difficult to prove" (144), Newton shows a nascent rhetorical awareness that discovery is different than proof, and that proof requires its own set of arguments.

Pardies' second letter (of 11 May) accepted all of Newton's added details and elaborations, but still denied his conclusions. Pardies claimed that alternate hypotheses explained the results equally well. Thus, although Pardies apologized at the end for calling Newton's conclusions hypotheses, Pardies still called them theories and considered them no more firm than the hypotheses of other people. More substantively, he treated Newton's claims as hypotheses by arguing there was no necessary link between the empirical evidence and the general claim (*Correspondence* 1:156–59; *Transactions* 7:5012).

Newton's reply to the second letter takes up the direct challenge of the concept of hypotheses and admits three alternative hypotheses which he considers legitimate (*Correspondence* 1:163–71; *Transactions* 7:5014–18; translation from Baddam 1:375–79). He even goes so far as to suggest ways of amending Hooke's, Descartes's, and Grimaldi's hypotheses to be consistent with his results. Thus he argues that there is no end to hypotheses: "since numerous hypotheses can be devised, which shall seem to overcome new difficulties." Newton claims that his doctrine is different in kind—for he has reduced the issues to empirical ones. This claim may not be precisely accurate, for we have already seen how much his early work was speculative, how his later experiments were driven by concerns arising from speculation, and how these early speculations creep into the article presentation. We will see later how he backtracks on these points to maintain his nonhypothesizing stance. Yet we can see the drift of a rhetorical strategy which attempts as far as feasible to reduce all questions to empirical issues.

Following this strategy, Newton takes his problem in this particular letter to translate the issues raised by Pardies into concrete empirical issues to be determined by experiment. "To lay aside all hypotheses" he considers the substantive force of the disagreement: "the whole force of the objections will lie in this, that colours may be lengthened out by some certain diffusion beyond the hole, which does not come from the unequal refraction of light or of the independent paths of light." Having redefined the issue so, Newton then recounts the experimentum crucis from the original article in greater detail and more concretely, with greater explanation of the meaning of the event. Moreover, he points to a procedural detail which Pardies may not have been aware of and which would lead to different results and different interpretation.

Whether or not crucial experiments are philosophically a valid and certain procedure, and whether or not they actually prove to be persuasive in the majority of disputes, in this particular case reduction of theoretical issues to empirical ones determined by a crucial experiment, elaborated adequately for all parties to share an understanding of the event, turned out to be a useful rhetorical strategy. Pardies replied soon thereafter,

> I am quite satisfied with Mr. Newton's new answer to me. The last scruple which I had, about the Experimentum Crucis, is fully removed. And I now clearly perceive by his figure what I did not before understand. When the experiment was performed after

Two: *Literary and Social Forms in Early Modern Science*

his manner, every thing succeeded, and I have nothing further to desire. (*Correspondence* 1:205–6; *Transactions* 7:5018; translation from Baddam 1:379.)

Such a successful outcome would certainly reinforce Newton's faith in the rhetorical power of detailed experimental accounts.

Answering Hooke

Robert Hooke's critique, although written almost immediately after Newton's paper was first read before the Royal Society, was by far the most difficult, penetrating, and challenging. In strategy the critique resembles Pardies' second letter, characterizing Newton's claims as hypothesis, no superior to a number of equally plausible hypotheses; however, Hooke scrutinizes in greater specificity Newton's corporeal assumptions, his own alternative wave hypothesis, and detailed points of divergence between the two. Following in order the numbered list of claims in Newton's paper, Hooke deems some of Newton's claims consistent with his own hypothesis, but offers explanations of some of the results to demonstrate that his hypothesis is of greater explanatory power than Newton's. Moreover, he denies that the experimentum crucis is indeed crucial in distinguishing between the two hypotheses, whereas Pardies only expressed some procedural uncertainties about the experiment. I leave out of discussion of Hooke's critique and Newton's response, technical issues concerning telescopes.

To answer, Newton adopts a strategy similar to the one he chose for Pardies' second letter: denying that his claims are hypothetical, discrediting hypotheses as a mode of investigation, then reducing the issues to empirical ones, and finally reestablishing the experimentum crucis. However, because of the intensity and specificity of Hooke's challenge, Newton must work harder and add new twists to the argument to achieve the same effects.

First, because Hooke more pointedly identifies the speculative remains of prior hypothesizing—the corpuscularity argument near the end of the article—Newton must distance himself from his comments. A simple denial of hypothesizing is not enough. He argues that this entire late passage was couched by a "perhaps" which identifies its hypothetical character and sets it apart from the main body of his more solid findings, which are discussed in terms independent of the alleged hypothesis (*Correspondence* 1:171–72; *Transactions* 7:5086). His invocation of the "perhaps" is a weak argument, for the word in the original article is proceeded by "it can be no longer disputed" and followed by related claims in equally urging language: "which in effect is to call it" and "we have as good reason to believe" (*Correspondence* 1:100; *Transactions* 7:3085). Moreover, the article continues with another more detailed set of questions based on the corpuscularity assumption before breaking off the discussion as entering into conjecture. Newton's defense that in the body of the article he avoided terms based on corpuscular assumptions seems equally suspect. The terms he says avoided all concern with the issue of perception, which is not a topic in the article, although he does discuss the issue in his notebooks. The public article gave him no occasion to use the assumption-laden vocabulary.

Whatever the strength or weakness of the defenses, Newton's face-saving and backpedaling aims to separate his main claims from anything he cannot identify as experimentally grounded. In all future presentations of the optical findings he was to avoid any language that would raise the specter of corporeality.[19]

But damage was done to Newton's position, and Newton felt it necessary to reconcile his doctrine with those details Hooke claimed were better accounted for by his own hypothesis. This was particularly important since Hooke had claimed that his gave a better account of color dispersion in layered plates, which seemed back then and still seems now, much easier to explain as a wave phenomenon. Newton explained how secondary wave phenomena arose by movement of corpuscles through the ether, as stones thrown into ponds create waves.

This attempt to reconcile wave phenomena with a particle account became the basis of Newton's explanation of rings in Book 2 of the *Opticks*, which we will not examine here.[20] In the current context, however, two points are significant. First, the discussion of rings, and thus the necessity of discussing wave phenomena, particularly in the cumbersome way Newton had to in order to reconcile it with his other conclusions, was separated out from the basic theory of refraction and colors. Once again he establishes clarity around an issue by distancing it from

---

19. Hooke began a response to Newton's answer in an unfinished letter (Newton *Correspondence* 1:198–205). It is uncertain whether Newton or Lord Brouncker, the current president of the Royal Society, ever saw-the letter. In it, however, Hooke beards Newton at some length for having relied on the corpuscular hypothesis in the "New Theory" article. By ostensibly excusing himself for the mistake in attributing the corpuscular hypothesis to the article, Hooke introduces extensive textual evidence to show how the hypothesis appears to be taken. These apparent references were, of course, the cause of his "mistake."

20. In "Uneasily Fitful Reflections on Fits of Easy Transmission," Richard Westfall provides an enlightening account of Newton's corpuscular explanation of wave phenomena.

Two: Literary and Social Forms in Early Modern Science

all troublesome matters. Second, the "fits of easy reflection" theory that derives from this formulation never achieves the crispness of presentation that the theory of refraction and colors does. That is, Newton is never able to reduce the material to a closely linked network of generalizations and empirical results of compelling character. In Book 2 there remains a large explanatory and hypothetical middle between claims and results, with consequences for the structuring of the argument.

In this answer to Hooke, despite having to address the comparison of hypotheses, Newton must distance his main claims from the discussion of rings, lest his whole set of claims be tainted with the brush of uncertainty. So he introduces the comparison of hypotheses with a denial of responsibility for what follows: "But supposing I had propounded that Hypothesis . . ." And he ends the comparison by disowning hypothetical discussion as unnecessary for his doctrine: "But whatever be the advantages or disadvantages of this Hypothesis, I hope I may be excused from taking it up, since I do not think it needful to explicate my Doctrine by any Hypothesis at all" (*Correspondence* 1:174–77; *Transactions* 7:5087–91). Between these two disclaimers, Newton uses an analogy with sound phenomena to suggest that Hooke's hypothesis is consistent with his doctrine and provides a plausible alternative, up to a point. That point is when the analogy reveals a patent absurdity.

Newton uses the breakdown of the analogy to discredit hypothetical discussion and move on to his experimental discussion: "You see therefore, how much it is besides the business in hand to dispute about Hypotheses. For which reason I shall now in the last place proceed to abstract the difficulties in the Animadversor's discourse, and without having regard to any Hypothesis, consider them in general terms" (177, 5091).

Newton uses here what strikes modern ears as strange locutions to talk about empirical results in contrast to hypotheses. To us terms like *abstracting* and *general* seem associated with theories and hypotheses, instead of being opposed to them. Newton has also used similar language both earlier in this response and in his second response to Pardies, so it cannot be dismissed as a chance usage. What Newton seems to be meaning here and in similar contexts is that concrete implications of a general character can be abstracted, or pulled out of hypotheses and treated empirically separate from the larger explanatory systems of the hypotheses. These generalities are in the form of empirical claims, and are thus open to empirical tests. Thus although certain generalizations may have their origin in corporeal hypotheses, they can be cast in empirical terms and tested in ways that do not explicitly invoke corporeality.²¹ He continues in this passage immediately to propose a series of empirical questions, that are the result of this process of abstracting:

> 1. Whether the unequal Refractions, made without respect to any inequality of incidence, be caused by the different refrangibility of the several Rays; or by the splitting, breaking or dissipating the same Ray into diverging parts?
> 2. Whether there be more than two sorts of Colours?
> 3. Whether Whiteness be a mixture of all Colours? (178; 5092)

As in the second response to Pardies, the strategy is to reduce all issues to empirical ones. The effect of this, however, as Newton's language is beginning to recognize, is to create another level of claim, in between the large explanatory hypothesis and the specific empirical result. This claim is a generalization based on results and is to be held specifically responsible to empirical results. The premium is to be placed, in fact, on establishing as strict a link between the result and the claim as possible. Newton will work out further implications of this middle level empirical generalization for the form of his argument in the later exchange with Huygens, but here he already begins an amplification and reorganization of his materials around general empirical questions and their answers.

With respect to the first proposed question, Newton adds experimental and interpretive details to support his claim that the unequal refractions were not caused by factors other than differing refrangibilities. He recounts a number of experiments directly relevant to the claim but not reported in the original article and spells out their direct meaning with respect to the claim. The added detail of interpretation is of equal importance to the added detail of the account.

With respect to the second question concerning Hooke's claim of only two fundamental colors, Newton gives two arguments. The first argument is not an empirical reduction; quite the contrary, it is a theoretical argument of why Hooke's results could not be precisely as he reports them. Newton recognizes the weakness of this first form of argument

21. In modern terms, this is Newton's first attempted solution to the problem of translatability and gaining some measure of intertheory agreement on empirical grounds. The continuing difficulties Newton encountered in gaining agreement suggests the difficulties in translation between theoretical systems. On the other hand, the success of his later solution of building mutual understanding and experience from first principles on up suggests the possibilities of intertheory agreement based on carefully constructed empirical grounds.

by putting greater weight on his second, as revealed in his transition to the second point: "But, supposing that all Colours might according to this experiment, be produced out of two by mixture; yet it follows not, that these two are the only Original colours" (180; 5095).

The second argument is indeed the more essential, developing a fundamental distinction between simple and compound colors. Although Newton had made the interpretive distinction in the original article, he had not applied it to the kinds of experiments Hooke, and later Huygens, discuss. Only by making the application of this distinction to such experiments intelligible and persuasive could Newton establish publically what he believed was the proper interpretation of the results. The manner of Newton's handling of this point is particularly important, for he finds it necessary to make this point repeatedly, and each time he finds a crisper way of articulating it. The final way he finds, in the Huygens exchange, will provide a structural model for Book 1 of the *Opticks*.

In articulating his position more clearly, he also refines the particulars of his claims, in response to forceful objections concerning the kind of light accounted for by his theory. The clarifying of the definitions of simple and compound colors, makes clearer the issue of the character of various forms of white light, including the sun's. As Alan Shapiro elucidates in "The Evolving Structure of Newton's Theory of White Light and Color," the challenges to his earlier formulations, by himself and others, forced him to withdraw from broader claims about all white light and about the sun's light. Even the final narrowed definitions of the *Opticks* contain some irregularities which troubled Newton, but which he was unable to resolve (see also Shapiro, "Experiment and Mathematics," and Westfall, "Development"). Interestingly, his reformulations are often accompanied by statements that he is only clarifying or elaborating an earlier formulation, rather than that he is retracting a position. Whether this is simply a face-saving ploy or a reflection of the psychological experience of being forced to a more precise statement is difficult to disambiguate. The terms of a claim can be refined through an agonistic struggle, and the refinements may have substantive consequences for the claim. By clarifying terms you can clarify them to yourself as well as to others, and that clarification may make distinctions visible that were not visible before. Is a prior obscurity rightly perceived as error? In any event, the refined claim is an improvement in being more defensible, given the current means of argument and use of empirical evidence. By allying itself more closely to the available empirical evidence, the improved claim is actively relying on passive constraints for its force, as Fleck argues is characteristic of modern science.

Newton recognizes the difficulty of communicating the distinction between simple and compound colors in the sentence introducing the substantive discussion: "But, because I suspect by some circumstances, that the distinction might not be rightly apprehended, I shall once more declare it, and further explain it by examples." He begins the substantive discussion by defining the two terms directly and contiguously so to make evident the contrast: "That colour is primary or original, which cannot by any art be changed, and whose Rays are not alike refrangible: and that compounded, which is changeable into other colours, and whose rays are not alike refrangible" (180; 5095). In the original article this distinction had been made descriptively over two substantial paragraphs, rather than as tightly contrasted definitions. Newton now continues to provide empirical prismatic tests to distinguish between the two kinds of colors. He then offers an experimental confirmation of the distinction by proposing that if two objects of apparently the same color are so tested and found to be distinguishable, then the colors must be of two kinds.

In this manner Newton has presented a general issue, reduced it to an empirical question, then organized the experimental material so as to present a direct line of reasoning tying the empirical to the general.

The last query presents no such complicated problems of tying claim to evidence, for Hooke had presented a rather straightforward empirical challenge: "Methinks, all the coloured bodies in the world compounded together should not make a white body, and I should be glad to see an experiment of that kind done" (Birch 14; Newton *Correspondence* 1:114). After reiterating an experiment already reported and examining the difficulties which Hooke would have in explaining away that result, Newton offers at least twenty-one other experiments and observations to the same point.

After this inductive pummelling of Hooke, Newton simply reasserts the importance and soundness of the experimentum crucis, without adding new substantive discussion.

Concluding the First Round

In dealing with the queries and objections of Moray, Pardies, and Hooke, Newton has learned to reorganize his discussion to argue specific claims or positions. Moreover, he has found it most useful to tie his discussions as closely as possible to empirical results. Even reasoning processes are to be supported by empirical procedures at each step.

As a conclusion to his response to this first set of challenges, Newton sends Oldenburg a set of experiments which he considers appropriate for testing what he now calls his theory. Although this list of experiments was published on 15 July, 1672, along with his response to the second Pardies letter and before his response to Hooke, it was written almost a month after he had finished all the other responses (see table, p. 100). His continued use of the term *theory* (first used in responding to Pardies) in place of the original term *doctrine* indicates his recognition that he is offering a higher level of claim beyond simple descriptions of experimental fact, yet still to be distinguished from hypothesis. He wishes, moreover, to remove discussion of his theory from confutation among a variety of opinions, by having all interested parties suspend all objections deriving from hypotheses and convincing themselves by "Experiments concluding positively and directly," as he claims to have done (218; 4044).

To that end, he reduces the relevant issues to a series of eight experimental questions. Although he provides no answer to these questions here and gives no indication that he has indeed satisfied himself on these issues, we know from his other writings that he already has confirming results for all of these issues. By leaving the questions apparently open, he takes the stance of letting the facts speak for themselves. However sure he is of the facts, nonetheless, Newton will come to distrust even this rhetorical strategy, for future controversy was to convince him that people won't read the experimental facts correctly unless he reads the facts to them. Indeed he has already begun to have second thoughts about the elliptical approach he has taken. In a letter of 8 July 1672, Newton writes to Oldenburg:

> Touching the Theory of Colours I am apt to believe that some of the experiments may seem obscure by reason of the brevity wherewith I writ them wch should have been described more largely & explained with schemes if they had been then intended for the publick. (212)

The Second and Decisive Round

The first round of controversy had set all the wheels of a new style in motion, but the style had not yet found its settled form. For that, Christian Huygens' challenge the following year was necessary. Huygens originally had been favorably impressed with Newton's article, but had an increasing number of questions with time. In his fourth

Between Books and Articles

letter to Newton, transmitted by Oldenburg on 18 January, 1673, and published in the *Transactions* of 21 July, Huygens offers serious criticism. He takes his lead from Newton's reduction of the controversy to empirical issues by pressing one of Hooke's earlier empirical proposals:

> I have seen, how Mr. Newton endeavours to maintain his new Theory concerning Colours. Me Thinks, that the most important Objection, which is to be made against him by way of Quaerre, is that, whether there be more than two sorts of Colours. (255; 6086)

Huygens then claims that yellow and blue can combine to form all other colors including white and proposes an experiment (which he admits to not having yet done, for the thought came to him just as he was writing).

Whether two colors might combine to form all others was not a question Newton thought essential to his findings and is not in his list of eight queries, nor is it considered anywhere in Newton's writings except in response to Hooke and Huygens. As we have seen, Newton considered experiments demonstrating this flawed because they did not distinguish between compound and simple colors. Once that distinction is accepted, the importance of such experiments for the validity of his theory evaporates. Yet because such proposals were such a sticking point with his opponents, because they did not see the import of the distinction, Newton had to lead his readers through the distinction.[22]

Newton's first response to Huygens was written on 3 April, 1673, but not published until 6 October due to an editorial mistake. It begins with a methodological point, suggesting that using compound colors to compound again will only lead to confusion and that one must begin with simple colors. Thus he insists on his distinction as a necessary methodological consideration even prior to its interpretive use. That is, while he still leaves the empirical confirmation to the reader, he recognizes more and more how that empirical experience must be led and con-

---

22. It also was apparently an unpleasant challenge that Newton first attempted to evade by withdrawing from the Royal Society. In a letter of 8 March 1673, to Oldenburg, Newton first suggested that Huygen's critique needed no response, since it was part of a private correspondence of Oldenburg and Huygens (despite the well-established practice by that time of reading the private correspondence at Royal Society meetings and printing it in the *Transactions*). However, after promising to respond if Oldenburg pressed the issue, Newton requested to resign from the society, for he lived too far from London to take advantage of the meetings (262). Oldenburg answered on 13 March by excusing Newton from paying dues (263). The matter was then dropped, and within three weeks Newton sent his first reply to Huygens' critique.

Two: Literary and Social Forms in Early Modern Science

strained so as to produce the proper results. Experiments must be done properly, embedding the proper assumptions and in the proper order. This further implies, so that the assumptions do not appear simply a priori, that the experiments establishing assumptions must be done first. In this case the experiments establishing the distinction between compound and simple colors logically must precede experiments on the production of other colors. The later experiments are to be constrained by the conclusions of the earlier.

Starting as Newton did with an immediate sense of the concrete and self-evident facticity of his findings, Newton has been discovering that empirical experience is a variable thing. His readers did not immediately understand that certain claims implied certain prior completed experiments nor that other experiments follow by immediate implication (as in the Moray case). Nor did they always perform the experiment in the way Newton had, which led to varying results and disagreements (as in the Pardies case). Nor did they even do the experiments Newton proposes (as is evident in Hooke's critique and following letters). Nor did they even do experiments they themselves thought up (as in this first letter from Huygens). Since Newton himself was convinced that anyone who went out and did the proper experiments could not doubt the concrete truth of his doctrine, it is not odd that he would get increasingly frustrated with what he might consider the readers' obtuseness in getting the experiments right. On the other hand, he is realizing that he must provide much more detailed instructions—of logical procedure, sequencing, and interpretation, as well as of apparatus and procedure—in order that they get the experience right. In Newton's words to Huygens we find both his attempt to challenge better experimental procedure from his critics and his increasing disillusion with the certainty of this happening:

> This, I confess, will prove a tedious and difficult task to do it as it ought to be done; but I could not be satisfied, till I had gone through it. However, I only propound it, and leave every man to his own method. (264; 6108)

Before discussing concrete experiments in answer to Huygens, Newton argues by analogy about how implausible Huygens' hypotheses are, but again immediately disowns this analogical discussion as being irrelevant to his purpose in exhibiting concrete phenomena—as he had done with Pardies and Hooke. Newton then discusses how Huygen's experiments out to be done properly and how they ought to be interpreted. In passing, he elaborates some of his own experiments, which, although producing results resembling Huygens', point to substantially different conclusions. Thus he ends with an experimental challenge as to what Huygens must do, and even that result, if obtained, Newton promises to show is not what it appears:

> If therefore M. Hugens would conclude anything, he must shew, how White may be produced out of two uncompounded colours; wch when he hath done, I will further tell him why he can conclude nothing from that. (265; 6110)

Newton further goes on to claim he had tried that experiment and had not been able to get the results that Huygens would wish for.

In a letter 10 June (published 6 October), Huygens responds by backing off from further disagreement, except to bite at the bait that Newton had offered:

> I list not to dispute. But what means it, I pray, that he saith; Though I should shew him, that White could be produced if only two Un-compounded colors, yet I could conclude nothing from that. Yet he hath affirm'd in p. 3083 of the Transactions, that to compose the White, all primitive colors are necessary. (288; 6112)

At this point the discussion is now focussed on a single interpretive issue, which is Newton's task to make clear and unquestionable. Newton in his answer of 23 June (published 21 July) explains his position three times, ending with an argument in the form of a compelling mathematical derivation.

Newton had been thinking about such a format since having completed the first round of controversy, which ended in him reducing his claim to a series of empirical queries. Shortly after receiving the list of queries from Newton on 7 July 1672, Oldenburg had requested Newton in a letter of 16 July to elaborate on the appropriate experiments. On 21 September Newton replied to this request belatedly:

> I drew up a series of such Expts on designe to reduce ye Theory of colours to Propositions & prove each Proposition from one or more of those Expts by the assistance of common notices set down in the form of Definitions & Axioms in imitation of the Method by wch Mathematitians are wont to prove their doctrines. And that occasioned my suspension of an answer, in hopes my next should have conteined the said designe. But before it was finished falling upon some other business, of wch I have my hands full, I was obliged to lay it aside. (237–38)

In this mathematical form of proof Newton sees a way of compelling assent and ending controversy.

Two: Literary and Social Forms in Early Modern Science

In presenting his answer to Huygens on this highly focused issue, he has his first opportunity to display this new rhetorical strategy. Nonetheless, it takes Newton three levels of presentation in this one letter to reach the mathematical form he seeks. That is, he presents his main point in three different ways before the issue can be turned to one of mathematical argumentation from first principles and supporting statements. He starts with a direct answer to the issue at hand cast in general form. He then turns the answer into a general position which he supports by experiment and then uses to analyze Huygens' proposed results. Only then does he derive his conclusions from first principles. That is, he must lead Huygens from a hostile theoretical position, focused on a particular point in contention, through an alternative answer instantiated in an experiment, to a reconception of the original experiment. Only then can the exact meaning and full implications of the original experiment be made accessible by placing it within a rigorously drawn new system. The procedure seems to be to compel the hostile Huygens to take Newton's system seriously in its own terms instead of seeing it just as a proposal competitive with the Cartesian one held by Huygens.

The first statement of Newton's response is presented in a few general sentences of direct answer. The issue is why Huygens could not conclude anything from the compounding of white from two colors. Newton answers because "such a white would . . . have different properties from the white . . . of ye Sun's immediate light, of ye ordinary objects of our senses, & of all white Phaenomena that have hitherto falln under my observation" (291; 6087). Moreover, those differences of property would support his theory, for they would reveal how ordinary whites are produced by more than two colors.

To explain this difference more precisely, Newton must shift to the second level of his argument. This shift is well marked by a transitional sentence: "But to let you understand . . . I shall lay down this position" (291; 6088). This shift of argumentative level is accompanied by a change in discourse focus, organizational pattern, and graphic layout. The position he offers is italicized and separated from the surrounding text. It becomes the central focus of the following three paragraphs, organized as experimental demonstration, deduction of consequences, and application to the Huygens' experiment.

The sentence style is also particularly interesting here, in light of Newton's expressed intent in developing a mathematical type of argument. Earlier in recounting experiments, Newton had most frequently adopted a first-person past-tense narrative, although for the demonstration experiment at the end of the "New Theory" article he had adopted general imperative instructions; e.g., "In a darkened room, make a hole in the shut of a window . . ." (100; 3085). Here, although he claims to have done this experiment and many like ones, and although he is using it in support of a claim rather than as a demonstration, he again uses the second-person imperative mode. This casts the responsibility for doing the experiment back on the reader, as he had been trying to do with his queries. The instructions, however, have the advantage over the queries of leading the reader more strongly and precisely.

Newton takes charge even further by commanding not only the actions but also the interpretive process, as is done in a geometrical demonstration: "Let α represent an oblong piece of white paper" (291; 6088). Newton had of course used such language of mental command when engaged in geometric derivations and analyses of optical phenomena during his lectures, but here this is being applied directly to the experiment. This strategy of interpretive command further has the advantages of increasing the appearance of generality to the claims and lending the universal force of geometry. Moreover, it then presents the results of the experiment in the precise form and mode for the continuation of a geometric argument. With no change of tone, the second paragraph deduces conclusions, which are then immediately applied to Huygens' proposed experiment, which is treated as an abstract geometrical problem, since Newton does not consider Huygens' results plausible. Moreover, this hypothetical geometrical problem is described in the exact same style of the actual experiment ("suppose that A represents . . ."). Even the same diagram and reference letters used to describe the actual experiment are reused for the hypothetical.

Newton has succeeded in integrating an actual experiment into a general geometrically styled argument. Doing so, he has eliminated the need for the interpretive arguments he has needed earlier to make clear the import of the experiments. Plus he has found a way of totally divorcing his claims from his explanatory hypotheses, which he kept finding himself tempted to discuss and then having to disown as irrelevant. That is, the experiments don't find a meaning in any external explanatory scheme, but only within the scheme in which they are serving as cogs. Moreover, since the language of presentation is so tightly linked into that immediate scheme, no loose linguistic ends suggest switching to any analogical or explanatory mode of discourse. The geometrical precision ensures that its own boundaries are maintained. And finally, the geometrical argument in support of general propositions removes the local and direct confrontation with specific opponents. The text is addressed to a general proposition rather than against Hooke or Huygens.

Two: Literary and Social Forms in Early Modern Science

To complete this translation into a geometrical argument, Newton follows this three-paragraph demonstration with a brief return to personal confrontation to indicate that the points being made in this abstracted form are exactly the same points he had made in earlier presentations. Specific references and appeals to comparison attempt to establish that this is the position he has been maintaining all along, although not quite in this general form. This cross-reference also serves as a personal character defense against Huygens' comment that Newton "maintains his doctrine with some concern." Newton here suggests that the whole problem has been the lack of the readers' comprehension, and he has only been explaining previous answers to people who were not able to see his points.

Having succeeded in translating the point at contention into a geometrically styled argument of the lemma sort, and having established and elaborated that lemma, Newton has changed the level of discourse. Now he can begin to lay out his whole system in this general mathematical form, thereby indicating the precise meaning of the current claim. Again he recognizes the transition through a single sentence similar in syntax and phrasing to the transitional sentence cited earlier:

> However, since there seems to have happened some misunderstanding between us, I shall indeavor to explain my self a little further in these things according to the following method. (292; 6089)

This last level of mathematicization of the argument is further recognized by the organization and labelling of the parts: five numbered items under the italicized, separated, centered heading *Definitions* and nine numbered items under *Propositions*. Each definition consists of a single naming statement: e.g., "I call that Light homogeneal, similar or uniform whose rays are equally refrangible" (292; 6090). Similarly, each proposition statement consists of a single-sentence claim followed by one or more sentences of proof. The proof is sometimes experimental, as after the first two propositions. And sometimes the proof is deductive: e.g., "by Def. 1. & 3. & Prop. 2. & 3" (293; 6091).

Round Three: Reducing Disagreement to Error through System

Newton had now satisfactorily solved how to present his optical findings in a compelling manner within a critical forum of competing researchers. The remaining exchanges of letters required no

Between Books and Articles

rethinking or reformulation of argument, only a reiteration of existing statements. In this last set of exchanges with Francis Line and his students, Newton heavily cross-references his previous statements, using them as an articulated, coherent system which, when properly read, can answer all relevant questions and problems. In this constant pointing back to previous experiments and arguments, Newton displays increasing irritation with the inability of some readers to carry out proper experiments, to make appropriate judgments, or even to read his original text correctly. These developments strengthen Newton's rhetorical strategy of leading the readers very carefully down an intellectual and experiential path, controlling both the reasoning and experience of the reader. In what modern literary theory would call a closed text, Newton does the thinking and experimenting for the reader, with the reader needing only to comprehend each step as he is presented with it.

This last round of correspondence was initiated by a letter of Francis Line to Oldenburg at the end of 1674, doubting Newton's account of the first experiment in the "New Theory" article. Oldenburg replies, under Newton's instructions, by referring to Newton's second answer to Pardies (328; *Transactions* 9:219). In a second letter, Line persists in claiming that his own results differ from Newton's and questions specific lines from Newton's earlier papers. A supporting letter follows from one of Line's students, John Gascoines. Newton responds by giving increasingly detailed and directive instructions, heavily interspersed with exact-page cross—references. Newton also cautions about specific possible errors. For example,

> 1. Then, he is to get a Prism with an angle about 60 to 65 degrees, N. 80, p. 3077, and p. 3086. If the angle be about 63 degrees, as that was which I made use of N. 80. p. 3077, he will find all things succeed exactly as I described them there. But it be bigger or less, as 30, 40, 50, 70 degrees, the refraction will be accordingly bigger or less, and consequently the Image longer or shorter. . . . But he must be sure to place the prism so, that the refraction be made by the two planes which comprehend this angle. I could almost suspect, by considering some circumstances in Mr. Linus's Letter, that his error was in this point, he expecting the Image should become as long by a little refraction as by a great one; which yet being too gross an error to be suspected of any Optician, I say nothing of it, but only hint this to Mr. Gascoin, that he may examine all things. (419; *Transactions* 10:560)

The only slightly veiled irritation of the last sentence reinforces the

Two: Literary and Social Forms in Early Modern Science

impression, given by the simplified and directive instructions, that Newton by now is impatient with what he perceives as experimental and intellectual imcompetence. This impatience abates only slightly in the next exchange, when Anthony Lucas takes over from Gascoines. Lucas grants the substance of Newton's last answer, but raises a new issue, over the exact proportions of measurements resulting from the experiment in question. Lucas then provides an account of some other experiments which he claims contradicts Newton's theories. Newton, praising Lucas for being serious enough to actually do the experiments and taking some care over them, reciprocates by reporting fresh measurements to suggest how the quantitative results can be reconciled. Newton, however, simply dismisses Lucas' new experiments as beside the point and based on misunderstandings. Newton points back to his already published experimentum crucis as definitive.

Most interestingly, however, Newton here mentions for the first time in any letter for publication his completed book on the subject. The mention is to establish he already has considered and explained the kinds of experiments Lucas reports. While arguing here that only the experimentum crucis is important, Newton is yet coming to recognize the persuasive force of the entire system to answer all objections and to demonstrate how all related phenomena are to be accounted for.

> Had I thought more requisite, I could have added more [experiments]: For before I wrote my first Letter to you about Colours, I had taken much pains in trying Experiments about them, and written a Tractate on that subject, wherin I had set down at large the principal of the Experiments I had tried; amongst which there happened to be the principal of those Experiments which Mr. Lucas has now sent me. (174; 703)

Having worked out a full system of claims, representations, and arguments, and having a plethora of experiments, observations, and phenomena reconciled to that system, Newton reduces disagreement to error—errors in reading and errors in conceiving, carrying out, and interpreting experiments.[23] In further correspondence not published in the *Transactions*, Newton with increasing impatience identifies Lucas'

---

23. Although Newtonian system gained authority in England, it did not do so in continental Europe, where a different conceptual/empirical/rhetorical/social climate reigned. There the objections excluded in England through Newton's narrowing of issues and experience remained alive, as described in Henry Guerlac, *Newton on the Continent*. The rhetorical interchange between Newtonian England and the continent is explored in part in Schaffer, but interesting questions remain to be studied concerning the interaction of the two distinctive rhetorical systems.

"mistakes" against the authority of his entire theory (see, for example, *Correspondence* 2:254–60, 262–63). Newton, finding Lucas incorrigible, finally breaks off entirely in a letter to John Aubrey who had taken over Oldenburg's role as intermediary and editor.

> Mr. Aubrey
> I understand you have a letter from Mr. Lucas for me. Pray forbear to send me anything more of that nature. (*Correspondence* 2:269)

The Juggernaut as Persuasion: Book 1 of the *Opticks*

Newton was never again to publish optical results in a journal, nor was he to publish anything else in the *Transactions* or any other journal, except for a minor piece in 1701 on a scale of temperatures. He was to present his major physical findings only within the complete and comprehensive argumentative systems of the *Opticks* and the *Principia*. Moreover, not wishing to rekindle any of the controversies (or misunderstandings, as he saw them), he was not to publish the *Opticks* until 1704, even though in 1677–78 he was on the verge of publishing an earlier version based on the controversy correspondence until a fire in his rooms destroyed the manuscript, and even though he had essentially completed the final version by around 1694.

That final version totally scraps the expository structure and much of the content of the previously completed book of his optical lectures and adopts the argumentative structure that we have seen developing in the correspondence published in the *Transactions*. The book, in the manner of a Euclidean tract, moves from definition to axiom to propositions. The propositions, supported by experimental proofs, are sequentially arranged to create an ironclad deductive argument, as revealed by the organization, the hierarchical ordering of claims, the internal numbering system, and the graphic layout. The beginning of the analytical table of contents prepared by Duane H. D. Roller for the 1931 reissue serve as sufficient example of the structure and organization.

Two: Literary and Social Forms in Early Modern Science

ANALYTICAL TABLE OF CONTENTS*

THE FIRST BOOK OF OPTICKS

Book I, Part I

Purpose of this book 1

DEFINITIONS
1. Ray 1–2
2. Refrangibility 2
3. Reflexibility 3
4. Angle of incidence 3
5. Angle of reflexion, angle of refraction 3
6. Sines of incidence, reflexion, and refraction 3
7. Simple, homogeneal, similar light; compound, heterogeneal, dissimilar light 4
8. Primary, homogeneal, simple colours; heterogeneal, compound colours . . 4

AXIOMS
1. Angles of reflexion, refraction, and incidence, are coplanar 5
2. Angles of reflexion and incidence are equal 5
3. Path of refracted ray is reversible . . 5
4. Angle of refraction is less than angle of incidence for ray entering denser medium 5
5. Ratio of sine of incidence to sine of refraction either constant or very nearly constant—examples—use of this law to trace ray paths 5–10
6. Plane and spherical reflecting and refracting surfaces produce point images of point objects—computation of object position for plane and spherical mirrors, refracting surfaces, and lenses 10–14
7. Converging rays produce real images—the eye as a lens, its normal operation, effect of spectacles, of jaundice, of old age 14–17
8. Diverging rays produce image at the place from which they diverge—purpose of lens systems 18–19
These axioms are a summary of previous knowledge 19–20

*Prepared by Duane H. D. Roller.

PROPOSITIONS
Prop. I. Theor. I. Lights differing in colour also differ in degrees of refrangibility 20
Exper. 1. When a half-red half-blue paper is viewed through a prism, the halves are separated—blue light is the more refrangible 20–23
Exper. 2. A lens produces blue and red images at different distances . 23–26
Prop. II. Theor. II. Sunlight consists of rays differently refrangible 26
Exper. 3. A beam of sunlight is passed through a prism and an image or spectrum formed—position of minimum deviation is used in this and following experiments—detailed description of spectrum—effect of changing beam size—water-filled prism—experimental results disagree with accepted laws of optics . . 26–33
Exper. 4. Eye placed in various parts of spectrum—light from clouds used 33–34
Exper. 5. Two crossed prisms used—spectrum consists of overlapping circular images—penumbra of spectrum is due to finite beam size and may be eliminated by use of lens—separation of blue and red light originally incident at same place on prism 34–45
Exper. 6. Aperture used to select light of a single colour from spectrum—experiments with such light . 45–48
Exper. 7. Half of a paper or thread is illuminated by red light from the spectrum of one beam, the other half is illuminated by violet light from the spectrum of another beam—the paper or thread is then viewed through a prism as in Exper. 1.—two spectrums lying in a right line are viewed through a prism—violet of one spectrum and red of another are superim-

Between Books and Articles

The complete system is presented as a logical and empirical juggernaut, with every step in the reasoning backed up with carefully described experimental experiences precisely related to the formal proposition. As Newton states in the opening sentence of the text:

> My Design in this book is not to explain the properties of light by Hypotheses, but to propose and prove them by Reason and Experiments.

The reader is moved down a path of reasoning and vicarious, virtual experience through the experiments reported. The placement of the experimental descriptions within the developing framework also makes it more likely that the experiments will be understood, performed, and interpreted in the manner intended by Newton, if the reader wishes to move from the virtual, literary experience to the laboratory.

This control of reason and experience within a tightly developing network of claims, experimental representations, and deductions is well illustrated in his elaboration of "Prop. II. Theor. II. The light of the Sun consists of Rays differently Refrangible" through experiments numbers 3 through 10. The announcement of the theorem is immediately followed by the subheadings "The PROOF by Experiments. Exper. 3." The text proper begins "In a very dark Chamber, at a round Hole, about one third Part of an Inch broad, made in the Shut of a Window, I placed a Glass Prism" (26).

Experiment 3 is a much more detailed account of the experiment described at the beginning of the "New Theory" article, resulting in the elongated image. Here, however, the experiment is detached from any discovery account. It is presented only to establish the result. Both the methods of obtaining the results and the results themselves are told in far greater detail and precision than in any previous presentation. For example, the description of the solar image, which in the article was only a sentence long, here is given almost a page. Not only is the immediate image described, but all the variations that occurred as Newton rotated the prism. Not only does this description answer possible questions about what occurred, but it recreates the experience with sufficient narrative intensity for the reader to imagine the event. Throughout this and other experiments, Newton emphasizes the care he took, the places where mistakes might occur and which therefore required even greater care, and the many variations and trials he ran in order to avoid error and anticipate all disagreements.

Further, to establish the result as important, Newton presents a full geometrical derivation of what the results should have been given traditional optics. Thus the reader is carefully held in tow, to see what New-

Two: Literary and Social Forms in Early Modern Science

ton wants him to see in detail, to be silenced on all possible objections, and to find the proper meanings in the experimental experience.

Newton is next careful not to require that the readers find too much meaning in the experiments. He marks the steps of the argument very carefully, nowhere leaving a gap in reasoning that a critical reader might use to undermine the argument. He in fact calls to the reader's attention the limits of the conclusions that can be drawn from each experiment. After a description of experiment 4, he comments,

> So then, by these two experiments it appears, that in Equal Incidences there is considerable inequality of Refractions. But whence this inequality arises . . . does not appear by these experiments, but will appear by those that follow. (34)

Experiment 5 is then prefaced by some comments on its design to indicate how it is aimed at demonstrating a specific point not demonstrated by the previous experiments. The experiment is then described in the language of a geometrical demonstration referring to a schematic diagram. "Illustration. Let S [Fig. 14, 15] represent the sun, F the hole in the window, ABC the first Prism, DH the second Prism . . ." (35).

From the time of his student notebooks Newton had used schematic diagrams to display his experiments and analyses, but here the incorporation of the experiments into a geometrical argument, and the consequent easy movement from experimental description to geometrical analysis, often in reference to the same diagram, gives these representations a special function within the argument. They are treated as both real and ideal, combining experience and reasoning in a step-wise construction of reality. Indeed, immediately following the presentation of experiment 5, a schematized analysis of the results occurred, using prior experimental results and geometrical derivations and assumptions. The reader is again carried one more step into a carefully constructed perception of an ideal/real world. By this point Newton had a practical sense of the modern concept that every observation was theory laden. He wanted to make sure that his experiments were seen through the proper theory loading.

After Newton has marched his reader through almost forty pages of narration and discussion, through all the steps of experience and reasoning creating a tactile and ideal proof of the theorem, he then sums up the argument to this point. The summation is not just a series of claims, however. It is a series of experiences that reveals a coherent world, a felt vision of a world that we have all shared, turned around, and shared again:

> Now seeing that in all this variety of Experiments, whether the Trial be made in Light reflected, and that either from natural

> bodies, as in the first and second Experiment, or specular, as in the ninth; or in light refracted, and that either before the unequally refracted Rays are by diverging separated from one another, and losing their whiteness which they have altogether, appear severally of several Colours, as in the fifth Experiment; or after they are separated from one another, and appear colour'd as in the sixth, seventh, and eighth Experiments; or in Light trajected through parallel Superficies, destroying each others Effects, as in the tenth Experiment. . . . It's manifest that the Sun's Light is an heterogeneous Mixture of Rays, some of which are constantly more refrangible than others, as was proposed. (62–63)

Newton has vicariously given us that same concrete feeling of holding the phenomenon in our hands and turning it over and over again.

Through this juggernaut of a system, Newton has been able to create an authority and certainty of argument that seems to go against the tendency of the period to find in empirical experiences only uncertainty and probabilities. Such tentativeness is evident in Hooke's and Huygens' insistence of maintaining alternative hypotheses in the correspondence examined here, and in Huygens' own work on optics, *Treatise on Light*. In the preface to that work, Huygens states that the empirical evidence he presents cannot produce certainty, although "It is always possible to obtain thereby to a degree of probability which very often is scarcely less than complete proof."

The persuasive historical accounts of the rise of uncertainty and probability by Hacking, B. Shapiro, Dear, and Paradis set off by contrast just what a powerful tour de force of argument Newton has created. In this sense Newton seems very much a man against his times, although his solution was to remake his times. Never satisfied with uncertainty in argument, once he shed his professorial authority, he sought authority through establishing his credentials as a proper Baconian investigator in the "New Theory." When that failed he moved toward the compelling claim, supported first through structured experiment, and then embedded in a massive system built from fundamentals.

The Effects of Compulsion

The controlled experience Newton created in the *Opticks*, moving the reader from first principles to a fully articulated and fully imagined system has a remarkable literary effect, as noted by

Two: Literary and Social Forms in Early Modern Science

many readers, both scientists and nonscientists. Marjorie Hope Nicholson's book *Newton Demands the Muse* documents the mighty force of the *Opticks* on the eighteenth-century literary imagination; almost all of the literary impact of the volume came from the first book. Albert Einstein in an introduction to a modern edition of the *Opticks* attests to the imaginative force of the work, which he sees as a portrait of Newton's mind:

> He stands before us strong, certain, and alone: his joy in creation and his minute precision are evident in every word and every figure. . . . It alone can afford us the enjoyment of a look at the personal activity of this unique man. (lix–lx)

Einstein was not alone in commenting on this experience of felt thought in reading this book. But this experience of the reader must not be taken naively as the actual fact of the writer. As the evidence reviewed in this chapter indicates, the book is far from the spontaneous workings of the creative mind. The book is a hard-won literary achievement forged through some trying literary wars. The texts that are closer to the spontaneous outpourings of Newton's mind, such as his student notebook, have hardly the compelling presence.

The compelling effect of Book 1 of the *Opticks* is rather evidence of how well, totally, and precisely Newton has gained control of the reader's reasoning and perception, so that he can make the reader go through turn by turn exactly as he wishes. In modern literary theory such a text is called a closed text as opposed to an open one that allows the reader greater freedom in providing alternative interpretive procedures and meanings, and projecting personal considerations on to the text (Eco). In the closed text we read only the author; in Book 1 of the *Opticks*, Newton powerfully grabs hold of our reason and experiences until we have seen exactly what he wants us to have seen, in both the concrete and cognitive senses of the word.

With the writer so closely shaping our experience of reading, it is inevitable that the author's voice should be compellingly powerful and the authorial presence imposing. The author has taken over our minds and we become subservient to the powerful directions laid down by the guide and master. Of course, we hand over our wills only to the extent that other firmly held beliefs and experiences are not violated in ways that cannot be and are not reconciled to the emerging vision. As we have seen, Newton is quite careful to recognize and deal with those places where common beliefs and experiences would likely pull the reader out of sympathy with the closed text and thereby remove the reader from the cognitive compulsion.

In this respect, it is important that the text provide an account of the phenomenon that encompasses all contemporary experiences and satisfactorily addresses all contemporary issues. Forceful criticisms must be attended to with a compelling answer or with a revised claim for the closed text to maintain its compulsion. And it must be able to weather the continuing experiences, experiments, and thought of the readers. Compelling scientific texts are embedded in nature and in science. A compelling text, whose end is an authoritative representation of the world, is not simply a textual matter. The text can only create a formulation that serves as a resting point for thoughts and experience of reader and writer. In the current context, the text must appear to be "the right answer."

In Books 2 and 3, where Newton felt (and modern scientific belief agrees) that he had not gotten to the bottom of the issues, he could not create this kind of compulsion. But when the contemporarily satisfying answer combines with a compelling form of argument, an intellectual network is established that seems to spread the presence of the author over a vast and certain domain. And that domain becomes defined by the terms of that intellectual network, making it hard to escape and establish contrary claims. Powerful arguments and experiences must be mounted to break through the compulsions of the earlier system.

Newton's encounters with criticism and opposition, some of which were recounted here, in all instances show his personal conviction and desire to sweep away all objections as ill-founded, if not ignorant. But only in this kind of form did he find the strong vehicle that really would push opposition off the stage, demonstrate the power of his claims, and leave him and his claims in the center spotlight. In his success we can recognize his great effect on the scientific community to follow. It was not just Newton's findings that dominated eighteenth-century science; it was his presence.

And it was his mode of argumentation that also dominated. I. B. Cohen in his analysis of the Newtonian style, which he argues set the tone for the science that followed, focuses almost solely on the *Principia*. He dismisses the *Opticks* as not amenable to the kind of tight, logical system-building with empirical consequences that he finds characteristic of the Newtonian style (*Newtonian Revolution*, 13-14; 134-35). But the kind of closer inspection of the *Opticks* and its literary history that we have carried on here suggests how much the style of the *Principia* may owe to Newton's rhetorical struggles and solutions in trying to shape the optical work. The *Opticks*, to be sure, does not contain the radical split between deduction and induction, between logic and empiricism, between mathematics and physics as there is in the *Principia*. But the *Opticks* does attempt an empirical argument with the same kind of compulsion as the mathematical-deductive argument. The choice of separating out the empirical elements into the final book of the *Principia* is only another option in the same kind of literary problem.

On the form of scientific argument developing in the journals, the solutions reached in Book 1 of the *Opticks* seem to have had a more immediate and powerful impact than the more abstract machine of the *Principia*. As we have seen, the form of Book 1 of the *Opticks* was a direct response to the rhetorical situation and rhetorical problems created by the emergence of the journal. In its rhetorical solutions it served as a precursor of many of the later developments in the scientific article that we examined in the last chapter. It seems that it took the community as a whole over a century to discover what Newton worked out in about a decade, from his first notebooks to his answer to Huygens. Even the way-stop of the failed experiment of discovery narrative used in the opening section of the "New Theory" article seemed to foreshadow the reliance on discovery accounts a century later in the *Transactions*.

Certainly Newton's final rhetorical conclusions seem to match very closely with those realized in the *Transactions* article of 1800 and after: (1) That experimental methods and results must be spelled out explicitly and in detail, both to allow replication and comparison of results and to create a plausible virtual experience for readers; (2) That the discourse must be organized around a central claim or sequential series of claims, and the experimental accounts should be structurally and logically subordinated to those claims to serve as a form of experimental proof; (3) That the coordinated series of claims and articles, incorporated into a coherent system, becomes a mutually supporting network framing a way of working, viewing, and thinking, so that reliance on the network becomes an essential cognitive and argumentative resource. Serious arguments can only be cast within the closed system that realizes the mode of perception, activity, thinking, and interchange. Arguments that step outside the closed system are no longer considered properly scientific.

The framework that Newton developed and relied on was entirely his own and was the system codified in his books, whereas ultimately the scientific community was to develop a communally constructed framework. But this was to require inventing not only the modern apparatus of citation and embedding of others' ideas, not only developing forms of theoretical argument, but also the invention of complex synthetic genres that allow the emergence of codified beliefs without hindering the argumentative and negotiative processes that occur in the research front articles—genres such as review articles, forums, handbooks, and textbooks. Much of this integrative machinery was not developed until the nineteenth and twentieth centuries. The late arrival of integrative machinery makes Newton's awareness of the necessity of a coherent system to provide a powerful account of phenomena all the more remarkable and his solution all the more powerful a resource. His individually conceived system, without the more modern integrative apparatus, both drove science that followed him and created difficulties for integrating viewpoints, discoveries, and claims from outside the system. One suspects that there are important correlations between the breakdown of the Newtonian systems and the emergence of new rhetorical devices both for mounting oppositional arguments and for creating integrated communal theory. Certainly the emergence of integrative machinery allows for more flexibility and modification of the communal system, allowing for changes in argument without stepping outside or causing breakdowns of the system.[24]

That Newton's mode of argument was a model as well as a precursor for later developments in the journal article is more than likely given the omnipresence in the eighteenth century of editions of the *Opticks* and the other evidence of wide circulation, greater than that associated with the more difficult *Principia*.[25] However, the details of the path of literary influence have yet to be drawn out to support this claim.

This single example of an individual working with book and article modes of publication hardly resolves the issue of the relationship between book and journal publishing, but it does begin to suggest the complications, particularly in a time of transition. In this one case the book, which at first was conceived as an extension of an expository series of lectures, became—through contact with the more intimate argument of journals—an argumentative system, shaping consciousness, reason, and experience to compel readers down an incontrovertible path. It appears likely that such a rhetorical style came to reside most fully and permanently in the journals; books gradually moved to other functions, popularizing and codifying the results of such arguments.

Whatever books and articles have become, and whatever relationship between them has developed, the result has been the consequence of individual writers making assessments of their perceived rhetorical situations, choosing among available resources and adding a few new tricks of their own. Books and articles are all the products of writers writing.

---

24. For one early step in the development of this integrative machinery see Bazerman, "How Natural Philosophers Can Cooperate."
25. For discussions of the popularity of the *Opticks* see Cohen's preface to his edition of the book, and Nicholson, chap. 1.

5 LITERATE ACTS AND THE EMERGENT SOCIAL STRUCTURE OF SCIENCE

Elizabeth Eisenstein, in her monumental work, *The Printing Press as an Agent of Change,* details major events in the formation of literate culture, which in turn transformed politics, society, economics, and knowledge. That transformation, although fomented by a single technological invention, was realized only through a nexus of many innovations—linguistic and social as well as industrial. Similar lessons are to be found in Goody; Graff; Havelock; and Scribner and Cole. The history of scientific writing also reveals the many developments necessary to realize literate culture.

In the previous two chapters I have examined the emergence of a linguistic technology that has helped shape modern literate culture. I have associated this linguistic technology with the generic features of scientific experimental communication, which in our time has been associated with certain regularities of form. However, as the change and variation within the pages of the *Transactions* and of Newton's optical writings suggest, the technology and the genre are no simple, rule-determined set of inflexible procedures and forms. They rather represent continuing realizations of social activity within socially structured situations. Industrial, social, and linguistic inventions, such as the inventions of the printing press, the scientific society, and the scientific journal, helped shape the situations out of which the technology emerged and in which the new technology provided the means of social action. The linguistic inventions of this new communication technology, because they themselves embodied social actions, in turn set in motion changes within the structured social situation. Humanly made solutions addressed the immediately perceived problems and provided an environment influencing the perceived structure of future problems.

This chapter looks up from the pages of the texts examined in the previous chapters to observe more directly the interaction between linguistic technology and social structure. In examining how social situations structure communication events and how forms of communica-

Literate Acts and the Emergent Social Structure of Science

tion restructure society, this chapter will foregound sociological theory. Thus the literary analysis (contextualized in a social account) of the last two chapters will here give way to a sociological analysis (based on a literary account). I will be working largely within the view of social structure elaborated by R. K. Merton in *Social Theory and Social Structure*. As Stinchcombe points out in his commentary, in this view social structure lies within the individual's choices of socially structured alternatives. That is, individuals through perception of situation and available alternatives and in their choices make and remake social structure. Through microdecisions individuals both realize and create social macrostructure. In this chapter I argue that this Mertonian position is a contextualized, constructivist one.[1]

The First Editor of a Scientific Journal

In 1665, three years after he had been named secretary of the newly formed Royal Society, Henry Oldenburg founded the first scientific journal in English, *The Philosophical Transactions of the Royal Society of London*.[2] Although not a scientist himself, he saw his mission to advance science through increased communication. Already by the late 1650s he had started correspondence with a number of scientists, be-

---

1. By *constructivist* I mean simply the position that humans construct their own activities and knowledge. The constructivist position in the sociology of science has been associated with a critique of Mertonian social theory as falsely asserting that people behave according to preexisting abstract norms that seem to contradict the individual's immediate interests and actions (Knorr-Cetina, *The Manufacture of Knowledge*; Collins, *Changing Order*; Mulkay, *Science and the Sociology of Knowledge*; Barnes, *Scientific Knowledge and Sociological Theory*). I neither read Merton that way nor agree with the critique.

Sociological constructivists have favored microscopic studies of individuals' situated actions over studies of larger patterns of regularities in individuals' social behavior. The social belief and apparent social influence of such larger patterns has correspondingly provided a puzzle in constructivist accounts. Attempts to explain the status of apparent macrostructure and the mechanisms by which that apparent macrostructure may be generated from microactions are presented in Knorr-Cetina and Cicourel; and R. Collins. Such accounts are thoughtful, ingenious, and interesting, but would be aided by an understanding of the mechanisms linking microactions and macrostructure already implicit in the Mertonian theory they have largely rejected.

2. The French *Journal des Scavans* first appeared three months prior to the *Transactions*. Various authors still contest which nation shall have the honor of giving birth to the first scientific journal, with the crucial point hanging on the broader character of the French journal.

Two: Literary and Social Forms in Early Modern Science

coming a conduit for exchange of scientific information across Europe.[3] As his correspondence and skills as a correspondent grew, he began to see how increased sharing of information goaded working scientists to produce more and to reveal more of what they were doing. Conflict inevitably resulted as correspondents learned the opinions of others and as Oldenburg synthesized the findings of scientists working in the same area. Oldenburg, although becoming highly skilled at elaborate flattery and social graciousness, did not try to gloss over such differences, but rather encouraged their recognition. From the beginning he sensed that science needed to be agonistically structured, so that each player—seeing the moves of the others—makes countermoves attempting to defend his position and to eliminate his opponents from the field.[4] This is not the exposé of the dirty social underbelly of science—this is the plan for science. As long as such conflict was played out in the semiprivacy of correspondence, it did not lead to serious hostilities (M. B. Hall 187).

The role of correspondent in the persons of Oldenburg and, on the continent, Marin Mersenne, helped bring together a previously dispersed scientific community, which had communicated primarily through books. The slowness of book publication, the limited distribution, and the increasing popularity of vernaculars had kept the scientists' audiences and communicants limited. Moreover, books tend to present self-contained universes, accounts complete in themselves with little opportunity for response, except in the muffled comments of the unsatisfied reader. Communication through books minimizes confrontation, disagreement, discussion, synthesis, and sense of competition.

The reactive social dynamics encouraged by Oldenburg's correspondence were also encouraged more locally by the early scientific societies, the Royal Society of London, the Academia del Cimento, and the Académie des Sciences. Standing between the Royal Society and the rest of the scientific world, Secretary Oldenburg became the center of scientific communication, It is little wonder then, that Oldenburg, needing an additional source of income, created a journal of scientific information and found a ready market. The journal put Oldenburg even

---

3. Information on Oldenburg's life and works is to be found in the introductions to the nine volumes of *The Correspondence of Henry Oldenburg*, edited by M. B. Hall and R. Hall; in M. B. Hall, "Henry Oldenburg and the Art of Scientific Communication," *British Journal for the History of Science* 2 (1965): 277–90); and in R. Hall, "Henry Oldenburg," *Dictionary of Scientific Biography*.

4. Latour and Woolgar, chapter 6, expresses a similar imagery of scientific research as an agonistically structured game, where each move restructures the game.

more in the center of communications with his correspondence doubling in the first year and tripling again within three years.[5]

Although Oldenburg did not succeed in turning as much profit as he had hoped, he did succeed in turning himself into an editor, the first scientific editor. In the earliest issues he was still very much the correspondent, writing an extended newsletter of all items of interest that had come to his attention: a new book from the continent, a presentation he had witnessed at the meetings of the Royal Society, a report he had received from one of his correspondents. All was filtered through his voice as he selected and focused attention on those aspects he thought his readers might find most newsworthy. Some features of his writing do change from his previous correspondence: the information is selected to be of generally wider interest, and the long passages requesting information and continued correspondence vanish, although they remain in the private correspondence. Nonetheless, important stylistic features remain: the chatty informativeness; the assumption that the readers are knowledgeable about the subject at hand and are therefore only looking for the latest news, which they will largely know how to interpret; and the consistently complimentary tone, aimed at encouraging continued cooperation. In short, although personal business has been eliminated, Oldenburg still treats the readers as correspondents, people who write to him with information in return for the information he provides them.

Editor, Author, and Reader

However, the new social dynamics of a broadly circulated periodical soon necessitated changes in Oldenburg's relationship with his audience and authors. Within scientific correspondence even the distinction between author and reader had hardly been a sharp one. Whereas previously his correspondents both read and wrote letters, now only a small subclass of the readership contributed information for

---

5. By a count of letters written by him in the published correspondence, which includes all extant items, in the two years prior to his secretaryship, his letters numbered 14 in 1661 and 9 in 1662; in 1663 and 1664, his letters numbered 52 and 59, respectively. In the first year of editorship, 1665, the number jumped to 115; even more strikingly, through April, before the appearance of the first issue, his letters numbered only nine, with nine more added through June, with all remaining 97 letters being written in the second half of the year. In 1666 he wrote 114 extant letters and in 1667, 151. In 1668 the number jumped to 318, and continued at high levels for the next years.

Two: Literary and Social Forms in Early Modern Science

the benefit of the rest of the readership. The contributor becomes a more distinctive and important voice than the newscarrier. Accordingly, Oldenburg increasingly lets the contributors speak for themselves, turning them into authors. He rapidly increases the amount and length of quotations from his sources, until he soon prints entire letters with only a short editorial introduction. Eventually that editorial introduction vanishes as does the form of the letter, leaving freestanding authored articles. In changing from a correspondent, passing on the news through his own perception and personality, to an editor enabling authors to communicate directly with readers, Oldenburg seems to vanish from the pages of the journal, appearing only in the occasional editorial statement. Yet, while the editor is apparently nowhere, he is of course implicitly everywhere, in the appearance, content, style, and personality of the entire enterprise. An editor's voice is a composite voice, comprising all the voices that make up the journal. The quieter the apparent editorial voice, the stronger the corporate one.

In standing between the journal authors and journal readers, the editor helps define not only his own role, but the character of these other two roles. Oldenburg could not keep his journal afloat unless he had authors to fill up the pages. Although at first he could rely on the residual habits of correspondence, the new configuration of editor standing between authors and audience could no longer support the old motivations of information sharing and competitive reaction. Indeed, the new publicness would prove a serious irritant to potential authors. Oldenburg had to offer other lures, such as public exposure of ideas, priority, fame, cooperation of amateur fact-gatherers, and participation in a great universal undertaking. Competitiveness was recast in the threat that the competitor might win these rewards first (for example, see Oldenburg's *Correspondence* 2:439-43; 3:631-33; 4:331-33).

Once Oldenburg enticed a correspondent to share information to be published, he had to keep the contributor satisfied with the results to ensure continued contributions. This we can see in three areas: first, accurate reporting of the information being shared (adding to the pressures for increased use of the author's voice and placing limits on editorial modifications of submitted articles); second, ensuring contributors perceive the benefits they receive from publication (through praise in the editorial voice and in private correspondence with the contributors); and third, protecting the contributor from some of the less pleasant consequences of publication (primarily through ego-stroking and appeals to higher values in private correspondence surrounding an open controversy in the journal pages). These activities to maintain good relations with his contributors potentially conflict with his responsibility to the communal endeavor of science as embodied in the Royal Society. To resolve this conflict, Oldenburg removed editorial commentary on individual contributions from the pages of the journal, leaving the flattery for the letters. He now stroked his authors in private, not public.

Insofar as authors see the benefits of publication, they start writing for the audience, which has the power of granting recognition, instead of for the editor. The editor becomes an intermediary. Thus contributors write Oldenburg increasingly public, formal letters for publication rather than private communications to be digested by Oldenburg. By anticipating the editorial process, authors gained greater control of how their work would be presented. Thus letters came to have clearly marked expository sections, with private material gathered together in other deletable sections, through time reduced to a few prefatory personal comments. Eventually entirely public letters were written, accompanied by private letters of transmittal. Dropping the pretense of the letter form, authors began addressing readers directly in article form, transmitted with a private cover letter to the editor.

The role of the reader is less visible, the act of reading leaving little physical trace. We do know, however, that the early membership of the Royal Society and the readership of the *Transactions* were far wider than the collectivity of active virtuosi. During this early period society membership and journal readership were dominated by leisured gentry, neither professionally nor personally committed to orderly, extensive, systematic investigation. Rather, as members of a largely urban and educated class, they sought amusement and novelty. They were excited by the new philosophy but not necessarily critical or thoughtful in their appreciation. A few merchants and artisans from fields like mining and lens-grinding supplemented this primary readership, as did a few rural and colonial gentry (Hunter, *Sciences and Society* 70–80). In appealing to this nonprofessional, novelty-hungry audience, Oldenburg took for his domain the wide wonders of the world including earthquakes, medical monstrosities, language education, and foreign journeys.

Contributors to the early journals also wrote for this kind of audience, using the language of curiosity and wonder to create appreciation for new findings and inventions. Contributors used their texts to gain publicity and other forms of support for their work. Newton, for example, presented his optical findings in the *Transactions* to promote his completed book on the subject. More actively, contributors sought support for their investigations by requests for meteorological, oceanographic, naturalist, and anthropological data from travellers.

Two: Literary and Social Forms in Early Modern Science

Public Identities and Role Conflicts

The public presence of the journal and other forms of publicity, such as Sprat's *History of the Royal Society*, established a public identity for the journal, its contributors, the society, and its membership, as standing for a new movement in knowledge. Satires by Samuel Butler, Thomas Shadwell, and Aphra Behn relied for their effect on general public recognition of the social type of virtuoso.[6] For *Transactions* readers as well, the cast of characters and the enterprise started to take on social meanings. The *Transactions* became a point of contact for readers in small cities outside London and aided in the formation of local societies. Part of the purpose of these local societies was to make available copies of the *Transactions* and to imitate the activities reported therein (Hunter 81). Oldenburg, as the center of an increasingly organized communication system, took on a recognized scientific role and identity, even though he himself was not a contributing scientist. Finally, individual scientists, such as Boyle, Hooke, and Newton, became public figures through regular publications; in Newton's case his public presence was only on the rarest occasions supported by actual attendance at a Royal Society meeting.[7]

As public figures, natural philosophers were expected to live up to norms of genteel and politically responsible behavior. But their roles as natural investigators required rather odd behavior, such as looking at the moon and waterdrops, using peculiar contraptions like vacuum chambers and microscopes, and suggesting unorthodox opinions about taken-for-granted objects. Not only did they do this at public meetings, but they wrote about it in the journals so that anyone could read about it. These role tensions and violations provided grist for the satirists. (Sociological role theory emphasizes the importance of publicness as a key factor in role conflict; see Marwell and Hage; Merton, "The Role-Set"; Stinchcombe; and Stryker.)

At first, role conflicts were perceived more outside the nascent scien-

---

6. Samuel Butler, "Elephant in the Moon" and "On the Royal Society" in *Genuine Remains in Verse and Prose* (London, 1759); Aphra Behn, *The Emperor in the Moon* in *The Works of Aphra Behn*, vol. 3 (London: Heinemann, 1915); Thomas Shadwell, *The Virtuoso* (London, 1676). Shadwell's play in particular shows evidence of the author's extensive readings of the *Transactions* in search of satiric details.

7. Although becoming a member of the Royal Society in 1672, Newton did not attend his first meeting until 18 February 1675. His attendance remained sporadic even after he moved to London and was elected to the society's council at the end of the century. Only with his election as president did he begin regular attendance, after missing his first meeting as chief officer. Richard Westfall, *Never at Rest*: 267–b8, 476, 629.

Literate Acts and the Emergent Social Structure of Science

tific community than in it. Inside this community, members were recognizing a separate professional identity, establishing themselves as their own primary reference group (see Merton and Rossi, and Turner, "Role-Taking"). However, an emerging division within the readership of the *Transactions* soon led to new types of role conflict. Within the largely amateur, uncritical readership was a smaller circle of readers more actively concerned with the advance of knowledge. These would read critically, comparing what they read with what they believed and observed. Of course, critical reading occurs whenever a reader has a stake in the writer's topic, but now the critical reader could criticize in a public forum proximate to the original text. The journal facilitated not only criticism, but the public role of critic. Just as correspondence networks had served to increase the amount and immediacy of criticism, the journal made the critical activity public. And the answer also became public, casting the natural philosopher into the regular role of public defender of his work. The role of the third-party audience became important in the resolution of disputes.

This argumentative situation creates role conflict for the authors, who are caught between publicizing their own work in terms that would most appeal to the general reader and defending their work from the inner circle of specialized readers who have the power to criticize and therefore cast doubt upon work in public. Power begins to flow to a subclass of the readers, those best able to assess or criticize the work being presented, thereby affecting the general public impression of the work. If all potential critics are satisfied, no debate will ensue and one's work will appear unchallenged. Similarly, if an article avoids the domains of all potential critics, the work will again appear undisputed. However, if one makes claims in an area where others have interests and those claims unsettle those interests, challenge is likely. The article to be successful must then either disarm potential opposition or lay the groundwork for proper public defeat. Thus contributors' interests are best served by developing standards of public argument and adhering to them. The narrative of chapter 3 describing the emergence of the experimental article details both the pressures shaping standards of argument and the consequent standards as embodied in textual practices.

As the articles in the *Transactions* became more concerned with professional argument, other more popular journals (such as *Weekly Memorials for the Ingenious* and *The Athenian Mercury*) filled the gap between professional and popular audience (Hunter, *Science and Society* 55). Since the general audience was no longer the more powerful force for the authors, authors in the primary journals no longer served the needs of the general audience so well. Moreover, the serious natural scientists found the general audience interlopers. Several attempts were made to control the membership and increase the professionalism of the Royal Society (Crosland; Hunter and Wood; Hunter "Early Problems"; Stimson 147–51). Editors began to eliminate articles of insufficient professional interest and quality. In 1752 referees were introduced to maintain professional interests and quality further.

Thus the author's role conflicts in relating to two separate kinds of audiences in the same public forum led to separation of the two audiences (see Biddle and Thomas; Marwell and Hage; Merton, "Role-Set"; Stinchcombe; Stryker; and Turner, "Navy Disbursing Officer"). This social reconfiguration of the participants in the journal communication process led to further redefinition of roles, new conflicts, and new mediating mechanisms.

Exclusions and Gatekeeping

The reconfiguration relies on the social facts of recognition and authority, both externally and internally, of the Royal Society and its publications. Public recognition of the Royal Society as the primary social institution committed to inquiry increased the prestige of membership and publication in its journal and gave the society sufficient public capital to be exclusionary. Supporting this symbolic power of the Royal Society was the transfer of the *Transactions* from private ownership to the society in 1690, freeing it from private mercantile interests (Stimson 114), although it was not technically the official journal of the society until 1752. The editors (all secretaries of the society until 1751) could now look solely to the ideals of the society for guidance in shaping the journal. These goals now were to be achieved by exclusivity rather than inclusivity, turning the editor from a merchant of knowledge into a gatekeeper. At first keeping out information of only amateur interest, then keeping out work of amateur quality, the editor limited the potential audience and began to monitor the statements made among the professionals.

The exclusion of contributions, however, did put special burdens on the editor. First, the editor needed to establish sufficient authority to have his judgments respected as sound. Since this particular institution was founded on scientific inquiry, only the judgments of a respected natural philosopher would carry intellectual weight. An administrator secretary and editor from 1695 to 1713, and later as president of the society (Stimson 143).

Second, in order to retain authority and trust of the professional community, the editor must be perceived as fair and unbiased. However, since the editor has his own research interests and competences, he cannot remove himself from accusations of bias and/or selective incompetence. Moreover, insofar as the editor exercises authority by making judgments, he inevitably creates injured parties. No matter how much participation in the public discussion of a journal appears a desired good to members of a community, an atmosphere of unfairness and distrust, especially attributed to the chief interlocutor (the editor), will poison the atmosphere and destroy the communication.

Indeed, such a conflict took place in the early 1750s. At that time antiquarians' and historians' interests had become dominant in the society, and those interests were represented by the secretary/editor Cromwell Mortimer. John Hill took up a campaign of ridicule against the *Transactions*, pointing to the triviality and foolishness of many reports published therein. His criticism heightened after he was not elected to membership in the society. The response of the society to his satire was to take responsibility for the journal out of the single hands of the secretary editor and place it in those of a committee, which would review and select manuscripts to be published (Stimson 140-45).

Through this innovation, the Royal Society established the role of editorial board cum referee. The editorial function was maintained and strengthened by removal of the responsibilities from any one individual's hands. In order to maintain authority, the editor cannot be perceived as exercising it, but rather must take a distanced stance on all decisions which might be likely to be perceived as injurious to others. The invention of editorial boards to handle issues of general policy and of referees to handle issues concerning individual contribution not only helps the editor maintain authority and trust by assigning responsibility to other individuals, but it further allows the journal to establish a corporate identity, representing the field as a whole. Perceived scientific eminence of editorial board members and referees, as well as distribution among the various subcommunities of the larger scientific community, help maintain the authority of the journal as an institution through the appearance of fairness and generalized competence.[8]

---

8. Maintenance of the appearance of fairness is important to the maintenance of authority in bureaucratic settings; this generalization has been taken as almost axiomatic in the literature on bureaucracies since Weber. For a seminal discussion on the relationship between gatekeeping, critical criteria, and the maintenance of communal trust, see R. Merton and H. Zuckerman, "Institutionalized Patterns of Evaluation in

Two: Literary and Social Forms in Early Modern Science

Group Formation and Integration

Stringent gatekeeping only works when individuals so wish to enter gates that they are willing to satisfy the gatekeeper. The early motives to publish in the *Transactions*—publicity before mixed audiences, priority, possible cooperation of amateurs—were at best peripheral to the activity of carrying on natural investigations at that time. Even the lure to participate in the great universal undertaking, Bacon's Salomon's house, appealed more to ideals than to the realities of research. However, as the character of scientific communication changed from the late seventeenth century to today, publication became essential to research and integrated the working scientists into a communications network. Increasingly, one could only play the game by stepping onto the playing field, and stepping onto the playing field drew one into the social organization of the game players.

An early step in this process of group formation occurred when publication in the journal became a recognized identity-granting social activity. Presenting work before the Royal Society and being mentioned in the pages of the *Transactions* identified one as a natural philosopher. The success and prestige gained by the journal then accrued to whoever published therein. Perhaps more importantly, this prestige lent legitimacy to the work itself. It is one thing to mix chemicals in the back shed at the estate; it is another to be in contact with a secretive brotherhood of suspect alchemists; and it is quite another to participate in open demonstrations as part of a prestigious social institution.

Although at first criticism may have seemed a rather irritating by-product of public exposure, particularly within such a motivatedly critical crowd, this too became seen as a necessary, though unpleasant medicine. Statements acknowledging the usefulness of criticism appear in a variety of articles and letters in the seventeenth and eighteenth centuries, even from the notoriously intolerant Newton.[9] Only serious professional criticism could broaden the individual scientist's narrow view

---

Science," in *The Sociology of Science*. For other accounts of difficulties of early editors see Sherman Barnes and of modern editors see Fox, chapter 1.

9. Newton's grudging recognition of the benefits of communal cooperation and criticism can be seen in the closing lines of his article "New Theory of Light and Colours," and in his dubious compliment to Pardies: "In the observations of Rev. F. Pardies, one can hardly determine whether there is more of humanity and candour, in allowing my arguments their due weight, or penetration and genius in starting objections. And doubtless these are very proper qualifications in researches after truth" (*Philosophical Transactions* 6:4014. Translation from *Newton's Papers and Letters on Natural Philosophy* 106, ed., I. B. Cohen).

Literate Acts and the Emergent Social Structure of Science

and could separate personal conviction from universal truth; the professional forums of publication offered this criticism most readily and reliably.

Gradually researchers start to recognize the cooperative interlinking of their work. The shoulders of giants commonplace turns during the late seventeenth century from a resource in the ancients vs. moderns struggle to a recognition of one's near contemporaries (Merton, *On The Shoulders of Giants*). Informal and irregular recognitions of debt occur throughout the eighteenth century, and in the nineteenth century modern citation practices start to develop. Citations began only as a recognition of debt, but developed into a close interlinking of the current work with the on-going research and theory which formed a codified network of the literature.

In these ways, researchers recognized that their work meant more for being part of a socially legitimated, critical, socially interactive, and cumulative communal process centered on publication in socially recognized forums, screened by gatekeepers, facing public criticism, being cited by others, and being accepted into a codified literature. These elements form the core of most contemporary accounts of the current communication of Science (see, for example, Garvey; Meadows, *Communication*; Ziman, *Public Knowledge*). Group integration as represented in journal publication has become so much the hallmark of modern science that Kuhn takes it as the primary indicator of mature science.

Yet we must not idealize the integration as a simple vanishing of the individual into the group processes. This is the error of Salomon's house, science by bureaucracy, and the ill-fated French Royal Academy attempt to declare science from the outset to be an anonymous, joint endeavor (Hahn 26–28). Integration only worked as an integration of individuals who see personal interests and identity expressed through the group activity. The individual must not only identify with the community as a whole, but must see that his own contribution to the group endeavor will raise his own standing within the community, allowing him to contribute more fully.

Persuasion, Witnesses, and the Representation of Events

A fundamental change in group identification and individual assessment occurs when a contributor presents his work for the scrutiny of his peers as well as for their enlightenment. He no longer can adopt the pose and authority of the expert informing the uninitiated. He must rather establish the authority on communally accepted grounds beyond himself. Thus empiricism, which for Bacon was a mode of investigation, now becomes a mode of persuasion (Dear; Shapin, "Pump and Circumstance"; B. Shapiro; Hacking, *Rise*). To persuade someone of something you must show them what you have found. That is, an event in nature is not an empirical fact with scientific meaning until it is seen, identified, and labelled as having a particular meaning. Moreover, although it may be a fact to the person who first locates it, it is not a fact to other researchers until they have been satisfied that that event has occurred. Only by making the fact communal can one claim discovery of that fact for oneself and reap the rewards of it.

In the early Royal Society, persuasion of facticity was accomplished directly by public demonstration before the assembled members, then recorded in the notice published in the *Transactions*. The persuasion occurs at the public demonstration; the publication does not persuade, but rather only reports the fact of public persuasion. As the particulars of demonstrations become recognized as crucial to the outcome, not all members could witness all trials, so representative witnesses (sometimes of royal or other nonscientific status) came to stand in for the general membership. With time, as it became evident that one needed expertise to view and judge the event appropriately, witnessing was limited to recognized scientists. That is, as events become treated as more particular, and more difficult to interpret properly, witnessing became less and less a public matter. Finally, witnessing devolved on a single witness, the researcher himself. This had to do with the change in research from finding brute facts into inquiring into the meaning of difficult to understand facts—troublesome events had to be investigated by a series of observations and experiments that served as part of an intellectual path of inquiry for the researcher. This meant that persuasion depended not on the presentation of selected, displayed brute events to others, but on the symbolic representation of events in the published report.

How does one convince a critical audience that something happened when they didn't see it? One rhetorical strategy is to establish ethos; that is, that the author/observer is a credible witness, following all proper procedures thoughtfully and carefully. Newton attempts this in his early article "A New Theory of Light and Colours" where he first presents himself as a proper Baconian stumbling across a natural fact before then asserting the bulk of his results categorically. Similarly, in the latter half of the eighteenth century, writers commonly presented themselves as representative scientists by showing their reasonable path of inquiry. This strategy of establishing general credibility fails, as it did in the above examples, when other scientists get different results and come to different conclusions. Academic credentials today serve something of the same general function of lending credibility, but only in the most general union-card manner. That is, credentials permit one to present results, but the results must stand on other grounds (Cole and Cole).

With the failure of ethos as the primary means of validating results unwitnessed by others, the burden of persuasion fell on detailed accounts of each individual experiment—that is, on the representation: to establish proper procedure (that is, the experiment is done as any scientist might have done it), to specify all the conditions and procedures (that is, replication instructions), and to indicate how the experimental procedure answers potential objections. As findings and theory develop, consistency of results with other results aids in the persuasion. Anomalous findings raise more objections, requiring more vigorous counterarguments and powerful demonstrations. Seriously anomalous findings are also likely to undergo more serious attempts at replication than anticipated findings.

Consequently, while representation replaces immediate empirical experience of the audience/witness, the representation must appear experienceable by the audience. The representation must appear plausible to readers having expertise and experience similar to the author's, must seem so proper and controlled as to answer all objections and must offer an apparent replication recipe promising any trained scientist the possibility of experiencing the reported event. Although the replication instructions may in fact be incomplete, requiring additional craft knowledge to make the experiment work (Mulkay, *Science and the Sociology of Knowledge;* Collins, "Sociology of Scientific Knowledge"; Collins, *Changing Order*), the account must be consistent with replication procedures, whether or not the experiment is precisely replicated, for all future attempts at related findings serve as indirect replications. Thus authority now comes not from one's sources, nor from one's good person, nor even from a publicly witnessed fact, but from a representation of events, hewing closely enough to events and defining the events so carefully as to answer all critics, seem plausible to readers with extensive knowledge and experience with similar events, and to hold up against future attempts to create similar events.

As gatekeepers gain in power, restricting access to publication, the representation of empirical events becomes even more important. An editor or referee reading through a manuscript must judge plausibility and soundness solely on the written account. The longer term judgments based on consistency with future results cannot enter into the short-range publication decision; the writer must present the results so that they appear to have happened.

Authority deriving from the representation of events devalues the immediate standing of the individual, institutions, and traditional teachings. Within the network of scientific communication, even kings, nations, and sacred texts lose power before those representations of nature identified as empirical facts. Within the scientific article, authors adopt humility before the facts, putting their empirical findings and derivative generalizations in the central rhetorical positions. On the other hand, those individuals, institutions, and beliefs which have the power of facts behind them gain the authority of empiricism. This leads to a curious conflict. As science gained general social prestige, individual scientists took on the roles of public spokesmen, adopting the mantle of authority from science. This external role, representing science to the wider social and political worlds, was far from the humility before nature demanded internally in science. Similarly, social institutions and belief systems claiming to be based on empirical fact took on a power and attitude of power quite in contrast with the tentativeness required within scientific work. Even within science, an individual convinced of his empirical evidence may assume an arrogance with respect to his colleagues out of keeping with his "scientific" role as an inquirer after the facts of nature. When a scientist's sense of self grows from one of these public, nonscientific sources, his scientific credibility not uncommonly wanes.

Role Conflicts and Differentiated Audiences

Such conflicts between self-assertion and humility are classic conflicts within the role set. That is, an individual filling a status such as scientist or editor has a number of different role partners, with each set of role partnerships incorporating different norms and behaviors; insofar as the partnerships remain mutually invisible, one's behavior can respond only to the partnership at hand. But when the behavior becomes visible to other role partners, conflicts arise (Merton, "Role-Set"). A policeman can be a mean character on the beat, a good guy to school children, and a regulation-follower in the patrol house; however, a school child witnessing a drug bust or a police investigation unit looking into procedural violations on the beat presents the policeman with a conflict as to how to behave.

Scientific publication, by definition, is a public act, hard to keep secret from selected role partners. Moreover, journals provide a public forum and not just a public platform. Thus we would expect the public performance of journal publication to foment role conflicts and foster consequent mediating mechanisms. Further, since we have seen that the new institution of journal publication proliferates social roles, we would expect the opportunities for conflicts to increase with time. Finally, since the role behaviors we are most concerned with here are communicative behaviors, which are just where the conflicts are being publicly displayed, we would expect these conflicts to affect the writing.

As we have already seen, the changing social configurations of scientific communication created conflicts for contributors, who resolved these conflicts by addressing those segments where they perceived power to lie. At first power resided in the scientific correspondent for he controlled the return of useful news. With the journal, power began to rest with the readers who could grant recognition and spread of one's work. With the rise of criticism, power began to flow to the professional part of the readership who had to be satisfied to maintain credibility. The growth of gatekeeping placed the gatekeepers before all; and finally the development of cumulative science gave the last word to the readership of working scientists, for they held the key to incorporation.

However, these last three powerful partners did not displace each other: each retained power. To this day a successful publication must satisfy gatekeepers to get published, must defend itself against critics to maintain credibility, and must appear useful enough to readers to be cited and incorporated in future work. It is not easy to dance to all three masters, as evidenced by the many articles that get published and avoid criticism, yet never are cited, or articles that get published but then become the objects of controversy from which they do not emerge whole.

Conflict Mediation

The complex social configurations in contemporary scientific communication and community also present social complications for gatekeepers, critics, and readers, but for simplicity's sake the remaining analysis will primarily be from the perspective of the contributor. In particular, we will consider how four features of the contributor's role partnerships provide conflict-mediating opportunities. First, publication role relationships do not occur until near the end of the knowledge-production process, allowing extended areas of prepublication privacy and semiprivacy to develop problems, claims, arguments, and evidence. Second, the proliferation and differentiation of publication venues allow the contributor to limit his visibility to selected sets of gatekeepers, critics, and readers. Third, since communication with gatekeepers occurs chronologically prior to communication with critics and readers, and since the three role partnerships hold different rewards, the contributor may make strategic choices among the role partnerships. Fourth, contributing scientists usually fill the gatekeeper, critic, and reader roles. While this aids the contributor by creating certain uniformities in the audience—uniformities that the contributor himself shares—in the long run this creates more conflicting demands on the working scientist. Nonetheless, this integration of all the roles within the single working scientist allows an overriding identification with the entire enterprise of science. The manifold conflicts on the working scientist may then be finally mediated by a set of institutional ideals and goals that distance the scientist from particular conflicting roles and that absorb the various affronts and setbacks.

A closer look at these conflict-mediating processes reveals that many additional features of the social structure of contemporary science can be seen as responses to exigencies created by the communication system. To start, as scientific communication becomes liable to increasingly organized scrutiny by gatekeepers, critics, and research-motivated readers, the preparation of publishable statements retreats more deeply into private and semiprivate workspaces.[10] The primary empirical event (increasingly observable and interpretable only by the specialist) moves out of public sight into the experimenter's laboratory, with the public presentation becoming only the claim-maker's representation. This substitution of representation for presentation allows the claim-maker added selectivity and control—in planning and executing the empirical events (that is, experiments); in reporting only successful experiments and eliminating false leads, distractions, and unworkable experiments; and in presenting a cleaned-up account of the experiment, without bad trials, fuzzy data, or slips of the hand.

Such representational control does open the door for unconscious and conscious distortions, ranging from seeing only what one expects to see to outright fraud (Hanson argued first for observations being theory-laden; Shapin, "History of Science," reviews the studies demonstrating observations as interest-laden). This necessarily is a matter of concern, and procedures have developed to hold individual scientists

---

10. Here I am ignoring more recent issues that have arisen from the grant process which has brought gatekeeping in a new way into the early stages of work. See Greg Myers, "The Social Construction of Two Biologists' Proposals," for an illuminating study of how the funding process helps shape the direction and focus of proposals and work.

Literate Acts and the Emergent Social Structure of Science

accountable for what they report as happening in the laboratory. The extent and effectiveness of these procedures have from time to time come under question, particularly when major instances of fraud come to light. Here, however, we need note only that the systems of accountability are a result of the privacy of statement-making. The scientific community must assure itself that the writer of research is not a fiction writer, that the laboratory consists of more than a typewriter.

The scientist may also maintain a degree of privacy over work in progress by sharing early formulations of the work only with selected colleagues in informal settings—in the coffee lounge, in correspondence, or at closed seminars. These early exchanges help shape the ultimate public argument (for example, Latour and Woolgar, chap. 4). In some tightly structured specialties the less formal communication channels may be primary for the core group, with the published article only for the record and peripheral audiences. Once the informal communication in this tightly organized group passes the approval of the inner circle, then it is as good as published (Menzel; Price and Beaver; Crane). But not every informal communication passes that test to become approved, publishable material. Claims found faulty within the small group are unlikely to surface in reputable publications.

The emergence of validated claims from small research groups resembles the negotiation process that occurs among authors, referees, and editors before an article appears in the journal (Myers, "Texts as Knowledge Claims"). This semiprivate correspondence, shrouded by confidentiality, aims at transforming the private work into the most publicly acceptable form, although authors may not always see it that way and the semiprivacy raises questions about unintentional and intentional abuses (Mitroff and Chubin). Some accountability procedures have developed around the gatekeeping system to exercise control over the privacy.

Another kind of opportunity for conflict mediation through lessened visibility has been created by scientific specialization. The proliferation and specialization of scientific journals have preserved the publishing scientist from facing the judgment of the entire scientific community. In the evolving discussion of specialized research questions, local criteria for the acceptability and significance of work develop. These local criteria may be neither obvious nor superficially consistent with criteria of other specialties. On those limited occasions when specialized work strikes interspecialty issues clearly and forcefully enough to warrant more general presentation, contributors can seek publication in more widely distributed journals. Again, a negotiation process between author and gatekeepers may determine the level of generality at which a claim may be presented and the proper form and place for such a presentation (Myers, "Texts as Knowledge Claims").

Third, the differentiation of audience into three separate kinds of role partners offers the contributor strategic choices in appealing to partial audiences. An untenured junior researcher, needing publications more than public recognition, will likely be most concerned with meeting the criteria of the gatekeepers. Other researchers, humbled by an ideology of cumulative science rather than by the employment system, may be satisfied to contribute a careful, small piece of work, paying most attention to the critics. On the other hand, if one feels wider ambitions thwarted by entrenched gatekeepers and critics, one may attempt to bypass them by beginning a new journal or creating a less conventional channel of communication to the readers. Self-declared revolutionaries and communally declared crackpots may both follow this route; this dichotomous naming indicates the gamble of this procedure.

Finally, and most significantly for the social structure of science, all the communication roles of contributor, gatekeeper, critic, and reader may be taken at various times by a single scientist. Every scientist is trained to read the literature critically and habitually searches the literature for new findings to build on (see chapter 8). As careers develop, scientists then get to referee and perhaps edit journals. By adopting these various role perspectives, quite literally taking the part of the other in communication partnerships, the research scientist learns to understand, accept, and meet audience expectations and demands. Once you act as a referee, for example, you know better how to satisfy referees. This psycho-social integration into the entire process of scientific communication acts as accumulation of advantage that accrues to successful scientists just as much as the more tangible advantages of grants and large laboratories (Merton, "Matthew Effect"; Cole and Cole).

Role Unification and the Norms of Science

All these communicative roles were only gradually integrated into the single status of scientist. Prejournal critics included clerics, kings, and philosophers. The first editor was an administrative organizer rather than a working scientist. Readership was quite wide. Even the interim role of witness, later incorporated into the role of the scientist himself, was at first widely held, then more narrowly held by people of prestige derived from a variety of social institutions. Only the referee role, the last of the roles created in this process, was born requiring that it be filled by a working scientist.

Literate Acts and the Emergent Social Structure of Science

This gradual unification of roles results from empiricism replacing all other forms of authority in institutions concerned with natural knowledge. If authority lies in nature, those best capable of administering that authority are those who have the most intimate and rigorous contact with nature. At first, artisans and craftsmen held some authority because of their practical contact with nature, but this limited authority vanished with the rise of detailed, documented representations of nature replacing direct experience as the relevant form of knowledge. The intimate practical contact without the proper way to talk about it in public granted little prestige and no authority (Ochs; M. B. Hall, "Technology"). Not surprisingly, all the separate scientific roles, shared by the same set of individuals, became embued with similar norms and values. The shared value system of science was made possible by a common source of authority and a unified prestige system. Nobel Prize winners become editors and heads of labs and have their critical opinions taken most seriously. In fact the recognized quality of their work often leads to these other forms of authority long before prizes add worldly recognition and worldly authority to the previously established authority of empirically grounded research.

This unification of prestige, authority, and multiple roles in the single status of the scientist, however, presents the individual scientist with further role conflicts. Not only must the contributing scientist please a three-tiered audience, that same scientist when acting as reader, critic, editor, or referee must avoid irreparable breaches with those same individuals. In this situation conflicting role demands cannot be kept separate, as all the actors take all the roles. A contributor wanting findings to be accepted, but also having a critical role to fulfill, might hesitate alienating a significant potential reader or referee. Nor will the potential contributor accept without suspicion an editorial rejection that might be attributed to the editor's or referee's interests as potential contributors themselves. The possible conflicts and perceived violations are legion.

These conflicts become particularly omnipresent because the whole communicative system is based on conflict, a way of organizing the criticism that emerged with a public forum of communication. Critics are set against contributors, gatekeepers do make harsh choices, readers do select which material to build on and evaluate what they read—and the entire process brings the agonistic interactions into a form of public debate and discussion.

Very strong mediating devices are needed to hold this agonistic social structure together. Some of these devices are to create pockets of privacy within this rather public system. As we have seen, the editor was able to slough off some of the more sensitive conflictful choices to editorial boards and referees so that he might maintain good relations with contributors and readers. Similarly, anonymity surrounds refereeing to allow for "objective judgment." But these devices have only limited power. An editor must take responsibility for journal policy, assignment of referees, and thus the content of the journal. Anonymity in refereeing, often only a transparent veil, can at best hide only personalities and not intellectual commitments. The much stronger conflict mediating devices lie in the distancing values of science.[11] Commitment to organized criticism, communalism, universalism, and objectivity allow individuals to absorb individual strains, conflicts, and violations in the name of the communal endeavor. In this way the overall status of scientist is more than just an umbrella for the many roles taken by the individual; it is a crucial identity adopted by the scientist that allows him or her to rise above the conflicts and strains within particular roles adopted as part of this overall identity.

This overall integration of values and identity does not mean that all individuals equally identify with all parts of the system. Neophytes necessarily have limited experience and socialization. Individuals become alienated or remain marginal for many reasons. Cynicism, manipulation of the system, and fraud may appeal to individuals on the margin or individuals who are expected to fulfill demands beyond their legitimate means. Whole groups and scientific communities may develop other structures as they respond to different social/political/ or belief pressures. These qualifications notwithstanding, the general thrust of the development of the communication system of science has been to structure science in much the terms described by Merton.

The Social Construction of Social Structure

Thus a constructivist analysis of the social structure of scientific communication, examining actors' situated strategic micro-

---

11. Rose Laub Coser, "Role Distance, Sociological Ambivalence, and Transitional Status Systems," specifies Goffman's concept of role distance (from *Encounters*) as arising either in situations of ambivalence resulting from conflicting role expectations made by a single role partner or from transitions to new roles. The conflict situation I have described as occurring in scientific communication has elements of both situations arising from the complex multiple interactions with single role partners and from the expectation of critical skepticism which keeps creating distance between role partners. In this situation, adherence to the more abstract norms of science and identification with the generalized status and goals of scientist create role distance helping to resolve conflicts and ambivalences.

Literate Acts and the Emergent Social Structure of Science

choices, gives a picture of scientific structure consonant with more traditional macroanalyses.[12] This should hardly be surprising. What individuals who constructed the scientific community constructed was the scientific community.

Yet this inquiry has been more than tautological, for we have seen how the scientific community developed around the engendering and management of conflict. We have seen how the conflict-based interaction shaped the means of communication and its regularized channels. We have seen how the structuring of communication helped establish the role set of the scientist.[13] We have seen how norms of behavior and self-representation emerged out of the need to manage the conflicts and relieve the role tensions created by the structured activity of scientists. We have seen how commitments to a communal project beyond oneself help distance a scientist from personal strains and create the collectivity as a social fact.

Science, responding to its own dynamics and activities within its particular social circumstances winds up structured differently than other social systems, equally constructed out of their situations and activities, and developing their own appropriate symbolic systems. As a sociolinguistic system science has emerged through the socially contexted language choices of language users.

Finally, we gain an appreciation of how complex a social activity empiricism requires for its realization. It is not, as Swift's parody in *Gulliver's Travels* would have it, a group of men mutely gathering in a chamber and inarticulately pointing at one object and then another. Although perhaps some early members of the Royal Society might have had opinions not far removed from such parodies, the social realization of the empirical program soon pushed all participants to far more com-

---

12. Warren Handel, "Normative Expectations and the Emergence of Meaning as Solutions to Problems: Convergence of Structural and Interactionist Views," presents a similar analysis of the compatability of sociological frameworks by considering negotiated meanings as a means of resolving structured conflicts and thereby restructuring the perceived situation and the symbolic means of interaction. The protean restructuring of the sociolinguistic system embodied in scientific communication can best be seen in such a light. The evolving symbolic center of the interaction embodied in scientific texts constantly remakes social structure in ways that require renegotiation of what the scientific text should be.

13. Joseph Ben-David, *The Scientist's Role in Society,* also offers an account of the emergence of the role of scientist, but Ben-David's account concerns the broader social perception of what a scientist was, rather than what it meant to be a scientist within a scientific community. Ben-David provides an enlightening account of how the emergence of the public category of scientist shaped the possibilities of science in various periods. In this chapter, however, I have tried to provide an account of the emergence of the structured relations and activities of the scientist within the activity of science.

plex social behaviors. Yet this recognition of social complexity of human behaviors does not deny that the project is empiricist. Our contemporary Brobdingnagian microscopic examination of modern science need not convince us that it is a Grand Academy of Lagado, nor a petty world of Lilliput. The scientific community is what we have made of it.

PART THREE

TYPIFIED ACTIVITIES IN

TWENTIETH-CENTURY PHYSICS

6

THEORETICAL INTEGRATION IN EXPERIMENTAL REPORTS IN TWENTIETH-CENTURY PHYSICS

SPECTROSCOPIC ARTICLES IN *PHYSICAL REVIEW*, 1893–1980

The activity of twentieth-century physics is already well situated in developed institutions of social, communicative, and empirical practice that help shape the daily life and long term direction of the field. As Leslie White has pointed out, established cultures contain vectors shaping future developments, for every institution embodies a form of life and establishes the means for carrying forth that life.

In the last several chapters we have seen how by the end of the eighteenth century many features of the institutions of communication had emerged—in the regularized form of published communications, in the regularized ways of producing and receiving these forms, and in the elaborated social organization in which such communications took place. The practices of criticism and argument that had developed in agonistic competition over accounts of particular events and generalized patterns of events took on regular shapes. Particular literary forms, casting representations in certain detailed forms and necessitating certain practical empirical work for their production, were shaped for audiences adopting certain roles within the elaborating social system. Individual texts, realizing and further developing these literary forms, were produced and received by individuals enmeshed in common understandings and experience constituted by participation in the evolving community. Knowledge claims put forth in these texts were thus highly contextualized linguistic products, the printed trace of complex systematic activities.

Three: Typified Activities in Twentieth-Century Physics

The nineteenth century saw many new developments in institutions of communication, social organization, and the empirical practice of science in Europe and, nascently, the Americas. To carry this story into the nineteenth century is an overwhelmingly difficult task as disciplines proliferated and grew distant from each other, each developing its own set of institutions and practices. Communications forums increased as well and developed differing communications dynamics, with major consequences for literary form and social organization. I make no attempt here to construct any detailed, researched account of developments in the nineteenth century, which I leave entirely to future studies. Comparisons of eighteenth-century and current practices (see pages 78–79 and 126–27), however, strongly suggest that major changes occurred in the nineteenth century in the way scientific texts referred to and relied on each other. The emergence of modern citation practices is the most visible, but not necessarily the most fundamental, product of the development of implicit and explicit intertextuality in nineteenth-century scientific communication. Studies of nineteenth-century scientific writing would do well to take on the question of changing institutions of intertextuality.

Rather than take on the immense job of a comprehensive account of the complex social, linguistic, and research networks that draw disciplines more tightly together internally and separate them from each other, I will look at developments within a limited region once these differentiations have taken place.[1] In looking at the changing forms of the experimental report within twentieth-century physics, and more especially spectroscopy, I will be examining how the increasing prominence of an overriding and integrating theory helps reshape textual form and bind texts even more closely to each other. In the two chapters afterward, I will examine how the integrated discussion and communal endeavor of modern physics shapes the individual's activity in writing and reading texts.

---

1. This is not to suggest that important interdisciplinary links may not be forged over both phenomena and theories. Although different specialties may look at the same object or phenomenon from different perspectives and with different motivating questions, accounts created in one specialty may have strong consequences for another specialty, witness such a celebrated example as the implications of Watson and Crick's molecular biology discovery for genetics, because the account of the DNA molecule offered the mechanism for the carrying of genetic information. Similarly, a theory developed around a narrow question may turn out to have greater power that carries across the work of many specialties, witness the celebrated example of quantum theory's origin in certain specific problems in thermodynamics.

Theoretical Integration in Experimental Reports

Linguistic Code and Social Agon

Although here I discuss linguistic forms as evolving parts of the ongoing activity of a community, previous examinations of scientific language have tended to reify the highly elaborated linguistic forms of contemporary science into stable and independent textual structures. Linguistic studies of scientific language as a sublanguage (Kittredge and Lehrberger) or a special register (Crystal and Davy) consisting of particular lexical items (Savory; Hogben), syntactic forms (Huddleston; Lee; Gopnik) and organizational units (Meyer) treat scientific language as an independent system, to be learned as classical Latin or any codified school language is learned.[2] Indeed textbooks in scientific writing contain highly elaborated models of linguistic forms for students to follow. As a socializing and educational practice there may be some warrant for this attitude, despite significant pedagogical dangers in freezing forms and isolating them from practice (more of this in chapter 12). In any event there is some need for neophytes to be introduced to the current means of communication, to learn the ways of formulating statements appropriate to the community they wish to enter. Such an introduction both provides a repertoire and aids social acceptance of statements framed according to current habits.

However, such an approach to scientific language reduces its use to a matter of following prescriptions and avoiding prohibitions. Such a view isolates writing from the larger processes of formulation and interaction by making it merely an editing-for-propriety process, rather than a complex social event. Such a view hides the motive for writing, the larger part of the process of creating formulations, and the rhetorical import of these formulations.

Yet, in the last few chapters, we have seen how the forms of scientific representation emerged simultaneously and dialectically with the activity of science and the social structure of the scientific community. Features of the experimental article developed as part of an agonistic social activity, arguing over experienced events. The experience is shaped by the argument just as the arguments exploit the experience in a public linguistic forum.

Studies of scientific discourse coming from sociologists of science have indeed emphasized the agonistic force of language in the competi-

---

2. John Swales's analysis of article introductions offers a welcome exception to this general treatment of scientific writing as a disembodied code. Here and in consequent articles he considers the organization of article introductions as a solution to the rhetorical problem of establishing a place for one's work within a relevant literature.

tion over claims, power, and the satisfaction of interests. These studies have established that authors control the language and presentations of their papers so as to present their work in the most favorable light, so as to advance the acceptance of their own work, and to further their interests as scientists. Most aspects of the article, even the presentation of data, are open to forms of literary control, with the writers particularly concerned with persuading readers of validity and importance of their work.[3]

By representing scientific argument as an unbounded free-play of competing interests, however, these studies have erred in the opposite direction. They have ignored the historically evolving structure of scientific communications which has embodied and defined the evolving nature of the competition. While each participant in pursuit of individual goals may seek whatever resources are available and may bend the current rules and practices to personal advantage, those rules and practices and the recognized resources embody and shape the communal activity, evolve over time, and contain inherent goals and vectors.

Typically, most of the sociological studies of scientific discourse treat the previous literature as a persuasive resource, a validating set of scriptures to be effectively arrayed through references, but these studies do not consider how this prior literature helps define the current work. The sociological study of scientific texts, in an attempt to free itself of positivist historical whiggishness, which finds in scientific papers the march toward rational truth, has tended to cut itself off from the shaping effects of history even as it finds each separate moment indexically intertwined with a local sociohistorical context. Curiously, this leads to an assumed uniformity of freedom for the scientific writer, throughout history and in all situations, so that case materials from all times and across all disciplines are treated equally as sources for generalizations.[4]

---

3. Latour and Woolgar in *Laboratory Life* were the first to explicitly discuss the scientific text as making a move on an agonistic field, but also consistent with that view are Collins and Pinch, *The Social Construction of Extraordinary Science*; Gilbert, "Referencing as Persuasion," and "The Transformation of Research Findings into Scientific Knowledge"; Gilbert and Mulkay, *Opening Pandora's Box*; Woolgar, "Discovery: Logic and Sequence in a Scientific Text"; Gusfield, "The Literary Rhetoric of Science"; Knorr, "Producing and Reproducing Knowledge"; Knorr-Cetina, "Tinkering Toward Success," and *The Manufacture of Knowledge*; Latour, "Essai de Science-Fabrication"; Latour and Fabbri, "La Rhetorique de la Science"; Law and Williams, "Putting Facts Together"; Yearley, "Textual Persuasion."

4. There have been significant exceptions to this ahistoric tendency, most notably Martin Rudwick's exemplary detailed study *The Great Devonian Controversy*, which traces how evolving claims in an early nineteenth-century geological controversy were shaped by existing forums and forms of communication, the evolving state of the debate

Theory as a Textually Integrating Force

The following examination of experimental articles in physics since the late nineteenth century indicates how texts have become embedded in a web of common theory, a structuring force even more powerful than the web of citation and cross references (elaborated in the citation studies literature).[5] That common theory has become an extremely strong force in structuring articles and binding articles to each other. Acceptance of common theory not only creates common interests among the adherents, and a massive edifice to be elaborated by many practitioners, it binds together wide ranges of empirical experience, gathered by many different people at different moments engaged in different activities. The theory points them to certain kinds of experiences, suggests the appropriate means of designing and interpreting empirical events, and allows results to be harmonized with the results and ideas of others. Thus, over the period and within the range of texts examined below, theory has come to permeate writing in physics.

Just as the argumentative structure in Book 1 of the *Opticks* gave a coherence, force, and certainty of meaning and reference to Newton's claims, quantum theory helps place and stabilize claims and observations in contemporary spectroscopy. Thus one would expect that the discipline would find many ways to tie the texts in with the prevailing theory. Unlike Newton's presentation, however, the theoretical construct and its elaboration is the work of many hands. Thus the development of an integrated discourse cannot rely on a single Euclid-like exposition of a unified system from first principles. More elaborate and flexible linguistic means must be developed to permit communal construction of the unifying system.

This chapter, in particular, looks at the changing features of experimental reports appearing in the *Physical Review (PR)* from its founding in 1893 until 1980.[6] This period marks the rise of American physics from backwardness to world dominance (see Kevles), reflected by the journal's rise from a local, university organ to the primary international journal of physics.

---

and the evidence gathered and represented in the literature. Susan Cozzen's study "The Life History of a Knowledge Claim" examines the historical process by which texts become embedded in the literature of a field.

5. The literature on citation studies is reviewed in Cozzens, "Taking the Measure of Science."

6. Extensive background on the development of *Physical Review* appears in Merton and Zuckerman, "Patterns of Evaluation in Science" and in Physics Survey Committee, *Physics in Perspective IIB*.

Three: Typified Activities in Twentieth-Century Physics

Further, this period marks the virtual disappearance of the book as a way of presenting new results in physics. Early volumes of *PR* devoted as much as one-sixth of their pages to reviews of new books, including new contributions to the research front as well as textbooks. By 1910, however, new books were only listed, not reviewed; after a short revival of reviews in the 1920s, all mention of new books in physics vanished in the early 1930s. By that time research physics meant journal physics exclusively, with the article and shorter note (or letter) as the standard genres. In 1929 letters were added as a regular feature of *PR* until they were split into the separate journal, *Physical Review Letters*, in 1958. This study, however, will attend only to full articles, eliminating all texts placed in sections identifying them as notes, letters, minor contributions, or the like. One other regular feature of the journal from its founding through the 1950s was conference reports, including abstracts of delivered papers; these reports and abstracts also will not be studied here.

Finally, the period from 1893 to 1980 contains the introduction and establishment of the new physics and the enormous growth in the amount of physics research. Radioactivity was discovered in 1895; Einstein's first paper on relativity was published in 1905; Bohr's trilogy on the structure of hydrogen appeared in 1913; and the main features of quantum mechanics were settled with the publication of DeBroglie's and Schrödinger's equations in 1925 and 1926. The exponential growth of physics in this century has been demonstrated by Price; this growth can also be seen in the increase of equivalent words appearing annually in *PR*.[7]

Thus the period examined and the research site within American physics help highlight the impact of the development of an integrated and extensive professional community on the discourse of the field, although it may distort the international picture somewhat by hiding developments in nineteenth-century European physics. Some of the developments we will see in this chapter have likely been anticipated or at least prepared for in Europe. Further, differing events and relations within nineteenth-century European physics may have led to textual

---

7. Equivalent words are calculated by assuming the entire page to be filled with printed words with the size and spacing used throughout the main body of the article; this method helps incorporate changing use of equations, illustrations and other non-word features, while taking into account changing typographical presentation. In the first year of publication 190,000 equivalent words appeared in *Physical Review*; in 1900, 260,000; in 1910, 600,000; in 1920, 570,000; in 1930, 1,700,000; in 1940, 1,800,000; in 1950, 4,200,000; in 1960, 8,400,000; in 1970, 29,000,000; and in 1980, 30,000,000.

developments not reflected in the more recent American case. Such observations, however, await further research.

Methodological Problems and Selection of Materials

The attempt to characterize a large body of writing presents enormous problems, especially when the examination is carried out by a single researcher. The kind of analysis generally considered most revealing about the nature, organization, function, and style of a text is the traditional method of literary criticism: close analytical reading. The method is not only time-consuming, it is particularistic, revealing in detail the special qualities of individual texts. The method tends to militate against generalization and to produce masses of incommensurable findings. On the other hand, statistical methods, such as those adopted in computer studies of style, do provide comparable data open to generalization. Moreover, certain statistical comparisons were available for this study that were not available for the study of earlier *Philosophical Transactions* because the genre had by the end of the nineteenth century stabilized in many significant ways. The stabilization of the genre helps create countable and comparable features as well as providing a framework for the interpretation of the results of such counting. However, statistical counts provide only information about the most surface features of the text (at least at this stage of methodological development). My strategy to contend with this dilemma is to employ a mixture of methods—using statistics to indicate gross patterns or trends but using close analytical reading to explore the finer texture, the meaning and the implications of those trends. The statistics are to indicate that something is happening, and the close readings are to find out what that something is.

As implied earlier, the indicators and analytical readings are aimed at establishing gross trends in style and genre, as suited to the study of a historical body of articles not discussing the same immediate problem. Other analytical tools and different kinds of selections of articles would, of course, tell more about the detailed interplay among specific articles and authors as they use the conventions of style and genre revealed here to pursue individual interests, and/or to resolve particular issues of knowledge.

Given my limited resources, both the statistical and close reading analyses had to be carried out on limited selections of material, too limited to warrant the statistical designation of samples. I have tried to avoid making strong inferences where the numbers are small, but the entire endeavor must be granted some statistical charity until more comprehensive studies can be carried out.

For different levels of analysis, I have used three different selections of material. For the measure of article length, I have considered all articles through 1900 and every fifth year thereafter through 1950; since 1955, because of the increasing volume of annual publication, the data are limited to the first few issues, totalling 3,000–6,000 pages, of each fifth year.

For analysis of references, graphic features, organization, and mode of argument, I have examined a total of forty experimental articles reporting spectroscopy as a primary technique and appearing in 1893, 1900, and every ten years thereafter through 1980. If fewer than three appropriate articles appear in any year, as in 1900, articles from the next year are also included; if more than six appear, as in recent years, only those from the earliest months are used.

Finally, for sentence-level analysis, a subset of the spectroscopic articles is used, comprised of all the selections from 1893–95, 1920, 1950, and 1980—totalling seventeen in all. Appendix 1 gives the bibliographical citations for articles explicitly discussed, which will be identified in the text by year of publication and author's initials (e.g., 1893-EFN).

Given the variety, changes, and proliferation of specialties in physics over the life of the *PR*, it seemed advisable, except for the overall measure of article length, to limit the texts examined to a single specialty. Of all the specialties in physics, spectroscopy has been the most stable over the period examined. To stabilize the selection further, I have eliminated work based on the recent innovations of electron spectroscopy and the application of spectroscopic technique to the study of nuclear events, both of which have opened up some new directions for the field. I have also eliminated purely theoretical articles, for, in this specialty, they too are a phenomenon of the last half century, in the wake of quantum mechanics; the theoretical components of experimental articles will, however, remain part of the examination. Astronomical spectroscopy is a different field.

The major empirical discoveries of this narrowed specialty (what we might now call "the experimental study of the electromagnetic spectra of orbital events") were made before, or just at the time of, the founding of *PR*. Fraunhöfer lines were discovered in 1802, and through the middle of the century variations in lines for different substances were noted. Techniques and standards were refined until, in 1896, Zeeman

Theoretical Integration in Experimental Reports

discovered the fine-splitting of lines under a magnetic field. On the theoretical side, Kirchoff proposed in 1859 that absorption spectra were the same as emission spectra; between 1885 and 1890 equations were proposed to account for the distribution of lines, most notably by Balmer, Kayser, Runge, and Rydberg. Until the emergence of quantum theory, however, no comprehensive theory accounted for spectral lines, which by then had been observed for over a century.

The earliest articles on spectroscopy in *PR* already incorporated what were to remain the primary purposes of spectroscopic research: to measure the lines of different substances under different conditions, to account for the distribution of these lines, and to use the lines to help describe or understand unusual substances or phenomena. Thus, in the first two years, articles appeared reporting on the infra-red spectra of common substances, testing whether an equation predicted a set of lines, and using spectroscopy to investigate limelight. Since then techniques have changed (resonating lasers and electromagnetic counters tuned to narrow reception channels have replaced the prism or grating and photographic plate as measuring devices) and changes in surrounding knowledge have changed ideas of what lines would be interesting to study; but the basic tasks remain the same. Articles in 1980 still reported on the lines of various substances under various conditions, accounted for those lines by assigning starting and finishing quantum states, and used lines to measure and understand dense plasmas. This stability of basic activity simplifies the task of analyzing changes in language and modes of argumentation.

The limitation of material does, unfortunately, leave open several questions about the generality of the findings. First, the narrowing to experimental articles eliminates consideration of developments in the purely theoretical article, of increasing significance in recent decades. Second, without a wider cross-section of material we can only speculate on the extent and manner in which the writing in spectroscopy is typical of writing in the other specialties of physics. The stability of the specialty is in itself idiosyncratic in twentieth-century physics. Other specialties may have different intellectual or social structures, calling forth different kinds of argumentation; even the age or rapidity of change within a specialty may affect discourse patterns. On the other hand, given the stability of spectroscopy, the discourse changes may suggest the more general drift of the entire discipline, freed from the intricacies of specialty change. In any event, the problems in studying more rapidly changing specialties, many of which did not exist in anything like the modern form until recently, make such studies difficult, at least until maps of some simpler specialties are drawn to serve as comparative models. Finally, there is the problem of attempting to generalize from an American journal to all of international physics. In particular, the early features of articles in PR may be as much a consequence of the backwardness of American physics as of the general discourse patterns of international physics. Today, PR clearly represents the standard in international physics, but when this became established is not clearly known. Again, only a wider cross-section of material, including historical examination of European journals, will resolve this issue. Such comparisons may even reveal abiding differences in national style. The current study, nevertheless, as a first foray into the description and analysis of changes in the scientific article, will at least provide one reference point for later comparisons.

Results

ARTICLE LENGTH

A comparison of the lengths of PR articles through the years suggests, as a first approximation, some of the changes that have occurred (see figure 6.1). From 1893 until 1900, the average length of an article dropped from about 7,200 equivalent words to about 4,500, then immediately began to rise to a secondary peak of about 5,700 in 1920. The average then dropped to a bottom of about 4,600 words for ten years from 1925 to 1935, before beginning a sharp and steady rise continuing to the present, with a 1980 average of over 10,000 equivalent words. The splitting of the line in 1970 reflects the splitting of the journal into four sections: A, General Physics; B, Condensed Matter (Solid State); C, Nuclear; and D, Particles and Fields.

This graph contradicts the commonplace that in the nineteenth-century scientific writing was more expansive, but in this century articles have become increasingly compact under several pressures, not the least of which has been publication costs. The consistent expansion through the middle and latter part of this century confirms Abt's survey of astronomical journals from 1910 to 1980, and the more limited statistics on PR presented in the Bromley Report (Physics Survey Committee).

Figure 6.1, moreover, bears little relation to the major editorial events and policy changes of PR. When the journal changed sponsorship from Cornell University to the American Physical Society in 1913, an editorial claimed that recent more stringent editing had kept lengths down and made the sponsorship shift economically feasible; in fact, the major

Theoretical Integration in Experimental Reports

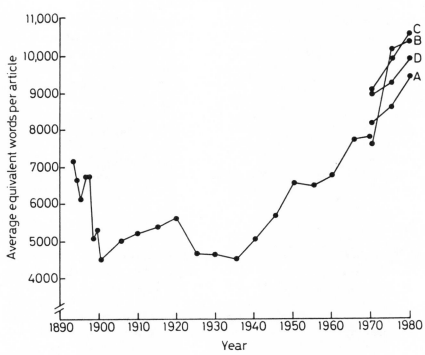

Figure 6.1. Average Length of Article in *Physical Review*

drop in article length had ended thirteen years previously, and article length was rising at the time. However, a decrease in total pages, from about 1,500 pages in 1910 to about 1,050 pages in 1915, had been achieved by a decrease in the number of articles published (from 104 to 83), and by a 25 percent increase in the number of words per page. Similarly, neither the page charge (instituted in 1930), nor the letters section (instituted in 1929), had any noticeable effect; nor did the splitting of letters into a separate journal in 1958; nor did the splitting of the journal into four sections in 1970. In the last two cases, the length simply continued an ongoing rapid rise, apparently moved by other forces.

Similarly, changes of editor seem to have had, at most, a marginal effect on article length. Turnovers of the editorship occurred in 1913, 1923, 1926, 1950, 1951, and 1975. The 1913 and 1975 turnovers do not correspond to any changes in the graph; the turnovers in the mid-1920s and early 1950s do correspond to temporary flattenings in the length curve, but such flattenings are only small adjustments to other, larger, longterm trends.

The data analyzed in the remainder of this chapter will suggest other, more substantial reasons for the length changes, related to intel-

Three: Typified Activities in Twentieth-Century Physics

lectual changes in the discipline. The lengthy articles of the mid-1890s will be seen to reflect a looseness of style, a focuslessness of argument, and a lack of compact technical vocabulary. By the turn of the century, articles will be seen to gain focus on particular issues of theory, becoming more selective in content and more purposeful in organization. The radical theories of the new physics will be shown to be associated with a more tentative, contemplative style, reevaluating and adjusting theories. Once the most confusing theoretical issues had been sorted out in the late 1920s, increasing length will be shown to be related to increasing knowledge and theoretical elaboration, with articles becoming more focused and compact, but relying on increasing amounts of background and contextual knowledge so that length and density rise together.

REFERENCES

A strong indicator of the reliance of a text on background and contextual knowledge is the use of explicit references to prior literature. The amount, pattern and function of references have changed significantly in the articles examined, suggesting the increasing embedding of arguments in the web of the literature of the field. Figure 6.2 presents the average number of sources referred to in the decade-by-decade selection of spectroscopy articles. Note the rapid decline over the first twenty years, and then the generally consistent rise until the present.

A detailed look at these references reveals what happened.[8] In the early years, references are used rather generally in the text of the article; they do not refer to a specific finding, nor identify a specific relation to the current work. Serving as a roll-call of previous work in the general area, references congregate at the beginning of the article, never to be raised in a significant way in the course of the argument—except perhaps in relation to methods and apparatus. For example, 1895-EM contains eleven references in the first quarter of the article, one reference in the second quarter, and none thereafter. In the same spirit, 1893-EFN, the first article of the premier issue, begins:

> Within a few years the study of obscure radiation has been greatly advanced by systematic inquiry into the laws of disper-

---

8. I have followed the procedure of examining references within the context of the entire article, as recommended by Chubin and Moitra. I use a fuller descriptive technique, rather than the kind of formal typology proposed by Chubin and Moitra or Moravcsik and Murugesan, although the description here does rest on concepts of reference use, as considered in both articles. The description also rests in part on ideas from H. Small, "Cited Documents as Concept Symbols," and S. Cozzens, "Life History."

Theoretical Integration in Experimental Reports

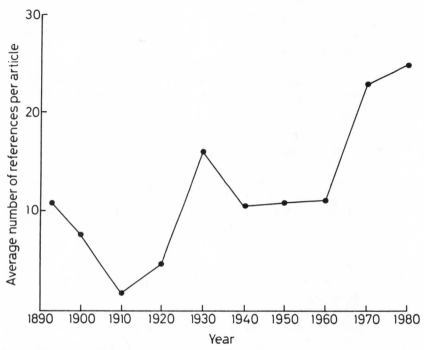

Figure 6.2. Average References per Spectroscopic Article

sion of the infra-red rays by Langley,[a] Rubens,[b] Rubens and Snow,[c] and others. Along with this advancement has come the more extended study of the absorption in this region. The absorption of atmospheric gases has been studied by Langley[a] and by Angstrom.[d] Angstrom[e] has made a study of the absorption of certain vapors in relation to the absorption of the same substances in the liquid state, and the absorption of a number of liquids and solids has been investigated by Rubens.[f]

The references here serve to establish a tradition the author is working in, but do little to define a specific context of knowledge, theory or problems that circumscribe the current task. The author only promises to do more of the same:

> In the present investigation, the object of which was to extend this line of research, the substances studied were . . .

The lack of concern with dating references, and the age of the references that are dated, further weaken the sense of a coherent, moving

Three: Typified Activities in Twentieth-Century Physics

research front. In both 1893–95 and 1900–1901, 52 percent of the references are undated, and only about 30 percent are dated six years or less from the article's publication.

By 1910, the number of references per article has decreased dramatically to only 1.5, and the few references are dated and of recent vintage, suggesting immediate relevance to the work at hand. In this spirit, 1920-CDC/DC begins:

> A knowledge of the relation between the spectrum of a substance and that of its isotope is important in that it may throw further light on the structure of the atom. Some work along this line has been done. Aronberg,[a] working with a grating spectrograph has reached the conclusion that the wave-length of the line λ 4058 is greater by 0.0043 A. for lead of radioactive origin than it is for ordinary lead. The work of Aronberg has been corroborated by Merton,[b] working with a Fabry and Perot *étalon*.

The passage continues with a discussion of the work of Duane and Shimizu, and of Siegbahn and Stenstrom, in the same spirit: these are specific findings of concrete relevance for the current investigation. Furthermore, all four references are less than four years old.

Even as the number of references per article has increased over the last sixty years, the specific relevance for the work at hand and the lengthy discussion have increased, with the result that new work appears increasingly embedded in the literature.[9] For example, in 1980-KHF et al., the extensive discussion of results is structured around comparison with the results and models presented elsewhere in the literature:

> The strong 'structures' on the lines resemble those predicted by Oks and Sholin.[a] As described there, one typically finds a valley with one hill on each side. However, the strongest 'structures' are not at spectral positions corresponding to the plasma resonance but rather at positions between $1/4\, \omega_{pe}$ and $1/2\, \omega_{pe}$. The calculations of Oks and Sholin predict similar structures not only at the resonance frequency but also at some harmonics and subharmonics, i.e., at $1/2\, n\omega_{pe}$ with $1 \leq n \leq 8$; $n\, /\, = 7$ for $H\alpha$. The predicted positions are marked in Fig. A. Because of the

---

9. The historical depth of the references did increase in the World War II period, with only 53 percent of references six or fewer years old in 1940 and only 37 percent in 1950, indicating the disturbing effect of the war on research. More recently there has been a like stretching out of references, with 40 percent six or fewer years in 1970, and 61 percent in 1980, indicating perhaps the maturity or lack of "heat" in the field.

uncertainty in the density determination and therefore the value of ω_{pe}, we cannot decide at present whether the observed line contour corresponds indeed to the model of Oks and Sholin.

Note the great length of the discussion, the specificity of the summary, the quantitative comparison (through the figure) between the reference and the work at hand, the attempt to evaluate the correspondence, and a discussion of the difficulties in carrying out the comparison. The work of Oks and Sholin is made an integral part of the intellectual content of the new article. References, as well, have tended to spread throughout the article, so that every stage of the argument relies on the work of others. 1980-SJR, for example, uses fifteen references in the first quarter of the article, eleven in the second quarter, eleven in the third, and three in the last.

Analysis of references then suggests a loose cognitive structure in the early years, with one piece of work claiming only general connection with earlier work. In the early part of the twentieth century, tighter standards of relevance developed, bringing work into greater coordination. Throughout the remainder of the century both the amount of relevant work for each article and the integration of references into the argument have increased. More references are being discussed in greater detail at more junctures throughout the article. This increasing discussion of sources is a factor in the growing length of the contemporary article, just as the deletion of the loose roll-call of forebears at the beginning of early articles was a factor in the decrease of length at the turn of the century.

SENTENCE LENGTH AND SYNTAX

Sentence length, on the other hand, has remained fairly stable: in 1893–95 it averaged 27.6 words per sentence; in 1920, 28.3; in 1950, 25.3; and in 1980, 23.7.[10] Sentences have also tended to remain generally simple in structure, averaging (in traditional grammatical terms) about 70 percent simple sentences, and 30 percent complex sentences, in all four time periods. Similarly, the types of phrases used to expand simple sentences, and the number of clauses used to develop complex sentences,

---

10. The sentence length, syntax, and word choice data were obtained from all the selected articles in 1893–95, 1920, 1950, and 1980. From each of the articles three to five passages for analysis were chosen, representing whichever of the following sections of the argument were present: introduction, theory, experimental, results, discussion/conclusion. The passages began at the beginning of each of the sections and ended either at the first sentence break after two hundred words were reached, or at the end of the section if it was under two hundred words in length.

Three: Typified Activities in Twentieth-Century Physics

show no significant changes over the period. These three levels of sentence stability suggest that neither changes in article length nor perceived changes in the "difficulty" of reading can be attributed to changes in sentence patterns or sentence style.[11]

The only significant syntactical change found is in the types of subordinate clauses used in complex sentences. The percentage of relative clauses decreases regularly and significantly through the period (1893–95, 54 percent of subordinate clauses; 1920, 47 percent; 1950, 37 percent; and 1980, 17 percent).[12] Such relative clauses simply modify a noun already present in the main clause, adding information or precision but not adding to intellectual complexity, as in this example from 1980-RAR et al.:

> The spectra thus obtained were found to be identical except for slight variations in relative peak intensities, which were attributed to lamp fluctuations and variations of the analyzer transmission.

Although the relative clause tells us more about the causes of the variations, the primary statement of the sentence (the essential identity of spectra) remains unaffected. On the other hand, noun clauses (presenting facts, claims, or observations that serve as nouns in the main clause), and subordinate clauses establishing temporal or causal relationships (using subordinating conjunctions such as "when," "because," or "if"), both increase regularly and consistently in percentage throughout the period. The percentage of noun clauses increases from 15 in 1893–95 to 33 in 1980, and the percentage of temporal and causal clauses rises from 31 to 50. Noun clauses can keep two thoughts in the air at the same time, as in 1980-KHF et al.:

> The analysis of the continuum intensity and of the optical thickness of the plasma column as well as the Schlieren measurements showed that plasmas with electron densities between 5×10^{17} and 7×10^{19} cm^{-3} can be reproduced rather reproducibly.

---

11. The data support neither of two related folk beliefs concerning contemporary scientific style: an increase of sentence complexity resulting from an influx of German-speaking scientists, and a loss of syntactic control resulting from the general loss of command of the English language. If anything, the data show a limited consistency with what is believed to be a general simplification and shortening of the English sentence in America over this century.

12. The data were limited to two-clause sentences to control for more complex syntactical relationships established in sentences of three or more clauses.

Theoretical Integration in Experimental Reports

Similarly, the temporal and causal subordination puts two ideas or events in relation to one another, as in 1980-SJR:

> As the electric field was applied, the oscillator was simultaneously returned to within 10Hz of the shifted point of maximum slope.

Thus changes in subordinate clause types suggest increasing intellectual complexity, even while sentence length and syntactical complexity remain about the same.

WORD CHOICE

This tendency to expand intellectual complexity within unchanging linguistic complexity becomes more pronounced when we examine word choice. Most important are the words that fill the two main syntactic positions in the sentence: the subject and verb of the main clause. These two positions usually define the main meaning elements around which the rest of the sentence revolves, unless the main claim is hidden behind an empty phrase such as "there are" or "one can say that." Such empty phrases appear in only about 5 percent of the sentences examined.

Throughout the period, 70–79 percent of main clause subjects have been either names of objects (that is, apparatus, observed features, or objects presumed to exist in nature) or names of abstractions (that is, processes, qualities, or generalized terms), but the balance between the two has shifted from virtual equality in 1893–95 (36 percent objects, 34 percent abstractions) to a 1:3 ratio in 1980 (19 percent objects; 57 percent abstractions). That is to say, recent sentences are centered less on concrete descriptions and more on topics of theoretical significance. Thus the opening sentences of 1893-EM use the following concrete grammatical subjects: "fact," "substance," "plates," "turmalin." The opening sentences of 1980-RAR et al., on the other hand, use more abstract subjects: "excitation," "correlation," "ionization," "autoionization." The increasing abstraction of sentence subjects reinforces the impression of increasing content.

The main verb also has been conveying more substantial content over the years as the percentage of substantive active verbs has been increasing (from 16 percent in 1893–95 to 35 percent in 1980) and the percentage of reporting verbs has been decreasing (from 10 percent to 3 percent). Passive verbs and forms of the verb "to be" have remained equally important throughout the period, with passives accounting for almost half of all main clause verbs, and "to be" for about one quarter. The decrease in reporting verbs (for example, "Smith reports . . .") and

Three: Typified Activities in Twentieth-Century Physics

increases in active verbs (for example, "temperature increases . . .") suggest that the finding or theory has increasingly been brought into the central grammatical position, while the publishing scientists have been given a back seat, thus adding density to the discussion and integrating source material into the continuity of the argument. The following two examples highlight this stylistic change. The opening section of 1895-EIN presents some findings with the aid of reporting verbs:

> In 1885, Messrs Siemens and Halske of Berlin published the results of measurements for the purpose of showing the superiority of the silver-grey surface obtained by treating filaments of glow-lamps by bringing the same to incandescence in an atmosphere consisting of volatile hydro-carbons. In the following year Mr Mortimer Evans described comparisons of the radiation from bright and black incandescent lamp filaments in which the superiority of the former was very clearly demonstrated.

In this chronological narrative, the point of theoretical interest remains obscure, as do the significances of the various details. What we most learn are the doings of scientists. In 1980-KF et al., two sentences pointedly summarize a large body of research with specific purpose for the work at hand by making the point of interest the grammatical subjects, and the relevance of those subjects the verbs (the first active, the second passive). The scientists have vanished to the footnotes.

> Laser techniques provide both an efficient population of highly excited states as well as a resolution frequently only limited by the radiative width of the excited state. Thus, Doppler-free two-photon spectroscopy,[a] quantum-beat spectroscopy,[b] level crossing,[c] rf resonance[d] and microwave resonance techniques[e] have been used for studies in sequences of D states, especially, but also P, F, and G states.

Thus changes in main clause verbs and nouns have made sentences more directed toward the argument, more active and denser.

A more general inspection of the vocabulary also indicates increases in the density of information and the theoretical meaning—that is, the embedding of meaning within particular bodies of knowledge and theory. These increases are evidenced by growth in the percentage of words having technical meanings (in 1893–95, 15 percent; in 1920, 14 percent; in 1950, 29 percent; in 1980, 32 percent). Consider the two passages quoted just above. In the passage from 1895-EIN, the first term with technical meaning is almost thirty words in, and most of the technical terms are not far removed from their then-common usage: "fila-

Theoretical Integration in Experimental Reports

ments," "glow-lamps," "incandescence," "atmosphere," "volatile," "hydro-carbons," "radiation." Only one term, "hydro-carbons," does not have a closely related common-use meaning. The terms do gain some specificity of meaning from the technical context, such as "filament," meaning not just a thin fiber, but one through which electric current is passed to produce heat and/or light. The terms also gain meaning from the accumulated work to perfect the incandescent lamp, and from existing electrical and chemical theory. The passage from 1980-KF et al., however, contains a higher number of technical terms, with meanings further removed from ordinary use. Not only do terms like "laser," "Doppler-free," "photon," "spectroscopy," "quantum-beat," "rf resonance," and "microwave" have their origin in scientific theory and practice, they incorporate large amounts of scientific knowledge in their definitions. In order to understand the terms with appropriate precision one must have substantial understanding of current physical theory and knowledge. Even terms with common-use meanings have highly specific, content-laden meanings in the context of the scientific article: "efficient population," "excited," "state," "radiative width," "level crossing," "sequences," "D, P, F and G states." Many of the meanings, in fact, derive rather directly from quantum theory.

One final lexical feature, the multiword noun phrase, has increased density and theoretical import. These phrases, sometimes hyphenated, combine words from common and technical vocabularies to create new terms of highly specific meaning. For example, the opening two paragraphs of 1980-KHF et al. contain such hybrids as "plasma spectroscopy," "electron densities," "free-bound continuum," "half-width," "line profiles," "mean particle-electric-field strength," "thermally excited longitudinal plasma waves," "collective wave field," "mean interparticle field," "current driven turbulence," and "thermal equilibrium." Such phrases are to be distinguished from ordinary adjective-noun clusters in that they modify not just by adding information, but by placing the object, event, or concept within a more specific framework of knowledge. An equivalent passage from 1893-EFN contains far fewer of these hybrids, and they tend to resemble more traditional nouns modified by adjectives: "atmospheric gases," "lamp-black," "potassium alum," "ammonium alum," "aluminum-iron alum," "fifty-volt Edison incandescent lamp." It should be remembered that from the time of Chaucer until the early part of this century, "alum" was a common term.

Three: Typified Activities in Twentieth-Century Physics

GRAPHIC FEATURES

Scientific articles contain, of course, more than running text: graphic features—drawings, graphs, tables, plates, and equations—interrupt the block of prose. They shift the argument into different symbolic media, but the decisions of when and where to employ them, how they should be designed and what information to include, are as much writing decisions as are word selection or organization. Here, as in other features already examined, we see the movement from early concreteness to recent abstraction, from early representations as ends in themselves and intelligible without extensive scientific knowledge, to recent issue-directed, interpretive arguments dependent on substantial disciplinary knowledge. To put it more concretely, a scan of articles of *PR*, series 1, volume 1, leaves a visual impression of detailed apparatus drawings and extensive tables of raw experimental data, while a scan of the journal of 1980 leaves a visual impression of extensive equations and schematized graphs.

Specifically, the decade-by-decade selection of spectroscopy articles contains, first, a decreasing use of apparatus drawings. Up to 1920, all but two of the selected articles had equipment illustrations—some realistic in representing the actual appearances of devices, others more schematic in representing only the essential optical features, but all directly representing the equipment employed. By 1930, however, fewer articles contained such illustrations, and those included tended to be abstract. Of the eleven articles examined from 1960, 1970 and 1980, only four had equipment diagrams, and all four were schematic representations of functions (functions being identified by word label), rather than representations of actual equipment.[13]

A more recent form of illustration is the schematic representation of quantum states and transitions hypothesized as present in the experiment at hand. Such illustrations first appeared in 1940 in one of four articles examined; in 1950 transition schematics appeared in two of six; in 1960, two of three; in 1970, one of three; and in 1980, two of five. Such diagrams, being specifications of quantum theory, are theory dependent, abstract, and interpretive (that is, at several removes from the raw data, and serving as explanations of those data).

Similarly, tables of results, originally presenting all results and often in raw form, become increasingly selective, summary, calculated, and focused with respect to theoretical importance. Tables become shorter

---

13. The detailed representation of novel apparatus has migrated to instrumentation journals, but the very separation of such materials from primary research reports signals that information about instrumentation advances is not considered of the same category as research findings.

Theoretical Integration in Experimental Reports

and by 1980 appear in only two of the five articles examined. The burden of data presentation has increasingly been placed on graphs, especially since 1950, even though graphs were always present in substantial numbers. All of the 1980 articles, for example, display their data through graphs. Graphs, in addition to displaying data, show trends and allow comparison with other data and with theoretical predictions displayed on the same or neighboring graphs. In fact, all five of the 1980 articles examined incorporate some comparative features in the graphs, and four out of five compare results, theoretical values, and other relevant curves extensively—through multiple curves on single graphs, multi-part graphs displaying different kinds of curves, and adjacent graphs (as many as eight at a time). The display of data has thus become more purposeful, interpretive, intellectually complex, and intertwined with the theoretical argument of the paper.

Finally, equations make more frequent and more prominent appearance in spectroscopic articles as the period progresses. The three articles examined from 1893 to 1895 contain no equations or mathematical expressions, while the five articles from 1980 contain forty-three lines of equations and expressions, not including those printed as part of the running text. The contrast would have been even more striking if theoretical articles were also considered. In the early years of *PR*, no purely theoretical article appeared on the topic of spectral lines; but since the establishment of quantum mechanics, they have abounded. It is not uncommon for recent theoretical articles to have twenty or more lines of equations and expressions per page. The appearance of equations is a clear indicator of the integration of theoretical explanation and prediction into the argument of the paper.

It is instructive to notice the difference in pattern of illustration change here from that observed in the *Transactions* in the earlier period. From 1665 to 1800, apparatus illustrations increased in number and detail as part of the article's increasing importance as a surrogate for first-person observation. Here, however, the verisimilitude of surrogate experience decreases as a significant rhetorical issue, to be replaced by the relation of the reported events to a more general theory. Authors seem less concerned to establish that the events occurred as reported than to show how these events fit with and elaborate the communally shared account of theory. When the community shares a generalized vision of the world, explicit connections to the abstractions carry more sense of veracity and more communally significant information than concrete representations of one-time events in the laboratory. Strong theories apparently can create stability of reliably reproducible events (see chapter 11) with greater force and generality than can concretely reported

Three: Typified Activities in Twentieth-Century Physics

events, for the generality allows application to a variety of circumstances, while the concrete event only encourages attempts at exact replication, with all the attendant difficulties. (See Collins, *Changing Order,* for a discussion of the difficulties of replication.)

ORGANIZATION, ARGUMENT, AND EPISTEMOLOGY

The features examined above strongly indicate the increasing abstraction, web of background information, density of knowledge, interpretation, and focused argumentation going into the *PR* article since 1893, but an examination of the structure of articles will reveal even more about the way discourse is intimately linked not only to knowledge and theory, but to epistemology—beliefs about what can be known, how it can be known, in what form it can be expressed and how it should be argued.

The analysis of organization and argument will examine three levels of data: (1) the self-identification of the article's structure as embodied in formal divisions and section headings; (2) the proportion of space devoted to the various parts of the argument; (3) the texts themselves, to extract the mode of argument and the logic of presentation.

Prior to 1950, only about half the articles had formal divisions with section titles; after 1950, section headings were a consistent feature of almost all articles. Moreover, section divisions became more complex after 1950; prior to 1950, those articles using subdivisions averaged 4.5 per article, while in 1950 and after, the average was 7.4. All articles in the decade-by-decade selection were examined for this feature.

Before 1930, those division headings that exist indicate that articles ended with results, with no conclusion or discussion sections, as though the results could stand alone and complete in their meaning. Before 1910, some articles contained conclusory sections, but only in the form of summaries of results. Starting in 1930, however, discussion and conclusion sections—sometimes so labelled, sometimes given more substantive titles—became increasingly common. This again is a clear indication that the articles have become issue-oriented rather than fact-presenting.

Similarly, with a single exception (1901-BEM, which later content analysis will show not to be anomalous), articles did not have explicit theory sections, although they appear with some frequency since then.

Early articles, then, basically have methods and results sections, sometimes with two or three methodological sections. More recent articles tend to have only one methodological section, but several discussion, conclusion, and theory sections. Moreover, in early articles those sections given original names tended to be methodological; for example, in 1910-EIN/EM, the first four of the five sections are methodological and are given specific descriptive names: "Determination of the Distribution of Energy in the Spectrum of the Comparison Flame," "Comparison of the Fluorescence Spectra with the Spectrum of the Standard Acetylene Flame," "The Correction for Slit-width," and "The Correction for Absorption." More recent articles, on the other hand, give methodological sections standard names (for example, "Experimental") and give original names to discussion and interpretation of results on occasion, as in 1980-RAR et al.:

I. Introduction
II. Experimental
III. Results
IV. Interpretation—A. Yb ($5p^64f^{14}6s^2$)—1. Autoionization, 2. Auger Decay, B. Ba($5p^66s^2$)—1. Autoionization, 2. Auger Decay.
V. Discussion
VI. Conclusions
Acknowledgment

These titling choices indicate that early authors considered methodological sections to present special problems and achievements, while more recent authors are inclined to call attention, and give specific designation, to the theoretical meaning of the data.

Finally, acknowledgments sections did not explicitly emerge until 1940 and were not a regular feature until 1960. The implications of this will be discussed later.

Analysis of the percentage of each article devoted to each part of the argument confirms and supplements previous findings. In the 1890s, the introduction and review of the literature sections were substantial, although, as indicated in earlier discussion of references, unfocused. By 1900, these parts had become more compact. Since then, the introductory material has expanded both proportionally and even more in absolute terms (as the size of articles has increased). Moreover, in recent years the introduction has been sometimes supplemented by presentations of background theory. Methods and apparatus sections have been generally decreasing in their proportional share of each paper. Results sections have always remained important, but, as noted earlier, the data display has tended to shift from tables to graphs. Tables still in use in recent years have tended to present conclusions, such as the identification of quantum-level transitions with specific spectral lines. Discussion and conclusion sections have taken increasingly large parts of the

Three: Typified Activities in Twentieth-Century Physics

articles, sometimes becoming so intertwined with the presentation of data that the results section takes on a discussion character. Finally, acknowledgments disappeared after the first few years, only to reemerge in a different form around 1920. The acknowledgments of the 1890s were personal testimonials to friends and mentors. 1895-EIN is filled with passing acknowledgments of the aid of the author's brother, such as this:

> . . . a method nearly the same as that described by E. F. Nichols in the first volume of this Review. Indeed in many of the measurements Mr Nichols did me great service, bringing to bear upon what was in many respects an operation of unusual delicacy the skill attained by long practice in similar research.

The acknowledgments that reappeared in the 1920s were more spare, sharing limited forms of credit and recognizing institutional dependencies. Even the acknowledgment of intellectual fellowship lost personal effusiveness. These trends have continued, as indicated by the two following examples, the first from 1920-GR, and the second from 1980-TFG et al.:

> The present investigation was suggested by Dr W. W. Coblentz who has shown continued interest in the problem. The apparatus was placed at my disposal and set up in the Randal Morgan Laboratory of Physics at the University of Pennsylvania. Suggestions have been made during the progress of the work by Dr Goodspeed and Dr Richards for which I wish to express my appreciation.

> We would like to acknowledge stimulating communications with R. Morgenstern in the course of this work. This work has been supported by the US Department of Energy, Office of Basic Energy Sciences.

An examination of the actual arguments presented in the spectroscopic articles gives a deeper insight into how the features already discussed are intertwined with significant intellectual and epistemological changes in the field. The remaining analysis consists of descriptive characterizations of selected articles, presented chronologically to suggest a rhetorical history of the field.

These descriptive characterizations reveal the substantive consequence of all the features examined through various indicators earlier in this paper. We see here presented the evolution of the kinds of argument that result from the mobilization of all the features examined. And we will see that the evolution of the argumentation has direct epistemologi-

Theoretical Integration in Experimental Reports

cal implications as the arguments become more theory-based and ultimately self-conscious about their constructed theoretical character. For instance, 1893-EFN employs a rhetoric based on an empiricist epistemology. Spectral lines and the substances that produce characteristic patterns are taken as unproblematic objects of nature. The main task of the article is to present measurements of these unproblematic objects. References to earlier work are only general because they only need suggest that others have identified and measured similar phenomena. The main problems are of methodological technique and are discussed in some detail. Results are presented in graphs and tables; the accompanying text only repeats the information presented graphically with no further interpretation, only further methodological comments. The conclusion consists only of a summary of results—that is, a third repetition of the findings.

1900-CJR shares the same empiricist stand, but presents its tasks, methods, and findings in closer relation to the work of others, thereby making the article more focused, concise, and aware of the concept of a "problem." The task described was to take a series of measurements already done, but with one change of circumstances to note the differences in results. The area of study is taken as a given, not requiring a roll-call of forebears; other work is referred to only as it bears directly on the current work. The apparatus is described as "about the same as that used by Foley," although a truncated description follows. Significantly, the author avoids discussing a methodological problem of possible distortion by referring to Foley's earlier treatment of the issue. In presenting results the author relies on prior literature by noting only those lines not reported in previous studies. Not only does this selective reporting of findings lend conciseness, it focuses attention on these new readings appearing under changed conditions, making the readings "problematic," something to be accounted for. The accounting is done in two ways: first, by associating them with an earlier set of predictions and, second, by attributing some lines to a specific element. In a final section the author discusses the conflicting observations of two previous workers and then describes some new observations "of some interest in this connection." He does not, however, draw the problem more sharply or propose a resolution; he only adds new observations. Thus, conflicts in the literature and comparisons of his own findings to other findings in the literature suggest topics for discussion, but the discussion remains concrete, only rising above the level of observation and measurement.

1901-BEM, anomalous by several of the previous measures (number of references, lines of equations, and presence of a theory section) is explicable when examined from the perspective of argument and epistemology. The article is nevertheless unusual for it attempts to move beyond empiricism to create a link between theoretical discussion and experiment, although the link is awkward and not very intimate. If 1900-CJR is a slight machine that rises a bit above ground by no great will of its own, 1901-BEM is a massive piece of equipment that struggles mightily but gets no higher than the other. 1901-BEM opens with a general theoretical discussion, beginning with a first principle and synthesizing much existing theory in textbook fashion, but without any indication of where the theory is heading, what problem is being addressed or what issues are at stake in the experiment. If not for the title and outline standing at the head of the article, the first five pages would give little clue that this was an experimental paper. The author does eventually apply the theory to the particulars of the experiment, but never defines a specific issue at stake. The theory serves only as a description of the experimental conditions. The presentations of apparatus, method, and results are not distinguished in any way from those of simple empirical work. Most significantly, the data presented are not selective concerning an issue at hand, but rather seem presented for their own sake. The discussion of results consists mostly of how method might have been improved. A few low-level generalizations are made in passing, and a conflict in the literature is discussed, but the data at hand are not adequate for a conclusive resolution. The conclusions section consists of a numbered list summarizing a miscellaneous collection of earlier observations, some of which are methodological.

Moving forward, 1910-HEI uses references to prior work to establish a problem, discusses relevant theory, proposes a solution, then discusses the limitations of the solution. In many respects, from the embedding of the problem in the literature and theory to the focus on problem solution and the recognition of the constructed and limited nature of the solution, this article foreshadows the intellectual structure, argument pattern, and epistemological stance of later work, except that in this case the problem is methodological and the solution is a new piece of apparatus, rather than the problem and solution being in theory. This parallel suggests the analogy between physical apparatus and intellectual apparatus. A piece of machinery (in this case, a photospectrometer) is clearly a human invention; if there are faults or limits to the apparatus, a study of existing machines and an understanding of their theory can lead to diagnosis of the problem and construction of an improved machine addressing the difficulty. Moreover, since the new machine is also a human construction, it can be assumed to have new limitations. It is not so easy to see symbolic representations of nature—intellectual constructions—in the same light; such perception is likely to come only after a science becomes organized around theory rather than around "empirical facts," and then gains some sophistication about that theory. Over the next period we will find theory moving to the center of arguments and an increasing awareness of the constructed nature of theory.

By 1920, a few articles present more substantial integration of theory into the argument. 1920-CDC/DC, although largely empiricist in manner, begins with a purpose of theoretical consequence:

> A knowledge of the relation between the spectrum of a substance and that of its isotope is important in that it may throw further light on the structure of the atom.

Although the consequences of the finding of this study are never explicitly discussed in terms of theory of the atom, the experimental design and results reported are directly relevant to this theoretical task. In this case, even though theory has not changed the structure of the argument, it has helped select and focus the contents.

1920-WD/RAP adopts a theory-driven task more fully. The opening paragraph, entitled "Object," identifies specific measurements important "for the purpose of testing certain relations deduced from theories of the structure of atoms and the mechanism of radiation." Theory testing becomes here an element of argumentative structure; after presenting apparatus, methods, and results, the article discusses how the data correspond to several current theories and to calculations from equations, although only in a general way. Some theories are supported, others questioned, and limited conclusions drawn based on theoretical interpretations of the data (for example, "It would seem in this case the electrons producing the lines did not come from exactly the same outer orbit").

In 1920, several purely theoretical articles relevant to spectroscopy also appeared, whereas none had appeared in 1893–95, 1900–1901, or 1910. Kemble readjusts an earlier theory of his to make it consistent with Bohr's theory of the atom; Baly tries to correct an earlier paper by adjusting its conclusions to new theories and findings; and Webster compares theories and results of quantum phenomena in the X-ray and visible light regions to draw conclusions about emitting mechanisms and to find some limitations to Bohr's theory.[14] This array of articles indicates that by 1920 Bohr's theory has cast the field into a more theoretical vein.

---

14. Edwin C. Kemble, "The Bohr Theory and the Approximate Harmonics in the Infra-Red Spectra of Diatomic Gases," 2:15:2, 95–109; E. C. C. Baly, "Light Absorption and Fluorescence," 2:15:1, 1–7; and David L. Webster, "Quantum Emission Phenomena in Radiation," 2:16:1, 31–40.

Three: Typified Activities in Twentieth-Century Physics

Not only does the empirical work gain more of a theoretical basis, but theory itself is unsettled, requiring testing, evaluating, readjusting, reconciling, and, in some cases, abandoning. The new situation calls forth new kinds of arguments in both experimental and theoretical papers.

By 1930, quantum mechanics had stabilized sufficiently to provide the grounds for empirical work without the theory itself being in question. 1930-SS takes on a task located and identified by theory, a task that appears from the discussion of references to be already commonplace: elucidation of the terms of the spectrum for selected elements. That is, measured spectral lines are being associated with specific electron transitions within the structure and fine structure of the atom. Thus, although the experimental description follows the typical empirical pattern, the topic of discussion in the results section is the classification of results to determine term values and to associate lines with transition intervals. These classifications and associations, rather than the raw measurements, are represented in the results tables. Thus, results are processed intellectually within concepts and operations derived from theory, and are expressed in a language also derived from theory. With the ground theory established, specific questions of elaboration and identification of mechanisms in specific circumstances can then become recognized questions in the literature. That is, theory helps organize the literature.

1930-SB takes a further step into theory by finding its problem in the literature ("there has been a great deal of speculation concerning the identity of the emitter") and presents an experiment testing one hypothesis. Since the ground theory has helped identify the problem, others can also be working on the same problem; therefore, the author must discuss the work of a colleague who published while his own work was still in progress. The article elaborates theory extensively, using the tools of quantum mechanics and discussing how the analysis varies from others proposed, as well as how it relates to experimental results in the literature. The author is well aware that he has organized his work around the concept of a problem, for he explicitly states in the acknowledgments, "Dr R. S.Milliken suggested this problem. . . ."

In 1930-SKA/JHW, awareness of the constructed nature of theory and language allows the authors to suggest a nomenclature innovation to allow better identification and analysis of a particular phenomenon. The distance between symbol and object becomes a resource of investigation. Thus, in addition to the usual features of a theory-located, problem-based article, this article devotes much space to explaining and justifying the proposed nomenclature convention. The results and discussion sections, moreover, become cases of the application of the new nomenclature.

Articles in 1940 and 1950 continue in the style of the theory-located, problem-based article, with the problem sometimes coming from the split between theory and data (for example, 1940-SM) and sometimes from disagreements in the literature (for example, 1950-RBH et al.). In 1950-WFH/TL a new style of argument appears that will be more fully developed in 1960-HA/AH: the modelling approach. Epistemologically, the modelling approach sees a split between nature and theory, theory being only a human construction, having no reasonable expectation of giving a complete and accurate account of nature. Under such an approach, a paper cannot propose a theory test, proving the truth or falsity of a claim, but can only propose a model that accounts for the data better than other available models. In terms of argumentative structure, a modelling article does not present a claim in the beginning to be explained, supported, and discussed in light of experimental data; instead, once the article locates the problem in relevant theory and presents appropriate data, only then does it offer its model or claim about what apparently occurred in the experiment. Results are first presented, then puzzled over. Only after the puzzlement is the provisionally best model presented.

Once the argument moves away from notions of absolute truth and error, the concept of fit between theory and data becomes more important. Consequently, 1970-NWJ/JPC finds its problem in the deteriorating quality of fit between one category of data and a new theory gaining acceptance because it improves fit with respect to other categories of data. The experiment is designed to find the cause of the discrepancy. The article ends by calling for new theory and experimental work.

1980-KF et al. compares the fit between two sets of experiments and two models. As knowledge has grown, theory elaborated, work proliferated, and individual problems have become located more and more specifically within the web of prior work, articles have become increasingly tentative about the certainty and epistemological status of their claims.

Discussion

What information people in a group convey to each other, the purposes for which they present that information, their means of persuading each other of the validity of their statements, the

Three: Typified Activities in Twentieth-Century Physics

uses others make of the statements, and the features of discourse they develop to realize these activities are all important aspects of a group's communal life, especially when a major activity of that group is to produce statements. The apparent function of the community of research physicists is to produce statements to be validated by that community as knowledge. The character of the statements presented for communal judgment embodies major (although not all) aspects of the community's social relations, and changes in the character of those statements represent changes in the social relations and social structure. Further, if, as in the case of *PR*, the changes in character of the statements are intertwined with cognitive changes of a discipline, discourse provides a concrete mechanism by which social behavior, social action, and social structure are related to cognitive structure.

Specifically, the discourse style in *PR* at the time of its founding suggests a group tied together by traditions of work, common objects of interest, common techniques, and personal apprenticeship loyalties. Its members engaged in a loosely organized mapping activity, confident of the solidity of the ground they were mapping, of the appropriateness of the tools and of a simple correlation between the ground and the map. Each contribution had only to identify the piece of ground, describe the tools, and present a piece of the map, with no particular need to demonstrate coherence within the piece or among the pieces. Much of the contribution of each article was methodological, so apparatus and methods were described at length, both to allay criticism and to make the innovations available for others. This situation, as noted earlier, may reflect more on the state of American physics at the time than on the general condition of international physics.

In the early part of this century, the spectroscopic community in America became more organized around its shared work. Members would scrutinize each other's work for patterns and would harness the work of others into the arguments of their own new work. They showed increasing effort to establish generalizations and coherence among the shared work and started to organize their work around theories, often casting empirical work in the form of theory-testing. They also felt obliged to argue for the theoretical significance of their work in order to anticipate the newly emerging criterion of significance.

Bohr's theory of atomic structure offered a single ground theory upon which spectroscopy could organize itself and its work. At first the full meaning, range of validity, and manner of application of the theory were in question. Physicists argued basic theory with each other: experimenting, deriving calculations from theories, comparing theories and data, examining the fuller implications of theories. Rather than being torn apart into mutually exclusive camps, the physicists seemed to be drawn more closely together as they had to examine, compare, rely on, discuss each other's work more closely in order to establish theoretical generalizations that would ultimately be validated by the entire discipline.

As quantum mechanics became established, it provided a coherent organizing principle for work and argument, but in each new contribution the publishing spectroscopist had to attend to the relationship between his own work and the general theory by locating his work in the theory, elaborating aspects of the theory, showing the theoretical meaning of results, and discussing theoretical consequences. The increasingly elaborated theory became a means by which his own work became tied to others' work, to which he more often referred. Problems, localized and suggested by theory, became shared. Theoretical significance, correctness, and consistency became major criteria. Attending to these criteria and tasks increased both article length and density of expression. In order to make a well-formulated statement to one's colleagues, one had to communicate more information.

As theory grew, it became apparent that it was a construction, separate from the nature it described. This awareness affected argument and social relations. Hard answers were not to be expected. The tentativeness of the "modelling" or "fit" type of arguments mitigated the confrontational conflict of theoretical dispute by recognizing that each contribution was only part of a process.

Concluding Thoughts

The evolution of the spectroscopic article over the past century in America reflects the growing knowledge and theoretical character of science and reveals some of the institutional consequences of these changes. The large-scale trends revealed here are consistent with the traditional view that science is a rational, cumulative, corporate enterprise, but point out that this enterprise is realized only through linguistic, rhetorical, and social choices, all with epistemological consequences.

This particular study highlights how a strong theory not only shapes the scientific activity, but becomes an important means of ordering social relations. A widely shared and elaborated theory can provide discrete and robust venues for individuals where they may formulate their own interests and carry forth their own work. In this sense a theory may allow a kind of bureaucratization of the scientific community, allowing individuals to sort themselves out into distinctive research roles according to rational principles generated by the theory.[15]

This, of course, differs from the classical bureaucracy where roles and tasks are established from the top down, although such bureaucracies may well exist within certain laboratories. Here, rather, roles and tasks are negotiated between individuals bidding to work on, modify, develop, elaborate, or apply part of the theory and employers, funders, editors, referees, critics, and audiences who grant the researcher various powers to continue, publicize, and gain acceptance for their work (see, for example, Myers "Social Construction" and "Texts").

This change from scientific entrepreneurship, where each individual stakes a private claim that recognizes few overt and lasting connections to the claims of others, where each claim is under threat from each other claim, to scientific bureaucracy, where competition is rather to attach yourself firmly to a powerful part of the communal apparatus, raises many new and intriguing possibilities for communal and individual pathologies, resulting in widening divisions between the abstractions of the theory and responsible empirical experience. Yet by organizing the experience of large numbers of individuals, pointing the individuals toward new kinds of experiences, providing means for comparing and coordinating varied results, and establishing topics and procedures for discussion, a strong theory can ground its generalizations on the empirical experience of an entire community. Whether the research program and the attendant social community pursuing that program thrive depends in part on whether that research program generates interesting venues for research—that is, places where the program can attach itself to accounts of empirical experiences. Furthermore, the program can continue to thrive only if the accounts created by empirical research coordinate well with the more general account offered by the theory. Otherwise, following Lakatos' analysis, the research program degenerates, offering little satisfaction for the interests of individual scientists. Few will fight for seats on a train going nowhere.

Spectroscopic Articles from *Physical Review* Discussed in This Chapter

1893-EFN. Ernest F. Nichols. "The Transmission Spectra of Certain Substances in the Infra-red." Series I, volume I, number 1, pp. 1–18 [hereafter I:I:1, 1–18].

---

15. These thoughts owe obvious debt to Max Weber's discussion of bureaucracy in *The Theory of Social and Economic Organization*.

Theoretical Integration in Experimental Reports

1895-EIN. Edward I. Nichols. "The Distribution of Energy in the Spectrum of the Glow-Lamp." I:II:4, 260–76.

1895-EM. Ernest Merritt. "On the Absorption of Certain Crystals in the Infrared as Dependent on the Direction of the Plane of Polarization." I:II:6, 424–41.

1900-CJR. Carl J. Rollefson. "Spectra of Mixes." I:XI:2, 101–4.

1901-BEM. B. E. Moore. "A Spectrophotometric Study of the Hydrolysis of Dilute Ferric Chloride Solutions." I:XII:3, 151–76.

1910-EIN/EM, E. I. Nichols and Ernest Merritt. "Studies in Luminescence: XI. The Distribution of Energy in Fluorescence Spectra." I.XXX:3, 328–46.

1910-HEI. Herbert E. Ives. "Scattered Light in Spectrophotometry and a New Form of Spectrophotometer." I:XXX:4, 446–52.

1920-GR. George Rosengarten. "The Effect of Temperature upon the Transmission of Infra-red Radiation Through Various Glasses." II:XVI:3, 173–78.

1920-CDC/DC. C. D. Cooksey and D. Cooksey. "The High Frequency Spectra of Lead Isotopes." II:XVI:4, 327–36.

1920-WD/RAP. William Duane and R. A. Patterson. "On the X-ray Spectra of Tungsten." II:XVI:6, 526–39.

1930-SB. Sydney Bloomenthal. "Vibrational Quantum Analysis and Isotope Effect for the Lead Oxide Band Spectra." II:XXXV:I, 34–45.

1930-SKA/JHW. Samual K. Allison and John H. Williams. "Experiments on the Reported Fine Structure and the Wave-length Separation of the Kβ Doublet in the Molybdenum X-ray Spectrum." II:XXXV:2, 149–54.

1930-SS. Stanley Smith. "An Extension of the Spectrum of Thallium 11." II:XXXV:3, 235–39.

1940-SM. S. Mrozowski. "Hyperfine Structure of the Quadrupole Line 2815A and of Some Other Lines of Ionized Mercury." II:LVII:3, 207–11.

1950-WFH/TL. W. F. Hornyak and T. Lauritsen. "The Beta-Decay of B^{12} and Li^8. II:LXXVII:2, 160–64.

1950-RBH et al. R. B. Holt, John M. Richardson, B. Howland, and B. T. McClure. "Recombination Spectrum and Electron Density Measurements in Neon Afterglow." II:LXXVII:2, 239–41.

1960-HA/AH. H. Arbell and A. Halperin. "Thermoluminescence of ZnS Single Crystals." II:CXVII:1, 45–52.

1970-NWJ/JPC. N. W. Jalufka and J. P. Craig. "Stark Broadening of Singly Ionized Nitrogen Lines." IIIA:I:2, 221–25.

1980-RAR et al. R. A. Rosenberg, S.-T. Lee, and D. A. Shirley. "Observations of a Collective Excitation in the Ejected-Electron Spectra of Yb and Ba." IIIA:XXI:1, 132–39.

1980-TFG et al. T. F. Gallagher, K. A. Safinya, and W. E. Cooke. "Energy Analysis of the Electrons Ejected in the Autoionization of the Ba $(6p_j\ 20s_{1/2})_s$ States." IIIA:XXI:1, 148–50.

1980-KHF et al. K. H. Finken, R. Buchwald, G. Bertschinger, and H.-J. Kunze. "Investigations of the Ha line in Dense Plasmas." IIIA:XXI:1, 200–206.

1980-KF et al. K. Friedriksson, H. Lundberg, and S. Svanberg. "Fine- and Hyperfine-Structure Investigation in the 5^2D–n^2F Series of Cesium." IIIA:XXI:1, 241–47.

1980-SJR. Stanley J. Rosenthal. "Differential Stark Effect in the Ground-State Hyperfine Structure of Gallium." IIIA:XXI:1, 248–52.

7 MAKING REFERENCE

EMPIRICAL CONTEXTS, CHOICES,

AND CONSTRAINTS IN

THE LITERARY CREATION OF

THE COMPTON EFFECT

The problem of reference haunts all studies of scientific language. How does language escape the narrow bounds of the linguistic code to say anything substantive about the natural world? How can language be anything other than an imaginative fiction, having anything to do with anything beyond the internal elaboration of the code? To anyone familiar with philosophy, sociology, linguistics, or literary theory I need hardly catalogue the way in which this issue persists, freighted with the frustration and acrimony of an irresolvable conflict over fundamental beliefs.

For just such reasons, once I became aware of the apparent intractability and acrimoniousness of the issue I tried to avoid it. I thought I could address some practical issues of writing without addressing the fundamental questions of the validity of science or belief in a natural world. However, I found I could not avoid the issue for several reasons. For one, most of the discussion over scientific language seemed driven by one position or the other—in classical rhetorical terms the discussion was epideictic, either to praise science for its truthful language or to blame it for the hubris of claiming a privileged path to knowledge. I found that every claim I made about scientific discourse was interpreted against this issue, even if there was no explicit relation and I had intended no implicit one. Rather than suffer the misunderstandings of imputed positions, it seemed wiser to address the issue full face. More substantively, I also found very early in the game that one could not contemplate any rhetorical system without taking into account the goals and objectives of the people using the system. Users of the scientific linguistic system seemed to believe their language was useful in gaining some control over the natural world, and many of their behaviors as writers and readers seemed constructed out of that belief. So at the very least I had to take the referentiality of science seriously as a communal assumption and as a question for investigation. I had to see how that belief shaped practices and to what extent the linguistic system lived up to its goals.

However, as I engaged the issue not purely as a textual matter, but within the complex matrix of social and individual practice, the issue no longer seemed to be so intractible nor rife with contradictions. My several discussions earlier and later in this book present my overall approach (see especially chapters 1, 2, and 11), but here I will examine how the developed system of scientific communication shapes the purposes, processes, and norms of statement production so as to make empirical experience a topic-, resource-, and constraint-shaping individual behavior. Specifically, the individual is placed within a communicative context that constantly encourages and demands that the individual at many junctures considers how empirical results either can advance the claim-making procedure or call for reconsideration of the claims and representations of phenomena. Through individual behavior and practice, the discourse is brought into increasingly close and precise exchange with the phenomena being examined. Through living people, the symbols of language come into contact with the world.

Language Moves People, People Move in the World

The first step out of the bind of the closed-system of language is to see that language is used by people and has an effect on them. The desire to understand and master that effect motivates the study of language and rhetoric. From the beginning of rhetoric, the ancients had no doubts that language could move people, both in their thoughts and actions. Even logic, as first developed, was not removed from human cognition. In the *Posterior Analytics,* Aristotle presents logic as a way for statement makers to move readers with greater certainty through arguments and for readers to monitor whether texts were moving them by compelling or by less certain means. Logic was a tool to help minds move in directions in accord with reason, rather than a way to enable reason to escape its human dwelling. Similarly, Plato's complaint about rhetoric was not that language moved minds, but that knowledge of rhetoric might be used to move people's minds falsely in service of unconsidered ends *(Gorgias).* That is, the speaker remains ignorant of or false to the true movement of his or her own mind, but rather speaks only to fulfill baser passions. The virtue of dialectic over rhetoric, as Plato argues in the *Phaedrus,* is that those engaged in dialectic find their words in their mutual search for truth, that the words are part of a motion upward rather than downward.

The study of rhetoric through the last twenty-five hundred years has never lost its concern for the connection between mental movement and words, sometimes the mental movement of the rhetor producing the words and more often the mental motion of the audience. Literary studies followed suit, sometimes more concerned with the overflow of the author's feelings into words that would then carry the reader, and sometimes more concerned with the emotive effects which the author could create through manipulation of form. Only in recent history has a consensus developed in the study of language and literatures, that texts could and should be considered independent of the human producers and consumers (to be discussed in chapter 11).

Much of the social sciences maintains this concern for how language moves people and how the movements of people are expressed in language. Sociology, political science, and psychology continue investigations into how groups and individuals are moved by linguistic symbols and even construct their realities out of their symbolic interactions. This concern for the effect of language on humans is shared by both cognitivists and behaviorists, although one group tends to see the human motions associated with language occurring in the mind, while the other is likely to find the motions in visible behavior, as a mother moves to a child crying and a consumer moves to a product embedded in a series of messages. Even Marxist social scientists understand language in relation to the large material forces that move individuals, whether they consider language as an epiphenomenon of superstructure, as a substantive part of the base, or in some more complex dialectic with social activity and structure.

Although many current studies of scientific discourse accept the rift between language and the natural world, they rely on this indwelling of language in humans. Through case studies they have demonstrated that scientific language is designed to move readers and derives from the various forces moving the authors. Yet they do not take the second step to see that mental motions influence behavior that occurs in the physical world. It is this second step, however, upon which the project of

Three: Typified Activities in Twentieth-Century Physics

empirical science is founded: to create symbolic accounts that will help us understand our daily concourse with the natural world of which we are part. These symbolic accounts can help us order our relations with this natural world, either by control or by reconciliation. And these symbolic accounts are created out of close concourse with that natural world, heightened and refined through the evolving procedures of empirical investigation.

The cases examined in the previous several chapters demonstrate how the institutions and institutional practices of scientific communication have developed in constant relation to empirical experience. Empirical work cannot be separated from the communications system which gives occasion for the work and within which the work will be represented. Scientists do not simply mutely walk into a laboratory, unconscious of concerns in the literature, with no words or thoughts in mind, and do an experiment, unengaged in any symbolic processing. Writing up results and engaging in professional debates cannot be totally separated from earlier events in the laboratory. As we have seen, the institutions of language developed around haggling over experience—the best way to represent it, how the representation can be held accountable to the experience, how experience can be strategically deployed in the debate over claims. Within the institutions of scientific communication, scientists discuss experience, use representations of experience in advancing of their arguments, and constrain their statements on the basis of their own representations of experience and the representations of others.[1] Moreover, and even more essentially, these linguistic representations are created in close relation to actual manipulation of objects to create the experience represented. Within the psychological and sociological manners of the community, these experiences are attended to in the language. (These ideas will be discussed more fully in chapter 11.)

With empirical experience given such a central role in the values, norms, expectations, procedures, and evaluations of the scientific community, a major and compelling way for an individual to pursue his or her interests is to cast claims in as close a relation as possible to empirical experience both as represented in the literature and as generated in new empirical work. An individual does well for him or herself, his or her social network, and for his or her claims, by doing good science; that is, by creating representations of some stability and power when held against the accumulated and future experience of the community.

---

1. Further accounts of how nature and empirical experience are used as argumentative resources appear in Bruno Latour, *Science in Action,* and in the various essays in Callon, Law, and Rip, eds., *Mapping the Dynamics of Science and Technology.*

Making Reference

The character and quality of reference in scientific language depend on the kind of work individuals do to create that reference empirically and to adjust constantly the representation to increasing experience. The institutions of scientific communication encourage that reference-creating work; embody practices, procedures, and forms for generating such reference-laden representation; and establish an agonistic social field against which these representations are held against the experiences of others. This communal structure encourages the production of scientific results whether the individual scientists are motivated by greed, vanity, commitment to a doctrine, faith in a private experience, or love of the game.

When we step into the middle of twentieth-century physics we see the game already highly elaborated. An individual scientist has structured opportunities, resources, and constraints out of which to construct claims and arguments that will move others within the same system to come to his view of experience. The scientist behaves normatively, creatively, and self-interestedly within a complex system.

This chapter examines how one creator of significant and successful scientific claims, Arthur Holly Compton, held himself and his claims accountable to empirical experience, even though he created his texts within a community, employed the communal language and concepts, and pursued communal and private interests. The investigation here examines how those texts are embedded in situated practices, through which meanings are created and embodied in the symbols. This study will consider how the author responds to the various social and natural difficulties to creating a stable, reliable, socially persuasive claim. Compton's responses will involve less fundamental rhetorical innovation than Newton's (as examined in chapter 4). Newton's improvisations helped invent the institutions of modern scientific communication; Compton is working within an already elaborated and stabilized system. That Compton's behavior may seem familiar and predictable is just the point. The developed system of scientific communication helps scientists to behave like scientists and do good science.

The Case of the Compton Effect

The case to be examined here is of Arthur Holly Compton's announcement of what is now called the Compton effect. In the standard history of early twentieth-century physics, the Compton effect is considered the first empirical verification of the quantum theory, although verification of the quantum theory was not his purpose in

Three: Typified Activities in Twentieth-Century Physics

designing his experiments or publishing his findings. Under current understanding, the Compton effect occurs when x-radiation is scattered by electrons. The target electron receives a quantum of energy from the incident radiation. In reaction the electron does not recoil as would be predicted by classical physics—in the direction and with the energy imparted by the absorbed energy (as would a billiard ball). Rather the electron recoils in a different direction and emits new radiation (of a lower energy than the incident radiation) in a third direction, so as to conserve momentum and energy. This discovery was announced in a May 1923 paper, "A Quantum Theory of the Scattering of X-rays by Light Elements."

The focus of the first part of this study is on the emergence of this paper out of Compton's reactions to the scientific conversation within the problem area of his work. The situation within the problem area offered Compton constraints and opportunities, out of which he made choices that shaped his contributions and reshaped the communal conversation. Compton's major discovery paper is embedded in and a response to historical forces. Yet the historical situation, the forces, and the response are all shot through with empirical experience.

This first part of the study is based on the primary record of published articles by Compton and his contemporaries and the secondary accounts of historians of this period, most notably Roger Stuewer's comprehensive history *The Compton Effect*.

To follow up the themes of constraint, opportunity, and choice in the greater detail, the second part of the study examines the emergence of a secondary paper by Compton entitled "Measurements of β-Rays Associated with Scattered X-Rays" (see appendix), written in the wake of the major discovery paper. This part of the study shows how Compton's smallest behaviors as a formulator of knowledge are shaped by his commitments as a scientist to empirical experience. The March 1925 secondary paper is chosen for examination because more extensive notes, drafts, and revisions of it are extant in Compton's notebooks than of any other of his articles. No draft material is available for the main discovery article.

Although Compton shared credit for the "Measurements of β-Rays" article with a junior author, Alfred W. Simon, Compton appears to be the actual writer and the shaping intelligence of the paper, while Simon assisted in the laboratory. All notes, draft, and revisions appear in Compton's third notebook in Compton's handwriting. Further, Simon, a graduate student when the paper was published, never pursued similar work except in collaboration with Compton (Cattell and Cattell, 897), while the paper fits closely with the topic and issues of Compton's continuing research. Finally, in the draft of the article, Compton unthinkingly refers to himself as the sole author.

For this part of the study I relied on the photocopy of Compton's notebooks at the Center for the History of Physics in New York; the original is deposited in the library of Washington University in St. Louis. The relevant materials from Compton's notebooks consist of about a dozen pages of notes on works by other authors, twenty-two pages of calculations and design sketches for a polyphase transformer, fourteen pages of analysis of photographic data, and seventeen pages of draft and revisions. Material relating to other work Compton was engaged in is interspersed, such as a draft of exam questions for a course Compton was teaching.

The Structured Situation in Which Arthur Holly Compton Worked

Arthur Holly Compton developed his claims within a situation structured at a number of levels, from the most general historical structuring of the scientific enterprise to the most immediate sequence of events occurring in the laboratory. These levels can be seen as nested within each other, each outer one providing a context for each inner one. Each outer level can, however, be seen as necessitating and depending on the inner levels for its historical realization and furtherance. All the outer contexts—of the scientific enterprise, the structuring of disciplines, the development of problem areas and emergence of specific problems, the shaping of an individual's research program, the arguments arising out of the public presentation of that program, and the designing of specific investigation—all point toward the most local context of the events happening in the laboratory, the designated empirical experience. The spot of time of this defined experience is both the greatest constraint and greatest resource for the scientist sitting down to write a specific paper. The scientific enterprise has been structured so that all the outer contexts keep pointing toward this spot of time for their resolution and fulfillment. The outer contexts are built on the representation, discussion, and accumulation of these spots of time, these spots of experience.

The largest frames for the creation of statements are the macroinstitutions of scientific community and communication, some of which were examined in previous chapters. The next context, of physics developing as a separate discipline with its own institutions and practices, has not been examined here. We will rather begin our account here with an examination of the established problem area within which Compton's work developed.

Constraints and Opportunities of the Problem Area

According to Stuewer's account, Compton's work grew out of the problem area of the nature of x-rays. In the twenty-one years between Roentgen's discovery of x-rays and the start of Compton's investigations, two competing theories developed to account for the properties of x-rays. The first, associated with Thomson and Barkla, described the x-ray as a wave-pulse phenomenon operating according to classical electromagnetic radiation theory. The second, developed slightly later and associated primarily with Bragg, held that x-rays and γ-rays were particles, neutral pairs comprised of α- and β-particles bound together electrically. Despite the publication of Einstein's light quantum hypothesis in 1905, quantum theory seemed to be ignored by those working in the x-ray problem area. Some attempts were made to provide nonquantum explanations of the photoelectric effect, which Einstein had claimed to explain.

This history of the problem area had several clear-cut effects on Compton's publications in the area. First, up until his 1922 review of the literature for the National Research Council, Compton employed arguments only from classical electrodynamics. When Compton finally turns to a quantum explanation, it is only because no other will fit the data. The consequences of this conversion to a quantum explanation for the structure of the argument in the main discovery paper will be discussed in the next section.

Second, the dispute between the adherents of the two nonquantum theories of the nature of x-rays centered around three types of empirical data resulting from x-ray scattering experiments that were anomalous in both theories: a forward-backward asymmetry in the secondary β-ray distribution, a forward-backward asymmetry in the secondary x-ray distribution, and a difference in hardness between the primary and secondary x-rays. The issue remained finding an appropriate theory or improvement on theory to fit these data. The argument of Compton's papers followed this pattern of proposing theory and evaluating data fit; moreover, these three kinds of data remained among Compton's primary data sources through the major discovery paper.

Finally, because the dispute over theories had narrowed to the issues concerning x-ray scattering, Compton tended to frame his problems in terms of explaining scattering incidents rather than identifying the nature of x-rays themselves. Although certain assumptions about x-rays are implicit throughout his work and made more explicit when he converts to quantum explanations, the problem is the scattering data, with the assumptions about x-rays only serving as part of a projected account of the scattering incidents.

The historical development of the problem area constrained Compton's work by providing the intellectual tools of classic electrodynamic theory which Compton necessarily began working with and by focusing attention on identified difficulties in the data; these difficulties provided the issues for discussion. At the same time these constraints provided the opportunities for Compton's work. They provided something to talk about and a way to talk about it—a puzzle and a method. Without the developed work in x-rays and classical electrodynamics there would be neither data difficulties to puzzle over, nor a theory against which the data would appear puzzling. There would be no occasion for a paper solving the puzzle.

Viewed as both constraint and opportunity the situation in the problem area is freighted with empirical experiences and imperatives. Classical electrodynamic theory is a generalization from the accumulated reported experience of phenomena considered relevant to the theory. Although anomalies, unreported phenomena, different selections of relevant phenomena, and alternate representations of the phenomena might exist or be possible, the theory was created to be consistent with certain classes of data and found to be continuingly consistent with ranges of new data. It was a useful generalization for the uses found for it. As a fairly robust theory, it enjoyed a substantial range of uses, generating continuing empirical contact.

Roentgen's empirical experience of unusual phenomena which he attributed to x-rays opened up the whole research area within which Compton worked. Although there were competing accounts of what these x-rays were, all the relevant researchers were able to produce these rays and observe curious phenomena in their laboratories. In particular three classes of data were regularly produced in the laboratory. When these results were first produced they provided challenges to the two popular accounts of x-rays. Thus they became interesting, were produced in a number of laboratories in the hope of understanding them better, and were the topic of professional discussion. For Compton these anomalous events became the precise research concern.

Constraining Choices of the Scientist's Research Program

Within a problem area the individual scientist's developing research program and theoretical commitments help determine the specific problems to be addressed and the kinds of answers sought. In the long range these choices amount to a line of inquiry and a process of scientific development; in the short range these choices determine how a scientific commitment is realized in specific hypotheses, lines of theoretical argument, and designed experiments, all of which may be reported on in resulting papers. In both long and short range the constraints are circumstantial as well as intellectual: where a scientist finds himself, surrounded by what ideas, and with what equipment and funding available for what projects.

Arthur Holly Compton's research program on x-ray scattering began—by Compton's own account and confirmed by Stuewer (96)—with data produced by Barkla which were not consistent with Thomson's classical electrodynamic x-ray scattering theory. In particular the data suggested that the absorption coefficient of the target material was dependent on the wavelength of the incident x-rays. Compton, deeply committed to classical electrodynamics, took on the task of reconciling the data with Thomson's theory. He first proposed alternative structures of the electron that might account for the variation in the absorption coefficient with the change of wavelength of incident radiation. Instead of considering the electron as a point charge, he proposed a large electron of a perfectly flexible shell, such that the radius would be of the order of the incident radiation allowing for diffraction as well as scattering (January 1918). When difficulties appeared with the flexible sphere, he proposed a ring electron, giving the electron magnetic properties (July 1919): this too presented difficulties. The form of his proposed solutions was clearly dictated by his perception of the problem.

Through this early period Compton was at Westinghouse Laboratories, without adequate equipment, working with crude experiments and secondary data. When he received a National Research Council fellowship to the Cavendish Laboratories to work with Bragg, he was able to devote himself to an investigation of the secondary radiation from the scattering (Stuewer, 137). From the intensity of the secondary radiation, he was able to distinguish two kinds of radiation, which he identified as scattered radiation (unchanged in wavelength) and fluorescent radiation (changed in wavelength). This change in wavelength of the fluorescent radiation would, he argued, account for the softening of intensity

Making Reference

of the secondary radiation (May 1921). This fluorescence hypothesis became the focus of his attention, even after he left Cavendish in 1920 to take a position at Washington University to be able to pursue his own line of research unconstrained by the concerns of Bragg's laboratory. He did, however, bring back with him a Bragg spectrometer which was to prove crucial in his ensuing work.

What specific consequences for the shape of the major discovery paper, "A Quantum Theory of the Scattering of X-Rays by Light Elements," did this earlier part of his research program have? First, the commitment to classical electrodynamics causes Compton to draw crisply the issue of choosing between classical and quantum theories; his conversion to quantum approach to the problem of scattering becomes the main justificatory task. The paper opens with a review of the problems arising from the classical Thomson theory; the review is detailed and lasts four paragraphs. In the stead of classical theory, he then derives a series of equations on quantum assumptions; he follows with a report of an experiment that provides confirming data. The latter section is in fact called "Experimental Test" and is followed by a short discussion confirming the validity of the quantum hypothesis.

The character of Compton's argument stands out more sharply if we compare it to Debye's paper proposing a similar quantum theory of x-ray scattering.[2] Debye's paper appeared before Compton's, but had been received by *Physikalische Zeitschrift* after Compton's paper had been received by *Physical Review,* so that Compton received priority for the theory. That particular aspect of priority, however, is less consequential now than then, for reasons to be discussed later. What makes the comparison important here is that Debye was not associated with the x-ray problem area, but rather was already deeply involved in the quantum theory and its elaboration. Consequently, the argument of Debye's paper is to present an extension of quantum theory that explains some data anomalous to electrodynamic theory. Rather than presenting the progress and general types of difficulties run into by classical theory, Debye points to specific data anomalies. The derivation of the equations then follows not as a proposed theory to be tested, but as a direct answer to the difficulties. For Debye the quantum theory already stands, and this is only one more demonstration of its power.

Thus the individual scientist's commitments and evolving research program will shape how he will define issues, create an argument, and develop his data, yet within that framework the scientist is committed to

---

2. P. Debye, "Zerstreuung von Roentgenstrahlen und Quantentheorie," *Physikalische Zeitschrift* 24 (1923): 161–66.

Three: Typified Activities in Twentieth-Century Physics

contend with the data derived from empirical experience and uses that data to further the investigation. Compton and Debye consider the same phenomenon from very different theoretical interests and frame different kinds of argument, yet they must contend with similar data that identify the peculiar character of the phenomenon. That they develop theoretically consistent accounts would not of course be necessitated by the phenomenon or the data, but that both kinds of discourse lead to similar conclusions adds persuasive force to both accounts. If the constraints of two robust research traditions meet the constraints of data to produce similar resolutions, the shared account carries the force of that much more scientific experience.

A second constraining effect of Compton's research program on the discovery paper is to be found in the data displayed as central in the paper. His increasing concern with the softening of the secondary radiation directly leads to the prominent role in the discovery paper (and several other related papers) for wavelength shift data on the secondary radiation. Not only do the data of wavelength shift and the data's analysis provide the chief substance of the empirical presentation in these papers, but the theoretical presentations are largely devoted to deriving the equations for calculation of the shift; consequently, the discussions and conclusions are devoted to matching equations and data of wavelength shifts. The comparisons between equations and data are, as well, presented in graph and tabular forms, to provide some of the more striking features of the papers. Key moments in this focusing on the wavelength shift were the move to Bragg's laboratory, returning to St. Louis with the Bragg spectrometer, then switching the use of the spectrometer from selecting the wavelengths of the primary radiation to measuring the distribution of the wavelengths of the secondary radiation.

Finally, we can see in Compton's earlier papers a series of reformulations of the problem with implications for the form of the appropriate solution. As a problem of reconciling data cast in terms of absorption coefficients to classical electrodynamic theory, Compton's early work looked to the structure of the target electron to determine scattering properties. As Compton began to reformulate the problem around secondary radiation, the work turned to the manner in which the secondary radiation was released, leading to hypotheses about Doppler-shifted fluorescence from scattered electrons. The next step was forumulating the problem more tightly in terms of wavelength shifts in the secondary radiation, which is indeed the formulation of the problem in the major discovery paper.

Before the major discovery paper could be written, however, it was necessary for Compton to draw together all the thinking and data on the subject and reformulate the existing theory and perception of the problem. If he was to abandon classical theory and turn to quantum solutions only as a last resort, he needed a comprehensive look at the subject to convince himself that the classical possibilities were exhausted. Two months before the discovery paper was written in December 1922, Compton published a lengthy (fifty-six page) review of the literature on "Secondary Radiations produced by X-rays." Compton undertook this review for the National Research Council as part of a special Committee on X-ray Spectra. Compton's monograph was the third and last part of the report of the committee. It was this institutional situation that gave Compton the opportunity to rethink and reformulate all the material in his problem area.

As Stuewer points out, writing this report helped Compton in four particular ways; each of these four ways affected what appeared on the pages of the discovery paper. First, in reviewing the data of the large electron hypothesis, he began to have serious doubts about the attempt to reconcile electrodynamic theory with scattering data. This was part of the process of cutting himself away from strictly classical explanations. Second, in examining secondary radiation he proposed a recoil electron hypothesis for the first time—that is, in addition to the fluorescent photoelectron, a second free electron results from the interaction through recoiling after scattering radiation. This, of course, is a key element of the theory presented in the discovery paper. Third, Compton presented new data which actually appears to be old data reinterpreted and reexamined to reveal a slight shift of wavelengths of the entire spectra between primary and secondary radiation. Previously he had mistakenly focused his attention on a grosser but less coherent wavelength shift. This subtler shift, noticed here for the first time, was the main phenomenon addressed in the consequent discovery paper. Finally, in a passage that appears to be a last minute addition, Compton offered a quantum explanation of the shift. Yet in the conclusion of the report, which may have been written before this insertion, Compton criticized any quantum explanation and reaffirmed classical electrodynamic theory. We see Compton clearly vacillating between two views; he resolved the vacillation by the clear choice represented by the discovery article (Stuewer, 193–211). Nonetheless, even after the quantum theory article was published, Compton continued to follow a secondary research program exploring x-ray reflection and diffraction, which did seem to follow classical electrodynamic theory. He made one choice for one set of problems and data, and another choice for another set of problems and data. After the moment of confusion he created a bifurcation in his work, with only limited cross-reference between the two parts.

Three: Typified Activities in Twentieth-Century Physics

Constraints of the Laboratory

With the onset of work for any particular research project, theoretical or empirical, another process of constraint begins. Up to this point constraints helped define a problem, the starting point of the inquiry, and some formal features of the likely answer—what the field is asking the scientist to do and what the scientist would like to do. Once one gets down to the actual pen and paper work of theory construction or experimental design, however, one becomes constrained by what mathematics, logic, and prior well-established theory allow one to say, by what available equipment can do, and by what data actually turn up. In this wrestling with recalcitrant mathematics, logic, technology, and nature one finds not what one would like to say, but what one can legitimately say. If earlier constraints helped shape the form of the statement, here constraints shape one's substantive theoretical innovations and the content of one's findings.

Of this stage, unfortunately, little remains on the public record; research activities occur in relative privacy, whether tinkering with equipment in the laboratory or tinkering with equations on the back of a cocktail napkin. But imagination and mechanical creativity are not unfettered. In addition to the constraints on the focus of attention and nature of the endeavor, discussed earlier, one runs up against the limited possibilities of mechanical and intellectual manipulations and the limitations of what is out there in nature as revealed by the marks on the photographic plate or the readings on the meter. It is these resistances—called passive by Fleck because they are not under the active control of human culture—that are brought forward into the public record, in the form of data tables, methodological articles, and theoretical derivations, but the process of getting to these hard places of resistance is largely obscured in scientific texts.

The darkness which hides this stage of the emergence of scientific statements has proved intriguing to psychologists, sociologists, and philosophers of science; their inquiries have led to the observation that the process of scientific inquiry is something other than the knowledge reflected in the public record (Medawar). Latour and Woolgar, in observing the private goings on in a biochemical laboratory, have noted how real substances get reduced to symbols through mechanical and intellectual manipulations and how symbolic formulations are tried and abandoned against the criterion of what will gain the most credibility in the agonistically structured (competitive) field. Part of the process of gaining credibility requires that one's results seem not to be tied to the specifics of one's lab work. Thus the final paper gives only a thin, highly transformed, highly selective account of the biological matter investigated. Knorr and Knorr, observing another biochemical laboratory, similarly note that all that survives in the final paper from the complex wanderings of motives, plannings, errors, speculations, and tinkering with machines is a data chart. The rest of the final article seems to be created on other grounds. Both these studies imply that this process of shedding away obscures both nature and scientific activity out of the social motivations and interests of the researchers.

If one remembers, however, that the private activity of the laboratory occurs within a context created by the public record, the eliminations and reductions that occur within the laboratory and between the laboratory and the final text—the shedding away—may be seen as the mechanism by which the specialized, highly focused data of the laboratory is fit into the broader constraints of the developing science. Brute nature is, of course, not constrained by science, but only limited aspects of nature are consequential at any moment in the discourse about nature called science. Just as a fiction writer may select details according to criteria of vividness, thematic consistency, and verisimilitude, the scientific writer must seek out and select data according to such criteria as consequentiality for the problem at hand, form appropriate to the theories in question, lack of contamination by uncontrolled factors, and anticipation of what the rest of the scientific community is likely to consider as compelling proof. That is, brute nature is symbolized and those symbols refined to meet specific purposes of discourse, a discourse that must address the literature, the audience, and the scientist's own thought as well as observed nature.

This author has little evidence about the private events that led to the writing of Compton's major discovery paper, but Compton's notebooks do provide material relating to a follow-up paper, "Measurements of β-Rays Associated with Scattered X-Rays," to be discussed below, along with an extensively revised draft.

Focused Choices at the Writing Desk

By the time the scientist gets to the actual writing up of theoretical and experimental findings, much of what will appear on the page has been determined by earlier constraints and choices. Thus the writing up of results may seem to be a perfunctory necessity, a painful obligation, but not an essential part of scientific discovery; by extension the entire writing process can seem epiphenomenal, rather than essential, to science. However, the analysis here suggests that the

Three: Typified Activities in Twentieth-Century Physics

text gets shaped over the long haul by the essential elements of science, which in turn can be well understood as parts of the process of scientific formulating. In this larger writing process, specific limited tasks of formulation are left for the overt work of draft writing and revision. In this writing-up stage, the scientist-writer must put the pieces of the argument together so as to make his purposes clear and so as to satisfy the criteria of judgment he anticipates will be imposed by his audience. Final wrestling with the applied theories, the continuity of the argument, and the data may lead to basic reformulations even at this stage. Even if no major changes occur, the author in controlling the words for the final formulation must manage the impression of the prior literature, the experimental design, the laboratory happenings, the data and its relation to the phenomenon investigated, the conclusions, and the conclusions' certainty. The scientist-writer must fine tune the language to reveal the proper levels of precision and uncertainty. Yet the writer must also project a hypothesized world in which his findings are true. That is, even while the literature, research program, problem formulation, experimental design, and data constrain the solution's formulation, all these earlier constraints are presented in the context of a formulation of the world that takes the findings for granted. Thus, for example, a scientist on the basis of a programmatic conviction bolstered by his most recent findings, in reporting those findings may dismiss work based on contrary programmatic convictions as irrelevant and insubstantial. To readers who do not share the author's clarity of vision, however, such a representation of the literature may appear worse than imprecise.[3] Such an example suggests the difficulty of managing a representation that is adequately precise for both author and audience. All this impression management must be done while attending to the stylistic conventions and preferences of the editor and audience. These conventions and preferences allow for convenient, intelligible communication which calls least attention to itself.[4]

Writing-up is not an instantaneous process; preparation of the drafts, revisions, and editorial revisions take some time. In the course of the drafts and revisions the final form of the article comes into shape. Although many of the writing choices happen in the author's head—we know only those sentences he writes down—the changes within the

---

3. I interpret in such a light the examples of apparent distortion in introductory sections of papers, cited by Gilbert and Mulkay in *Opening Pandora's Box*, chapter 3.

4. Such interest in the audience's convenience is the basis for the research reviewed in Ennis, "The Design and Presentation of Informational Material."

drafts and revisions reveal many of the concerns uppermost in the writer's mind at these later stages.

Rewriting in Reception

After the scientist has chosen the words that appear in the published text, the meaning of that text still must be reconstructed by the readers. The text takes on a revised meaning depending on where and how it becomes incorporated into an evolving science. Small has suggested that texts come to serve as specific concept indicators in later articles, and Cozzens has found evidence that with time references to an important article tend to become more compact and fixed in meaning ("Life History"). Messeri has likewise found that citations to seminal articles are replaced by a few key terms that come to represent the findings of those articles. This reduction and transformation of the meaning of an article depends on what happens to science after the article is published, so that the article may be seen to have a rather different set of foci and implications than intended by the original writer.

Stuewer's account (237–73) of the reception of Compton's article "A Quantum Theory of the Scattering of X-Rays by Light Elements" and a limited survey of the citation contexts of later references to that article reveal several striking features of the transformation of Compton's findings. At first the article became an object of controversy, attacked on both theoretical and empirical grounds; at the same time other scientists attempted to improve on Compton's theory. Compton in response ran further experiments and proposed his own improvements. The article gradually became accepted as fact and was cited as the basis for new work. Within a few years the article (along with several surrounding publications) came to have a limited meaning referring to empirical observation of what was coming to be called the Compton Effect. As acceptance and eponymity were granted, the discovery became retold in less specialized, less argumentative ways in order to inform wider publics about the newly accepted fact and to place the new fact in relation to other facts. Compton himself participated in this process by his speech before the American Association for the Advancement of Science in 1923, his 1925 article in *Scientific American*, his 1926 text *X-Rays and Electrons*, and his 1927 Nobel lecture.

Two particular reinterpretations are involved in the current view of the Compton Effect as an empirical verification of quantum theory. First, Compton's work is now seen as part of the research program of quantum theory, even though the article does not cite any of the prior

Three: Typified Activities in Twentieth-Century Physics

work in quantum theory and even though Compton came to his discovery out of problems in classical electrodynamics. And although the major discovery paper offers a quantum solution, the problems of assymetry of radiation and wavelength shift which it addresses are anomalies that arise from an electrodynamic point of view. Second, the current view of Compton's work neglects his theoretical concern in developing an account of x-ray scattering consistent with electrodynamic theory in favor of an empirical result that was originally subordinate to theoretical issues. This interpretive shift began quite early. Compton's article appeared in volume 21 of *Physical Review*. Of the citations that appeared through volume 25 of that journal (a span of two years), excluding self-citations, nine appear in contexts that refer to his theory, and only one is concerned primarily with his empirical results. Of the citations in volumes 26 through 29, however, two are primarily theoretical, three are empirical, and one is mixed.[5] Given the progress of quantum theory during that period and since, and the consequent change of the importance of Compton's work, such a reinterpretation makes sense as part of the historically changing codification of the literature of a scientific field. But such reinterpretations based on current scientific belief in effect rewrite the original article.

One Paper Begets Another: "Theory" Begets "Measurements"

Almost immediately upon publication, Compton's discovery underwent a series of challenges, which Compton answered by carrying out further experiments and publishing the results, disconfirming the challenges and refining the theory. It was in this context of challenge and response, of elaboration and bolstering, that Compton pursued the work that would lead to the "Measurements" paper. The more evidence of the most varied kind he could find, the more likely he would be to gain acceptance of his original discovery claims.

At about the same time as Compton had published "A Quantum Theory of the Scattering of X-Rays by Light Elements" in May 1923, C. T. R. Wilson (and slightly later W. Boethe) identified, in cloud chamber experi-

---

5. I drew the citations from *A Citation Index for Physics: 1920–1929*; incidentally, Compton's "Quantum Theory" article was the most cited article in physics during the decade.

Theory citations: 22, 283; 23, 122; 23, 135; 23, 316; 24, 179; 24, 591; 25, 314; 25, 444; 25, 723; 26, 435; 28, 875.

Experiment citations: 25, 193; 26, 299; 26, 657; 29, 758.

Mixed citations: 26, 691.

ments on X-ray scattering, secondary β-ray tracks substantially shorter than photo-electron tracks. Compton immediately saw that these shorter tracks could represent the recoil electrons he hypothesized in the quantum theory article. He wrote a letter dated August 4, 1923, to that effect to *Nature*, which published the letter in the issue of September 22, 1923. Although at that time Compton continued to be mostly concerned with data revealing wavelength shifts, which data he kept gathering during the following year, he clearly understood how the cloud chamber findings filled out his work. He assimilated the cloud chamber findings into his consequent papers, often in lengthy discussions indicating how they supported his theory.

Wilson's and Bothe's data, however, only offered a rough correspondence to Compton's theory, as Compton noted: "They have shown that the direction of these rays is right, and that their range is of the proper order of magnitude" ("Measurements" 307). The roughness Compton ascribes to the use of insufficiently hard and too heterogeneous x-rays. The "Measurements" article can thus be seen as Compton's attempt to tie down the connection between his theory and the cloud chamber tracks more firmly and precisely by redoing other people's experiments in a way more appropriate to his programmatic purposes. He would then obtain support for his theory from a kind of data not at all available when the theory was first framed; such data, confirming the predictive power of the theory, is rather persuasive.

In this way we can see the "Measurements" paper motivated and shaped in specific ways by Compton's theoretical program, discoveries by other scientists as reported in the literature, the desire for closer measurement of the phenomenon, and Compton's persuasive intentions. Contextual factors provide pressures and offer opportunities to gather fuller, more precise, and more focused data about the observed radiation—confirming and adding detail to the representation of nature embodied in Compton's theory. His social interest in establishing the proposed phenomenon leads Compton to search actively for passive constraints of new and more precise kinds; criticisms in the literature actively push him again to seek passive constraints that make his formulation more likely; finally, new techniques, actively created (although embodying passive constraints in what they can accomplish and in the results they produce), provide opportunities for closer looks at

---

6. C. T. R. Wilson, "Investigations on X-Rays and β-rays by the Cloud Method. Part I.—X-Rays," *Proceedings of the Royal Society*, 104 (1923): 1–24; W. Bothe, "Über eine neu Sekundärstrahlung der Röntgenstrahlen," *Zeitschrift für Physik* 16 (1923): 319–20, and 20 (1924): 237–55.

the purported phenomenon, adding new passive constraints to the formulation.

Specifically, this set of forces and opportunities led Compton to design experiments and write a paper reporting those experiments:

1. adopting Wilson's cloud expansion apparatus;
2. referring to and discussing his own quantum theory of scattering;
3. employing higher energy (shorter wavelength) incident radiation than Wilson and Bothe;
4. designing and employing a method of obtaining more homogeneous incident radiation than Wilson and Bothe (consequently reporting data for higher energy, more homogeneous data);
5. developing theoretical predictions about aspects of the recoil electrons measurable through the Wilson apparatus;
6. and discussing the correspondence between the theoretical predictions and the experimental data.

These effects of the rhetorical situation correspond to the major structural features of the resulting paper. Compton, indeed, alludes to these effects when he describes the paper at the end of the opening paragraph:

> The present paper describes stereoscopic photographs of these new rays which we have recently made by Wilson's cloud expansion method. In taking the pictures, sufficiently hard x-rays were used to make possible a more quantitative study of the properties of these rays. (307)

Within the stylized terms of the field, the paper describes constraints imposed by the results of more precise measurements. By showing that Compton's theory is in conformity with ever-increasing passive constraints, the article seeks to establish factlike status for Compton's claims.

Another aspect of the rhetorical context consisted of one particular challenge to Compton's quantum theory of scattering. Bohr, Kramers, and Slater claimed that at the particle level the laws of conservation applied only statistically.[7] Compton's theory required event by event application of the conservation laws; up to that point, however, Compton had established the recoil phenomenon only on an aggregate basis through measurement of radiation wavelengths. Wilson's cloud photographs provided a way of capturing and measuring single incidents and were,

---

[7] N. Bohr, H. A. Kramers, and J. C. Slater, "The Quantum Theory of Radiation," *Philosophical Magazine*, 47 (1924): 785–802.

Making Reference

therefore, the ideal means of refuting Bohr, Kramers, and Slater. The full and explicit refutation of the statistical argument was to be made by Compton and Simon in a subsequent article—"Directed Quanta of Scattered X-rays," which appeared in *Physical Review* six months after the "Measurements" article—but the desire to refute the challenge remains an implicit shaping force on the earlier article. The effect can be seen in the emphasis given in both the abstract and the full paper on conclusions and evidence that the scattering occurs on an event by event basis, with each event maintaining conservation of momentum and energy. This emphasis is in fact increased in revision. Again, attack on the formulation provides pressure to seek and reveal passive constraints, consonant with the original formulation.

Laboratory Decisions, Events, and Results

The effects of the rhetorical situation are first realized in Compton's laboratory decisions before their full implications in the text are realized. The laboratory decisions, such as the use of the Wilson cloud apparatus, designing a more precise control over the incident radiation, the design of the scattering experiment, the choice of which plates to use as data, and the particular measurements taken from those plates, all have an effect on the final article, both in terms of the procedures described and the data reported. The first three decisions are design decisions based on the characteristics of the phenomenon investigated and the properties of the equipment, as both have been revealed through previous investigations. The Wilson apparatus, for example, is used only because it has earlier revealed tracks that Compton can identify with recoil electrons. Compton goes to great lengths to make design decisions that will permit observation with the desired precision; twenty-two pages of his notebooks are devoted to designs for a polyphase transformer that will provide him with stable enough voltage to provide homogeneous incident radiation of calculable energies. The experimenter can choose from among available technologies, but those technologies suffer many passive constraints. The experimenter cannot use impossible machines, nor can he make machines do what they cannot do (Notebook 3, 20–41).[8] The latter two decisions—the choice of

8. Latour and Woolgar, citing Bachelard, discuss laboratory equipment as a reification of theory. This idea is intriguing, but it must be kept in mind that no matter how fully suggested by theory, the equipment must accord with the functioning of nature to work; in this way the equipment is as much a test of theory as reification of theory.

Three: Typified Activities in Twentieth-Century Physics

plates and the choice of measurements to take from the plates—depend on what happens in the laboratory, on what turns up on the plates. Once the experimenter sets up the conditions of the experiment, what turns up is beyond his control. Only afterward can the experimenter reassert control through selection and manipulation.

In the final article Compton reports that he is using data from "the best 14 of a series of 30 plates taken," but the notebooks show him making calculations for 14 numbered plates running from number 15 to number 47 (Notebook 3, 49–52).[9] Assuming that plates 1 through 14 served as practice runs, that still leaves three plates unaccounted for, presumably so bad that they do not even count as plates. Although Compton gives no overt definition of what makes the selected plates "best," the sixteen deleted plates worst, and the three not plates at all, his notebooks offer two clues about his criteria of selection. First, he tends to select the higher number plates; in fact he records measurements for plates 38, 39, 40, 43, 45, 46, and 47. This suggests that Compton and Simon were still gaining the technical skill to produce plates that clearly revealed the tracks they were interested in. Second, at the bottom of the column of measurements for plate 38—which in fact was deleted from the article partway through the writing of the draft—the notation "uncertain because too crowded" appears. This notation reinforces the impression that selection was based on how clearly the plates represented and allowed distinctive counts of the data associated with the scattering phenomenon. That is, Compton and Simon were simply looking for clear and distinct tracks.

The tracks on the photographic plates are Compton and Simon's closest glimpse at the scattering phenomenon, and reproductions of some photographic plates are included in the final document to give the readers qualitative visual evidence. How those tracks are interpreted quantitatively, however, depends on a number of manipulations of measurement and calculation. The data tables in Compton's notebook, even in parts of the rough draft of the article, are filled with corrections. These corrections seem all to derive from two incorrect assumptions about the equipment which led to mistaken values for the potential of the x-ray tube and consequently for the energy of the incident radiation. The two causes for error—a warping in a frame and the effect of a condenser—are both carefully noted in the notebook and in the final article. Although on first glance all the corrections appear to be manipulation of the numer-

---

9. On the bottom right hand corner of page 51 there is a boxed-off set of data that is unlabelled that may represent a fifteenth plate; if so this would compensate for the apparent discrepancy caused by the later deletion of plate 38.

ical data after the fact, they really only serve to adjust the secondary numerical data to the actual event as occurring in the equipment and recorded on the plates. In addition, although Compton for the most part adheres to Knorr's observation that scientists tend not to report their wrong turnings and errors in the final report (Compton, for example, does not discuss what went wrong in the first fourteen parts nor in the later deleted ones), Compton is very careful to cover this error in both notes and text. His great care, and indeed the great detail with which he reveals this error in the article, suggests that this error is of a different order in that it comes after the laboratory event but seems to change the reality of what happened. To retain the integrity of the data, to make clear that he is constrained by the data and not fiddling with it, he must expose the error of calculation and measurement which leaves the reality of machinery and photographic plates untouched. Thus the representation of a certain class of error is necessary in the article to keep the relation between laboratory happenings and the report of those happenings as clean as possible. The purpose of exposing the error is not, as Medawar would like, to reveal the psychology of discovery.

The Writing-Up

The previous sections have examined some of the constraints and decisions that determined what the measurements article would look like, but still we do not have a text. Compton must sit down with blank paper in his notebook and create a string of words, equations, numbers, and graphics to fulfill the possibilities of the constraints. As part of that fulfillment he must represent nature at various levels of mediation: nature as perceived through the literature, as formulated in a problem and hypothesized answer, as inherent in the experimental design and the actual experimental happenings, as represented by the experimental data and the secondary calculations, as interpreted through discussions and conclusions. Thus the article, even while describing the forces that shaped it, is reconstructing views of nature at a number of levels of intellectual and physical mediation. By the convention and logic of the scientific report, however, all these representations must be weighed against the least mediated representation, the data—the photographs and numbers one carries away from the laboratory.

At this point of writing-up, the task of the scientist then becomes using language to create these various representations at a level of precision and completeness that adds no further confusion or lack of clarity at

Three: Typified Activities in Twentieth-Century Physics

any of the levels and that allows an intelligible comparison between the data and the other more mediated representations. When we look at Compton's draft and revisions of the article "Measurements of β-rays Associated with Scattered X-rays" (Notebook 3, 59–75), we see indications of just this concern for creating an adequately full and precise representation of nature at several levels of mediation. The larger part of the many changes and corrections he makes as he writes and revises manage the representation of the x-ray-electron interaction, the theory of that interaction, the experimental design and happenings, and the kinds of interpretations and conclusions that can be drawn on the basis of the data.

The following discussion of the drafts and revisions will first present the three major tactics of revision that Compton uses—postponing, extending, and fine tuning—and then will examine epistemological, phenomenological, and social issues raised by the draft and revisions. Line numbers refer to the final version, reproduced in the appendix to this chapter.

POSTPONING

Postponing is a structural decision made in the course of writing the draft. Four times Compton starts to raise major subjects, then decides he must first reveal some preliminary information. At the end of the opening paragraph in the draft, after only mentioning the photographs, he is about to present a set of reproductions with the phrase, "A typical series of these photographs is shown in figures . . ." Before completing the sentence, however, he strikes it out in order to insert a paragraph spelling out the cloud chamber, x-ray, and photographic equipment. Then in the third paragraph (line 28) he returns to presenting the reproductions of the plates. In the second case, after qualitatively discussing the photographs, Compton begins to raise a major theoretical issue with a new paragraph beginning, "One of the most important questions is whether . . ." He backs away from his direct assault, however, by striking the incomplete sentence and beginning a different paragraph introducing quantitative theory to be matched against empirical data (39). The quantitative material then continues as the main body of the paper. Although it is unclear what important question Compton has in mind, the discussion of all the major questions follows the quantitative presentation. The third case involves the presentation of the first data table. Some time after copying the first two columns of data Compton realized the errors in the potential and energy figures discussed earlier. He apparently then went back to check his equipment and recalculate his figures. He then corrected the figures in the first two columns and copied in the correct figures for column seven, which is calculated from the first two columns. Then in the draft immediately following the table he added a paragraph explaining the error (47–59). In the final paper, however, the table is postponed until after the explanation of the error. In the last case, Compton splits his first draft of the second table, which included data on both maximum range of R-tracks and the distribution of the ranges of the full set of tracks. The latter part of the original table appears later in the article in a slightly different array as table 3. The effect is to allow complete discussion of the issue of maximum range before raising the issue of relative distributions.

In all four cases the postponement is to allow the presentation of additional detailed information prior to the postponed material. In the first and third cases the additional material explains the equipment that produced the postponed data; in the second and fourth cases the inserted material is data logically prior to the postponed material.

EXTENDING

Extensions, giving more information about some item already under discussion, serve to clarify or make precise the item being discussed. For example, "primary beam" is changed to "primary x-ray beam" (5); "photographs" becomes "stereoscopic photographs"(11); "the x-ray tube, enclosed in a lead box" becomes "the Coolidge x-ray tube, enclosed in a heavy lead box" (19–20); and "$\tau + \sigma$" becomes "$\mu = \tau + \sigma$"(80). In a more extensive example, "To calculate the relative number to be expected, we have arranged this expression over the range of wave-lengths used in our experiments," grows in several steps into "To calculate the relative number of tracks for different relative wave-lengths to be expected, we have arranged this expression by a rough graphical method over the range of wave-lengths used in our experiments" (138–41).

In one case the addition serves to justify a statement. The phrase "in view of the fact that the photographs were stereoscopic" adds a reason to the original phrase which now follows, "it was possible to estimate . . ." (161).

In all the above cases the addition gives detail to the originally mentioned object or event, but in at least three cases the additions redefine the object of concern more precisely. "Track" becomes "length of a given track" (135); "40 tracks" becomes "the directions of 40 tracks" (159); and "short tracks . . . and long tracks" becomes "short tracks (type R) . . .

Three: Typified Activities in Twentieth-Century Physics

long tracks (type P)" (41). The last example involves a change in epistemic level (to be discussed below).

FINE TUNING

Word substitutions fine tune the language through more specific, correct, or appropriate phrasing. Compton achieves greater specificity by such changes as "an" to "the" (110), "the" to "its" (103), and "those" to "the quantity S" (125). More substantive specifications are made in such changes as "acquires" becoming "moves forward with" (109).

In some cases Compton is correcting an outright error, as when he miscopies an equation from a previous article (112), or he incorrectly calls an "expression" an "equation" (147). Elsewhere he must correct an inverted ratio (85), report that there was more than one "condenser" by making the word plural (52), and relabel a "scattering quantum" as a "scattered quantum" (151). More frequently the corrections are more subtle, as when measured values are described as "summarized" rather than "Shown on the following table" (117) or when "C. T. R. Wilson's datum" is changed to "C. T. R. Wilson's result" (119). A repeated subtle error needing frequent correction is referring directly to an object instead of the appropriate quality. Compton in the draft consistently refers to R and P and R/P when discussing the number of electrons but in the final version the notation is consistently changed to N_r, N_p, and N_r/N_p (42, table 1, 75, 83, 88, 96). Related are the wavering from "apparatus" to "chamber" back to "apparatus" (15), the change from "photoelectric absorption coefficient" to "true absorption coefficient" (43), and the revision of "amplitude" to "magnitude" (185).

The last category of fine tuning revisions corrects tactical errors of exposition and thereby modifies slightly the impression of what is being discussed. Compton first begins to describe the maximum frequency "required to" and then switches to "excited by the voltage" (122); a bit later Compton cites a finding "for the number" but then changes that to a finding "that the probability" (134); and a few lines later Compton starts a sentence, "This expression assumes that the electrons all . . ." then recasts the thought changing the subject of the assumption, "This expression assumes that the exciting primary beam . . ." (137). A more clearly consequential example occurs when Compton begins to discuss "the origin of the short" tracks but then changes the focus to "the origin of the two classes of β-rays" (40). Here he changes the topic from one phenomenon to two phenomena in order to prepare for an equation for the ratio of the two later in the sentence. The original singular focus, although not a factual or technical error, was a tactical error in not

Making Reference

providing for the continuity of the exposition; the writer must keep in mind what he will discuss in what order, and he must focus the discussion accordingly.

All three types of revision—postponing, extending, and fine tuning—indicate that the writer is moving through the imprecision and incompleteness of formulations to come to a more focused, accurate representation of what he did, saw, measured, and thought. The language of the original draft is in parts skimpy, fuzzy, misleading, and even wrong, but by struggling with the language the scientist writer can achieve a bit better fit between symbolization and experienced world.

CRITERIA OF ADEQUACY

The symbolic representation of nature is inevitably an approximation in an alien mode; absolute precision and completeness of formulation would be an endless task. Criteria are necessary for a writer to decide whether a linguistic representation is adequate. Compton's draft and revisions offer clues as to his criteria in the instances where he deletes detail or foregoes specificity. Compton seems to follow two criteria: what one can say and what one needs to say—that is, assessments of how finely one knows what one is discussing and of what level of distinction is necessary to carry the particular argument forward.

The rounding off Compton does in table 2 shows how these criteria are applied. In the original data tables in the notebook, the observed maximum ranges are all measured to the first decimal, but in the transfer of the table to the draft and the consequent revision three observed ranges are rounded off to the nearest integer, in accordance with a prior admission that the observed track lengths "could be estimated probably within 10 or 20 percent" (115–16). That is, the decimals give an appearance of greater accuracy than was probable. Two calculated values, as well, are rounded off to the nearest integer. On these calculated values no error range restrictions apply, but since the degree of statistical correspondence being demonstrated is quite broad (as large as ±3mm or 33 percent of the measured value), the decimals are unnecessary for the demonstration. Compton gives no greater statistical precision than he legitimately can or needs to.

Unneeded specificity is deleted in a number of cases, trivial and substantive. In trivial cases the specification has already been achieved elsewhere in the text as in the deletion of "x-ray" in "primary x-ray beam" (18). In more substantive examples the deleted material raises extraneous theory or inappropriately narrows the discussion. The expression V_c/h is eliminated after the phrase "maximum frequency" because the expression is not used in any of the ensuing calculations (121). The phrases "but radiates uniformly in all directions" (110) and "depending on the direction" (116) are similarly deleted for raising unnecessary qualifications. Another deletion, "mean of the experimentally" from the larger phrase comparing "calculated values with the mean of the experimentally observed relative ranges" (143), emphasizes that the data fit is independent of the voltage and therefore is valid for each of the cases individually rather than only in the mean. Thus the force of an entire set of data is strengthened by the removal of an unnecessarily narrowing qualifier.

The most interesting example of deletion occurs in the description of the photographic equipment (25-27). Compton twice tries to include phrases noting that the full aperture of the lens was employed, but he twice deletes this as unnecessary. Then he twice tries to give positive judgments about the quality of the lenses and plates—"which gave excellent defin . . ." and "very satisfactory." He deleted the first completely and removed the "very" from the second so that the text is left with only the comment that the plates "were found satisfactory." This judgment is all that is needed for the exposition of the experiment. Without a scale of excellence, the more effusive judgments, moreover, do not appear legitimately knowable or supportable to Compton; only the word *satisfactory* carries a criterion of adequacy to the task at hand. Compton's obvious technological pride in the laboratory accomplishment of capturing the scattering phenomenon on photograhic plates seems to motivate all four deleted phrases, but he recognizes that such feelings are extraneous to the argument.

Control of Theory, Persona, and Audience

In addition to controlling the more obvious representations of nature, Compton is careful to control the definition of the epistemic level of the discussion, the projection of his persona, and the relationship to the audience. These factors are important to maintain under control, because if improperly treated they could not only obscure the description of nature being proposed, but undermine the purpose of the discourse. By carefully identifying the epistemic level of discussion, Compton is able to identify exactly what he is representing and at what level of mediation. By controlling persona he is able to assert his individual ownership interests, identify where his judgment enters, and limit his intellectual risks, while still keeping attention on what the

Making Reference

data and theory suggest. By controlling the relationship to the audience, he serves the reader's convenience, helps the reader follow the argument, and submits himself to the audience's criteria of judgment, again while keeping focus of the article on the formulation and data; his most important task with respect to the audience is to maintain credibility, which is done by remaining responsible to and for the data.

EPISTEMIC LEVEL

As part of the process of adjusting language to necessary and possible levels of precision and completeness, Compton carefully assigns each statement to the appropriate epistemic level. That is, items can be represented at different levels of theoretical and empirical mediation. For example, near the beginning of the draft Compton shows uncertainty whether to discuss *rays* or *tracks*. *Rays* directly represents the purported object in nature, but *tracks* represents a manifestation of those rays as they pass through a cloud chamber to create vapor trails that are recorded on photographic plates. After a few equivocations and changes, Compton decides to discuss *rays* in the introduction and switch to *tracks* only after the photographic data are introduced. Thereafter the track terminology dominates the rest of the article. Thus Compton indicates that although rays are the object of interest, recorded tracks are all he has to observe and work with.

Even in the discussion of the purported object of nature there is recognition that the discussion is really about objects constructed in the literature. The opening sentence of the published article reads "In recently published papers, C. T. R. Wilson and W. Bothe have shown the existence of a new type of β-ray excited by hard x-rays." The word *new* is added in the draft, so its use is clearly a conscious choice. The word *new*, however, is only appropriate as meaning new in the literature, not new in nature.

Once the linguistic representation of an object is recognized as being a construction of the literature, then it is only appropriate that alternative terms should be used depending on the theoretical context invoked. Thus Compton changes "ray" to "quanta" (89) in accordance with the invocation of quantum theory a few lines earlier. Similarly, Compton begins to write "an [electron]" then corrects this to "a β-particle" (120) in accordance with an earlier switch in discussion from colliding objects to an analysis of ranges of particles. In both cases the changes are not compelled by technical accuracy, but they do help to maintain clear focus on the appropriate theoretical contexts.

Three: Typified Activities in Twentieth-Century Physics

AUTHORIAL PERSONA

Despite the familiar conjecture that scientists remove themselves from their writing so as to make their work appear less particular and so as to evade epistemological responsibility, Compton maintains an authorial presence in the article. The revisions in some ways enhance this presence and in other ways diminish it. The pattern is that authorial presence is decreased for the prior work, which is merged into the literature of the field, but authorial presence is increased for the current work, for which Compton and his co-worker Simon take responsibility as the thinkers, doers, and owners.

The merging of the individual into the collective of the literature for the scientist's prior work appears in a number of revisions involving self-citations. In the first paragraph of the draft, for example, Compton refers to his previous work "the quantum theory of X-ray scattering proposed by the [author]." Then Compton remembers that Simon is nominally coauthoring the article; he strikes out "the" and substitutes "one of us," to which he appends a footnote to his monograph for the National Research Council. But in the final version the entire phrase "proposed by one of us" is deleted (8–9), suggesting no credit in the text, and a citation to Debye is added to the footnote, sharing credit in the literature and emphasizing that the self-citation is part of a wider literature that is communal. Similar demotions of textual self-reference to footnotes occur at lines 101–3 and 128. In another case the self-reference is removed from the head of the sentence and given a less definitive verb; "Compton and Hubbard give for the . . ." becomes, "If the maximum range of the recoil electrons is S_m, Compton and Hubbard find . . ." (133–34). The most extreme case occurs in the last sentence, when Compton is stressing how well the current work fits with the findings of the literature. The phrase "strong confirmation of the assumptions used by one of us to explain . . ." is shortened by the deletion of the self-reference (187-88); moreover, the self-citing footnote is also eliminated, but a final phrase—the closing phrase of the article—is added: "on the basis of quantum theory" (180). Compton's earlier work is subsumed into a theory which is a fact of the literature transcending individual ownership.

In the previous example, however, even while self-citation is vanishing into the literature, strong reference remains to the authors as conceivers, doers, and owners of the current work. In all versions the last sentence opens with "Our results . . ."(187). Other first person usages remain through all versions to indicate the doing of the work (e.g., "photographs . . . which we have recently made" [11–12], "apparatus used in our work" [15], and "we used a mercury spark" [22]), responsibility for reporting the work (e.g., "In table 1 we have recorded the results" [47]), intellectual operations (e.g., "we have taken from his data" [78] and "the value of which we used" [81]), ownership of the data (e.g., "in our photographs [157]), the evaluation of the evidence (e.g., "In view, however, of the meager data as yet available on this point, we do not wish to emphasize this correspondence too strongly" [97–99]).

Three revisions, in addition, make the authors' role more explicit. The first two bring out the individual responsibility for the evidence. "Observed in the photographs" becomes "shown in our photographs" (115); "the experimental values" becomes "the observed lengths of the R tracks" (124). The third brings out the evaluative role; "can leave no reasonable doubt" becomes transformed to the more direct "we believe establishes" (83).

AUTHORIAL JUDGMENTS

Even where an author does not use first person to call attention to his evaluative role, he makes many evaluative judgments throughout the article through estimates of the reliability of various claims. Compton sharpens this evaluative role through revisions.

One set of judgments sharpened in revision assigns the way in which a relevant theory specifies a particular phenomenon. In the second sentence of the draft, radiation which has "been ascribed to photoelectrons" gets revised to radiation which has "been identified with photoelectrons," indicating a more specific association. A few lines later Compton flip-flops as to whether a particular interaction is "according to the predictions," "as predicted by," or "in accordance with the predictions of the quantum theory" (8); Compton winds up with the last, and weakest, assumption. As we shall see below, even the title of the article, characterizing the strength of the claim of the whole article, undergoes a similar weakening.

In the above examples the truth value of the claims was not questioned, but only the applicability to specific cases. But the larger set of revisions changes the certainty or character of a claim. "Fact" is weakened to "observation" (96); "suppose" is strengthened to "explained" (92); and a definite "are" wavers to "may be" then regroups to "are often" (68). "A satisfactory agreement" edges up to "a rather satisfactory agreement" (143–44); a "theory" is demoted to an "hypothesis" (154); and the direct identification of "are" weakens to the mediated explanation of "have tracks long enough to determine . . ." (157–58). Finally, in the last paragraph an inserted "about" (183) admits that the conclusions rest on approximate evidence.

Three: Typified Activities in Twentieth-Century Physics

The most direct judgments are made in the concluding section, and here we see the most adjustment of the strength of claims. In the third from the last paragraph, Compton begins to draw strong conclusions from the angles of ejection: "There can be no question but that the electrons ejected. . . ." But he then reconsiders and replaces this strong statement with a sentence about the calculation (173–74). In the next sentence he tries again: "There is undoubtedly . . ." But he also crosses this out and starts anew with a qualification: "In spite of some discrepancy at the largest angles, the R electrons ejected at small angles undoubtedly have greater energy than those . . ." In the final version, however, even this certainty is excessive, and a weaker judgment is passed to the reader who inspects the data charts: "It will be seen that the observed ranges . . . are . . . in substantial agreement with the theory" (174–77).

Again, in the next to the last paragraph, "thus constitutes a strong support of the . . ." becomes the weaker "is thus of special significance" (182). A judgment is again passed to the audience.

Despite these two weakenings the last sentence of the article is strengthened as much as it needs to be to assert the significance of the work. "Our results are thus in . . ." becomes "our results therefore afford a strong confirmation of . . ." (187). Compton thus urges no more than he has to, but does not evade responsibility for judgments. Elsewhere he calls attention to his judgments through italics in intermediate sets of conclusions (82–86 and 128–31).

AUDIENCE CONCERNS

The revisions show almost no concern with trying to urge the audience. The only persuasion seems to be that built into the article by the early constraints and early choices that shape the article. If one wishes to study persuasive intent one should look to those early decisions that position the work against previous work, that frame the problem to be addressed, and that determine the kind of evidence to be generated by the experiment; such modes of persuasion are in support of a theoretical position rather than in support of a particular set of results. The only overt attempt to urge the audience in the revisions is the addition of the word "heavy" in front of "lead box" (20) in the apparatus description to dispel criticism of contamination through inadequate shielding. All other revisions in anticipation of audience reaction have to do with the conventions and felicity of language: spelling and word form corrections, removing redundancies and excess commas, and rearranging word order and equations for easier reading. Many of these corrections occur between the completion of the revised draft and the publication of the final version. At that time certain small features are also made consistent with the journal style. *Centimeter* is spelled out, but *equation* is abbreviated; the degree symbol is substituted for the word, and the angstrom symbol is simplified by removal of the superior cycle.

Thus, although the audience is accommodated, it is not pushed. The reasons why the audience might want to believe the article are imbedded in the article's structure. A representation of the literature establishing and positioning a problem, an accurate understanding of existing knowledge, the drawing of a question sharply, the appropriateness of the research design, the fit of the results—these are what convince, but these are determined before the writing-up by the early constraints and decisions. The only thing the scientist as writer can control at the writing-up stage is the representation of these earlier constraints and choices. In the representation the scientist has some leeway, but the representations to be credible must still strike the audience as adequate accounts of actual situations. That audience has access to the same literature, has their own formulations of problems, knows what equipment is available and what the equipment can do, can inspect the author's equipment, and can replicate the author's experiment or run other experiments revealing the same phenomenon. In this light we can understand both Compton's throwing certain judgments to the reader under the assumption that the data are clear enough to speak for themselves within the theoretical context established by the article, and Compton's efforts in his revisions to make his descriptions as accurate and precise as needed for the argument. His credibility and persuasiveness depend finally on how close a fit his readers find between what he says and what is.

In order to maintain credibility Compton takes great care not to misrepresent his data. Not only is the first person maintained in contexts indicating his responsibility, the author takes explicit responsibility for miscalculations and errors, both through the section added prior to table 1 describing the sources of error and through another estimate of error (115–16). This latter discussion of error is difficult for Compton to formulate; he must make several revisions before he can make a reasonable and not misleading formulation of the probable errors. Finally, since the experimental error affecting the data was not discovered until Compton was part way through the draft, a number of corrections had to be made of figures in the text and in the first table.

Text as Object

Through all the constraints and choices we see the gradual emergence of a text—a literary object, separate from, although the

Three: Typified Activities in Twentieth-Century Physics

consequence of, all that went before. Particularly as the text takes shape in drafting and revision, we can see it take on the quality of an object, open to all the limitations and manipulations of language. But still the text is a linguistic object that takes on the overriding task of the representation of nature.

The act of revision itself treats language as an object. Certain revisions in particular call attention to the text as linguistic construction: the sharpening of the recognition of the obscuring effect of reproduction on photographs (33); the retrospective addition of a phrase because certain terms are needed in an equation on the next line (41); deletions in recognition of later repetitions (90 and 116).

Large organizational shifts call attention both to the manipulable quality of a text and to the gradual construction or emergence of the textual object. The splitting of table 2 indicates that Compton is developing an organizational sense of the article that he did not have as he started the draft. Similarly, he did not begin with the subtitles that mark the major divisions of the revised article in mind. The first subtitle in the draft, *"Number of Tracks,"* is clearly an afterthought, squeezed in between lines. But when he reaches the second set of data, Compton realizes that the organization does have major divisions, so he rather emphatically begins the next section with the title *"Ranges of the R Tracks"* on a separate line and centered. By the time he reaches the third of the ultimate divisions, he seems to have gotten used to the organizational structure, and he presents the title *"Angles of Ejection of R Tracks"* in a more subdued position, on the same line as the new paragraph. This is the position the subtitles take in the printed version.

If the subtitling indicates Compton's increasing awareness of the role of blocks of text, his titling of the whole article indicates his judgment of what the whole text does. The original title in the draft is "Measurements of β-rays Excited by Hard X-rays," but before publication the title was softened to "Measurements of β-rays Associated with Scattered X-rays." The changed title recognizes that the text is not so much concerned with the mechanisms of excitation as with the association of the rays through measurement and photographs of individual incidents. The text is limited to just an aspect of the phenomenon and just an aspect of Compton's thoughts and convictions about the phenomenon. A text is a limited object.

THE ABSTRACT

The article's abstract serves as one further step in turning the article into an object, for the abstract considers the article as a whole and then makes a representation of it. In this regard the point at which Compton decides to write the abstract is a good indicator of when he gains a grasp of the whole text. The draft of the abstract appears about two-thirds of the way through the draft of the main text, at a spot corresponding to line 142 of the published version. The earlier part of the abstract draft, in addition, contains the kinds of numerical errors that Compton was not aware of until he reached table 1 (59). These facts indicate that Compton probably began the abstract when he was part way into the article; he apparently turned to a blank page where he thought the main draft would end. He did not have a grasp of the whole when he began the article and had to wait until he saw what he had written before he wrote the abstract; nonetheless, he felt he needed to write the abstract before completing the article, in order to articulate his sense of the whole and to keep the later parts logically and structurally consistent.

Even in the abstract itself he seems to need to recapitulate the entire argument before summarizing the conclusion. He reduces the summary of the data to a one-sentence statement recounting the main topics: "Measurements were made of the maximum range, the relative number of different ranges, the relative number ejected at different angles, and the relative ranges of the R tracks ejected at different angles." This sentence does not find its way into the published abstract, but rather seems more for Compton's own benefit.

Furthermore, the draft of the abstract is not complete on the notebook pages allotted it, suggesting that Compton returned to the main article before finishing the abstract and did not leave enough blank space for the completion of the abstract. The abstract draft breaks off in midsentence at the bottom of a page; the next page continues with the main text in midsentence. If the abstract did get written in stages coordinated with the writing of the main text, that correlation would further emphasize the interaction between the gradual creation of the text and the growing perception and command of the text as an object.

The specific content of the abstract and its revisions further reveal Compton's perception of what kind of object the text is. The substantial discussions in the main text of the background literature and the experimental apparatus become only sketchy mentions via secondary phrases in the first few sentences of the abstract. The sentences are more concerned with the data and findings; the grammatical subjects are reserved for "photographs," "kinds of tracks," and "ratio." Moreover, the problem addressed in the paper, "a more quantitative study of the properties of these rays" (14), does not receive explicit mention in the abstract.

The first eight of the nine sentences of the abstract are devoted to reporting the findings in some statistical detail. The organization of sentences 3 through 8 follows exactly the structure of the body of the paper reporting the data and findings, with two sentences devoted to each of the topics announced in the subtitles of the paper. The conclusions are reported in the last sentence of the abstract; however, that sentence is very long, about eighty words, and manages to incorporate almost all the substance of the final two paragraphs of the full paper. The one-sentence summary in fact incorporates verbatim many of the key phrases of the full version.

The abstract, therefore, focuses on the outcome of the experiment rather than on the background, formulation of the problem, or the experimental design. Nor does the abstract try to recapture a coherent argument, which would require more emphasis on theory and context. The emphasis is entirely on what can be formulated about the out-there physical phenomenon as a result of the experiment.

The revisions of the abstract draft emphasize this focus. Specifying phrases are added about the observed phenomenon, and excess theory and reference to calculations are eliminated. Finally, the original terse summary of conclusions is greatly expanded to incorporate almost all the substance of the full conclusions, as previously noted.

Conclusions

This examination of the emergence of two of Compton's texts reveals that many forces, constraints, and choices shape the final textual object. A. H. Compton, as all authors do, chooses the words that go on the page and thereby creates a statement—a text, a linguistic object—that did not exist before. But Compton's choices are severely constrained by contextual forces, directed by procedures of scientific argumentation and motivated by his personal commitment to record his claims and data as accurately as he is able. Some of the contextual constraints are active (in Fleck's terminology) in that they reflect the structure of the scientific community, the thought style and expressive habits of the period, the social position and interests of the investigator within the scientific community, the research program and theory commitments of the scientist, and the nature of the challenges to prior formulations of theory.

Within this context Compton has some freedom in choosing what claims to advance, in formulating or reformulating those claims, and in designing experiments or other means of advancing those claims. It is at this point that Compton seems to have the most leeway to frame his

Making Reference

work strategically, positioning it against other claims and challenges. It is at this stage of basic positioning, I believe, that we should look for the locus of persuasive strategy rather than at the actual writing-up stage with its narrower manipulation of language. At this stage Compton decided what the real issues in the problem area were and how he could address them in the way most persuasive to his colleagues.

These strategic choices, nonetheless, were subject to constraints, but the constraints were passive. Compton could not violate the bulk of previously gathered data (although he could actively reinterpret or offer alternate explanations for the data.) He could not make equipment do what it could not do, and he could not control what data ultimately got recorded on the photographic plates. Moreover, given the canons of scientific argumentation which Compton observed, the center of the persuasive strategy was the active search for passive constraints. Compton bolstered his original discovery claim by developing a new source of data; he answered challenges by finding specific refuting data; and he advanced his own career by revealing more about the phenomenon and developing techniques for looking more intimately into nature.

Once the experiment has run its course, Compton could only choose to publish or not publish the results. Having chosen publication Compton is committed to presenting his theory and results as clearly, accurately, and precisely as the material and language allow. This precision, accuracy, and clarity in part serve the persuasive intention by identifying the tightness of fit among his claim, experimental procedures, and observed nature; in part they protect him from criticism of fuzziness or fraudulence (note particularly his careful revelations about the necessary recalculations to preserve the integrity of the data).

But the revisions are so careful on even such apparently inconsequential matters as his estimate of the quality of the photographic technique or the choice of "the" over "an" that they reveal a deeply internalized commitment to the best possible representation of the material within his theoretical, experimental, and linguistic scope.

Since there is no guarantee of an essential link between the objects of nature and the words and equations scientists formulate to describe those objects and their behavior, the nonfiction created by Compton, or any other scientist, cannot be taken as absolute, a transparent and congruent presentation of nature as it is. Compton, however, has worked to create orderly, significant experimental events that will produce results speaking to the issues before him and his colleagues. These issues are social, symbolic creations; scientific questions would not exist without scientists to find motives and ways to vex each other and nature with peculiarly human concerns of understanding and control. The repre-

Three: Typified Activities in Twentieth-Century Physics

sentations of results so as to speak to those issues are equally human constructions, which we again see Compton working at. Compton's text-constructing work creates strong bonds between the rhetorical tasks before him in the scientific forum and the empirical tasks he sets for himself in the laboratory. In revision work, Compton keeps those bonds as strong and untangled as possible by being as precisely explicit and detailed as the argument warrants about what those empirical experiences were and what abstractions he draws from them.

Compton creates a crispness of argument not only by detailed revisions of the representation of the experiment and the results, but also by his careful control of epistemic level, authorial voice, and authorial judgments. His persuasive ends can only be met if he maintains the confidence of his readership that his representation on all levels adheres to the current standards of scientific practice.

Although by the time Compton completes the text he treats it as a manipulable object, the text contains references constructed and maintained by Compton's active commitments throughout the constructive process. And although the reference is not absolute in the sense of a one-to-one correspondence with objects having self-evidently natural and unchangeable designations and divisions, the reference is more than a literary fiction. The scientist's hands, eyes, ears, and laboratory apparatus stand between the physical events and the symbolic representation. Compton is neither a fiction writer nor a mute mechanic. The experiments are worked out both in the library and the laboratory, and the writing occurs both over the lab bench and over the desk.

Compton's behavior as revealed here should not be surprising to anyone familiar with the practices of modern science. All the evidence here indicates he is acting just as a good scientist might be supposed to; in so doing has managed in this case to create a statement of some endurance and force within the canon of communally accepted scientific claims.

Of course, not all claims have such good fortune. Many are fleeting, many fail, many are of little force or interest; the majority of Compton's articles suffered such fates. This hardly means that these articles were less well written or that the authors were not acting as quite so good scientists. Which articles get identified as right and significant depends on many factors—including changing interests and future empirical experiences of the community, as well as luck. Moreover, the details of what it means to act as a good scientist change through time and from locale to locale, as each research community evolves around its own problems and emerging work.

Moreover, it would be wrong to hold up this single case as absolutely indicative of the scientific procedures even within Compton's particular time and community. Compton's work may be idiosyncratic in his explicit concern for language, for within twentieth-century physics his drafts and revisions seem unusually detailed. Further, the stakes involved in Compton's claims and reputation at this juncture were high, encouraging heightened care.

Yet within these cautions, Compton's strong drive to hold himself and his claims accountable to his and his colleagues' experiences suggests the mechanism which keeps reference alive and makes language capable of interacting with the physical world. In this one concrete case, we see in detail the kinds of material relations between word and world around which we have seen the larger institutions of communication developing. Further case studies of how people move in the world to create words and how words then move people to interact with the world will increase our understanding of the varieties, characters, and qualities of reference in language. Only if we imagine that people never lift their heads out of books can we accuse their words of being only bookish.

Appendix

MEASUREMENTS OF β-RAYS ASSOCIATED WITH SCATTERED X-RAYS

By Arthur H. Compton and Alfred W. Simon

Abstract

Stereoscopic photographs of beta-ray tracks excited by strongly filtered x-rays in moist air have been taken by the Wilson cloud expansion method. In accord with earlier observations by Wilson and Bothe, two distinct types of tracks are found, a longer and a shorter type, which we call P and R tracks, respectively. Using x-rays varying in effective wave-length from about 0.7 to 0.13 A, the ratio of the observed number of R to that of P tracks varies with decreasing wave-length from 0.10 to 72, while the ratio of the x-ray energy dissipated by scattering to that absorbed (photo-electrically) varies from 0.27 to 32. This correspondence indicates that about 1 R track is produced for every quantum of scattered x-radiation, assuming one P track is produced by each quantum of absorbed x-radiation. The *ranges* of the observed R tracks increase roughly as the 4th power of the frequency, the maximum length for 0.13 A being 2.4 cm at atmospheric pressure. About half of the tracks, however, had less than 0.2 of the maximum range. As to *angular distribution*, of 40 R tracks produced by very hard x-rays (111 kv), 13 were ejected at between 0 and 30° with the incident beam, 16 at between 30° and 60°, 11 at between 60° and 90° and none at a greater angle than 90°. The R electrons ejected at small angles were on the average of much greater range than those ejected at larger angles. These results agree closely in every detail with the theoretical predictions made by Compton and Hubbard, and the fact that in comparing observed and calculated values, no arbitrary constant is assumed, makes this evidence particularly strong that the assumptions of the theory are correct, and that whenever a quantum of x-radiation is scattered, an R electron is ejected which possesses a momentum which is the vector difference between that of the incident and that of the scattered x-ray quantum.

IN recently published papers, C. T. R. Wilson[1] and W. Bothe[2] have shown the existence of a new type of β-ray excited by hard x-rays. The range of these new rays is much shorter than that of those which have been identified with photo-electrons. Moreover, they are found to move in the direction of the primary x-ray beam, whereas the photo-electrons move nearly at right angles to this beam.[3] Wilson, and later Bothe,[4] have both ascribed these new β-rays to electrons which recoil from scattered x-ray quanta in accordance with the predictions of the quantum theory

---

[1] C. T. R. Wilson, Proc. Roy. Soc. A **104**, 1 (1923)
[2] W. Bothe, Zeits. f. Phys. **16**, 319 (1923)
[3] See, e.g., F. W. Bubb, Phys. Rev. **23**, 137 (1924)
[4] W. Bothe, Zeits. f. Phys. **20**, 237 (1923)

of x-ray scattering.[5] In support of this view, they have shown that the direction of these rays is right, and that their range is of the proper order of magnitude. The present paper describes stereoscopic photographs of these new rays which we have recently made by Wilson's cloud expansion method. In taking the pictures, sufficiently hard x-rays were used to make possible a more quantitative study of the properties of these rays.

The cloud expansion apparatus used in our work was patterned closely after Wilson's well-known instrument except that all parts other than the glass cloud chamber itself were made of brass. The timing was done by a single pendulum, which carried a slit past the primary beam and actuated the various levers through electric contacts. The Coolidge x-ray tube, enclosed in a heavy lead box, was excited by a transformer and kenotron rectifiers capable of supplying 280 peak kilovolts. For illumination we used a mercury spark, similar to that of Wilson, through which discharged a 0.1 microfarad condenser charged by a separate transformer and kenotron to about 40 kv. The photographs were made by an "Ontoscope" stereoscopic camera, equipped with Zeiss Tessar $f/4.5$ lenses of 5.5 cm. focal length. Eastman "Speedway" plates (45×107 mm) were found satisfactory.

A typical series of the photographs[6] obtained are reproduced in Plate I, (a) to (f), which show the progressive change in appearance of the tracks as the potential across the x-ray tube is increased from about 21 to about 111 kv.

Especially in view of the fact that the original photographs are stereoscopic, the negatives of course show much more detail than do the reproductions. These suffice to show, however, the two types of tracks, the growth of the short tracks with potential, and the fact that while the long tracks are most numerous for the soft x-rays, the short tracks are most in evidence when hard rays are used. These results are in complete accord with Wilson's observations.

Number of tracks. It has been shown[7] that if the above interpretation of the origin of the two classes of β rays is correct, the ratio of the number of short tracks (type R) to that of long tracks (type P) should be

$$N_R/N_P = \sigma/\tau \tag{1}$$

where σ is the scattering coefficient, and τ the true absorption coefficient of the x-rays in air; for σ is proportional to the number of scattered

[5] A. H. Compton, Bulletin Nat. Res. Council, No. 20, p. 19 (1922); and P. Debye, Phys. Zeits. (Apr. 15, 1923).

[6] These photographs were shown at the Toronto meeting of the British Association in August 1924.

[7] A. H. Compton and J. C. Hubbard, Phys. Rev. **23**, 448 (1924).

Three: Typified Activities in Twentieth-Century Physics

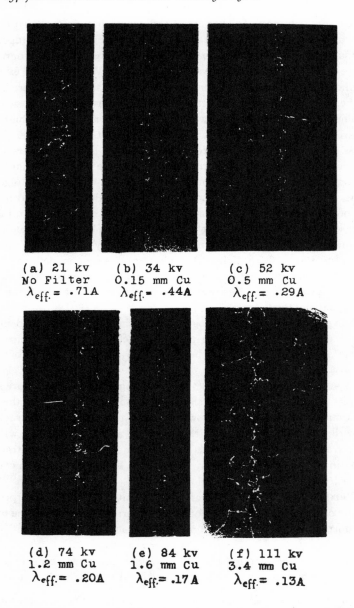

(a) 21 kv
No Filter
$\lambda_{eff.} = .71A$

(b) 34 kv
0.15 mm Cu
$\lambda_{eff.} = .44A$

(c) 52 kv
0.5 mm Cu
$\lambda_{eff.} = .29A$

(d) 74 kv
1.2 mm Cu
$\lambda_{eff.} = .20A$

(e) 84 kv
1.6 mm Cu
$\lambda_{eff.} = .17A$

(f) 111 kv
3.4 mm Cu
$\lambda_{eff.} = .13A$

Plate I. The x-rays pass from top to bottom. In addition to the copper filter, they traverse glass walls 4 mm thick. For the short waves the shorter (R) tracks increase rapidly in length and number. Thus while in (a) nearly all are P tracks, in (f) nearly all are R tracks.

quanta, and τ to the number of quanta spent in exciting photo-electrons, per centimeter path of the x-rays through the air.

In Table I we have recorded the results of the examination of the best 14 of a series of 30 plates taken at different potentials. The potentials given in column 1 of this table are based on measurements with a sphere gap. The potential measurements required corrections due to a slight warping of the frame holding the spheres, and to the lowering of the line voltage when the condenser was charged for the illuminating spark. The latter error was eliminated in the later photographs, at 34, 21, and 74 kv, and the former error was corrected by a subsequent measurement of the sphere gap distances, checked by a measurement of the lengths of the P tracks obtained at the lowest potential. The probable errors of potential measurements are thus unfortunately large, amounting to perhaps 10 percent in every case except that of 74 kv, which is probably accurate to within 5 per cent.

TABLE I
Number of tracks of types R and P.

Potential	Effective wave-length	Total tracks N	R tracks N_R	P tracks N_P	N_R/N_P	σ/τ
21kv	.71A	58	5	49	0.10	0.27
34	.44	24	10	11	0.9	1.2
52	.29	46	33	12	2.7	3.8
74	.20	84	74	8	9	10
84	.17	73	68	4	17	17
111	.13	79	72	1	72	32

The effective wave-lengths as given in column 2 are the centers of gravity of the spectral energy distribution curves after taking into account the effect of the filters employed. Because of the strong filtering, the band of wave-lengths present in each case is narrow, and the effective wave-length is known nearly as closely as the applied potential.

All the tracks originating in the path of the primary beam are recorded in column 3. Of these, the nature of some was uncertain. At the lower voltages it was difficult to distinguish the R tracks from the "sphere" tracks which Wilson has shown are often produced near the origin of a β-ray track by the fluorescent K rays from the oxygen or nitrogen atoms from which the ray is ejected. At the highest voltage the length of some of the R tracks is so great as to make it difficult to distinguish them from the P tracks. The numbers of R and P tracks shown in columns 4 and 5 are those of the tracks whose nature could be recognized with considerable certainty, the uncertain ones not being counted. This procedure probably makes the values of N_R/N_P in column 6 somewhat too small for the lower potentials and somewhat too great for the higher potentials.

The values of σ and τ given in column 7 are calculated from Hewlett's measurements[8] of the absorption of x-rays in oxygen and nitrogen. We have taken from his data the value of τ for 1 A to be 1.93 for air, and to vary as λ^3. The difference between the observed value of $\mu = \tau + \sigma$ and this value of τ gives the value of σ which we used.

The surprisingly close agreement between the observed values of N_P/N_R and the values of σ/τ we believe establishes the fact that the R tracks are associated with the scattering of x-rays. In view of the evidence that each truly absorbed quantum liberates a photo-electron or P track,[9] the equality of these ratios indicates that *for each quantum of scattered x-rays about one R track is produced.*

The fact that for the greater wave-lengths the ratio N_R/N_P seems to be smaller than σ/τ may mean that not all of the scattered quanta have R tracks associated with them. This would be in accord with the interpretation which has been given of the spectrum of scattered x-rays. The modified line has been explained by assuming the existence of a recoil electron, and the unmodified line as occurring when the scattering of a quantum results in no recoil electron. On this view the fact that the unmodified line is relatively stronger for the greater wave-lengths goes hand in hand with the observation that N_R/N_P is less than σ/τ for the greater wave-lengths. In view, however, of the meager data as yet available on this point, we do not wish to emphasize this correspondence too strongly.

Ranges of the R tracks. The range of the recoil electrons has been calculated on the basis of two alternative assumptions.[10] First, assuming that the electron recoils from a quantum scattered at a definite angle, its energy is found to be

$$E = h\nu \frac{2\alpha \cos^2\theta}{(1+\alpha)^2 - \alpha^2 \cos^2\theta}, \qquad (2)$$

where $\alpha = h\nu/mc^2$, and θ is the angle between the primary x-ray beam and the direction of the electron's motion. This energy is a maximum when $\theta = 0$, and is then,

$$E_m = h\nu \frac{2\alpha}{1+2\alpha}. \qquad (3)$$

[8] C. W. Hewlett, Phys. Rev. **17**, 284 (1921)
[9] See, e. g., A. H. Compton, Bull. Nat. Res. Council No. 20, p. 29, 1922
[10] See Compton and Hubbard, loc. cit.[7]

The second assumption is that the R electron moves forward with the momentum of the incident x-ray quantum. In this case the energy acquired is

$$E' = h\nu \cdot \tfrac{1}{2}\frac{a}{1+2a}(1-\tfrac{1}{4}a^2+ \cdots) . \qquad (4)$$

Eq. (3) was found to agree considerably better than Eq. (4) with Wilson's experimental results.

The lengths of the tracks shown on our photographs could be estimated probably within 10 or 20 per cent. These measured values, reduced to a final pressure of 1 atmosphere, are summarized in Table II. In column 2 are recorded the lengths of the longest tracks observed at each potential. S_m is the range calculated from Eq. 3, using C. T. R. Wilson's result[1] that the range of a β-particle in air is $V^2/44$ mm, where V is the potential in kilovolts required to give the particle its initial velocity, and the frequency ν employed is the maximum frequency excited by the voltage applied to the x-ray tube. S' is similarly calculated from Eq. (4).

TABLE II
Maximum lengths of R tracks.

Potential	Observed	Calc. (S_m)	Calc. (S')
21kv	0mm	0.06mm	0.004mm
34	0	0.3	0.02
52	2.5	1.8	0.1
74	6	6	0.4
88	9	12	0.7
111	24	25	1.5

It is evident that the observed lengths of the R tracks are not in accord with the quantity S' calculated from Eq. (4). They are, however, in very satisfactory agreement with the values of S_m given by Eq. (3). This result agrees with the conclusion drawn from Wilson's data,[11] but is now based upon more precise measurements. It follows that *the momentum acquired by an R particle* is not merely that of the incident quantum, but *is the vector difference between the momentum of the incident and that of the scattered quanta.*[12]

This conclusion is supported by a study of the relative number of tracks having different ranges. If the maximum range of the recoil electrons is S_m, Compton and Hubbard find[7] that the probability that the length of a given track will be S is proportional to

$$(2\sqrt{S/S_m}+\sqrt{S_m/S}-2) . \qquad (5)$$

[11] Compton and Hubbard, loc. cit.,[7] p. 449.
[12] That this is true for the β-rays excited by γ-rays has been shown in a similar manner by D. Skobeltzyn, Zeits. f. Phys. **28**, 278 (1924).

This expression assumes that the exciting primary beam has a definite wave-length. To calculate the relative number of tracks for different relative lengths to be expected, we have averaged this expression by a rough graphical method over the range of wave-lengths used in our experiments. These calculated values are given in the last column of Table III, for the relative ranges designated in column 1. A comparison of these

TABLE III
Relative lengths of R tracks.

Range of S/S_M	Per cent of R tracks within this range					Calc.
	52kv	74kv	88kv	111kv	Mean	
0- .2	44	66	60	54	56	53
.2- .4	34	20	26	32	28	22
.4- .6	19	8	4	8	10	14
.6- .8	0	3	5	3	3	8
.8-1.0	3	3	5	3	3	3

calculated values with the observed relative ranges shows a rather satisfactory agreement throughout. It will be noted further that the probabilities of tracks of different relative ranges is found to be about the same for x-rays excited at different potentials. This is in accord with the theoretical expression (5) for the probability, which is independent of the wave-length of the x-rays employed.

Angles of ejection of R tracks. On the view that the initial momentum of an R electron is the vector difference between the momenta of the incident and the scattered quantum, it is clear that these electrons should start at some angle between 0 and 90° with the primary beam. The probability that a given track will start between the angles θ_1 and θ_2 is on this hypothesis,[13]

$$\int_{\theta_1}^{\theta_2} P_\theta d\theta = 3ab \int_{\theta_1}^{\theta_2} \frac{a^2 \tan^4\theta + b^2}{(a \tan^2\theta + b)^4} \frac{\sin\theta}{\cos^3\theta} d\theta , \qquad (6)$$

where $\quad a = (1 + h\nu/mc^2)^2$, and $b = (1 + 2h\nu/mc^2)$.

In our photographs only those taken at 111 kilovolts have tracks long enough to determine the initial direction with sufficient accuracy to make a reliable test of this expression. In all, the directions of 40 tracks were estimated, with the results tabulated in the second column of Table IV. In view of the fact that the photographs were stereoscopic, it was possible to estimate the angles in a vertical plane roughly, though not closer perhaps than within 10 or 15°. The values in the third column are calculated from Eq. (6). It is especially to be noted that, in accord with the

---

[13] See Compton and Hubbard, loc. cit.,[7] Eq. (14).

theory, no R tracks are found which start at an angle greater than 90° with the primary x-ray beam. In view of the small number of tracks observed and the approximate character of the angular estimates, the agreement between the two sets of values is as close as could be expected.

A more searching test of the assumption that the R tracks are electrons which have recoiled from scattered quanta is a study of the relative ranges of the tracks starting at different angles. (See columns 4 and 5 of Table IV.) The calculated ranges in column 5 are based on Eq. (2) for

TABLE IV
Number and range of R tracks at different angles, for 111 kv x-rays.

Angle of emission	Per cent of total number (obs.)	(calc.)	Average range (obs.)	(calc.)
0°-30°	34	28	9 mm	11 mm
30°-60°	39	50	4	4
60°-90°	27	22	0.9	0.3

the energy at different angles. In this calculation the effective wave-length, as estimated in connection with Table I, is employed. It will be seen that the observed ranges of the tracks ejected at small angles are much greater than that of those ejected at large angles, in substantial agreement with the theory.

It is worth calling particular attention to the fact that in comparing the theoretical and experimental values in these tables, no arbitrary constants have been employed. The complete accord between the predictions of the theory and the observed number, range, and angles of emission of the R tracks is thus of especial significance.

The evidence is thus very strong that there is about one R track or recoil electron associated with each quantum of scattered radiation, and that this electron possesses, both in direction and magnitude, the vector difference of momentum between the incident and the scattered x-ray quantum. Our results therefore afford a strong confirmation of the assumptions used to explain the change in wave-length of x-rays due to scattering, on the basis of the quantum theory.

RYERSON PHYSICAL LABORATORY,
UNIVERSITY OF CHICAGO.
November 15, 1924.

Publications of A. H. Compton Discussed in This Chapter

Arthur H. Compton. "The Size and Shape of the Electron." *The Journal of the Washington Academy of Sciences* 8, (4 January 1918): 1–11.

———. "The Size and Shape of the Electron: I. The Scattering of High Frequency Radiation." *Physical Review* 14 (July 1919): 20–43.

———. "The Degradation of Gamma-Ray Energy." *Philosophical Magazine* 39 (May 1921): 749–69.

———. "Secondary Radiations Produced by X-rays." *Bulletin of the National Research Council* 4, 2 (October 1922).

———. "A Quantum Theory of the Scattering of X-rays by Light Elements." *Physical Review* 21 (May 1923): 483–502.

———. "Recoil of Electrons from Scattered X-rays." *Nature* 112 (22 September 1923): 435.

———. "Absorption Measurements of the Change of Wave-length Accompanying the Scattering of X-rays." *Philosophical Magazine* 46 (November 1923): 897–911.

———. "The Spectrum of Scattered X-rays." *Physical Review* 22 (November 1923): 409–13.

———. "Scattering of X-ray Quanta and the J Phenomena." *Nature* 113 (2 February 1924): 160–61.

——— and J. C. Hubbard. "The Recoil of Electrons from Scattered X-rays." *Physical Review* 23 (April 1924): 439–56.

——— and Y. H. Woo. "The Wave-length of Molybdenum K Rays when Scattered by Light Elements." *Proceedings of the National Academy of Sciences* 10 (June 1924): 271–73.

———. "The Scattering of X-rays." *The Journal of the Franklin Institute* 198 (July 1924): 57–72.

———. "A General Quantum Theory of the Wave-length of Scattered X-rays." *Physical Review* 24 (August 1924): 168–76.

———. "The Effect of a Surrounding Box on the Spectrum of Scattered X-rays." *Proceedings of the National Academy of Sciences* 11 (February 1925): 117–19.

——— and Alfred W. Simon. "Measurements of β-Rays Associated with Scattered X-rays." *Physical Review* 25 (March 1925): 306–13.

——— and Alfred W. Simon. "Directed Quanta of Scattered X-rays." *Physical Review* 26 (September 1925): 289–99.

———. "Light Waves or Light Bullets?" *Scientific American* 133 (October 1925): 246–47.

———. *X-rays and Electrons*. Princeton: Princeton University Press, 1926.

———. "X-rays as a Branch of Optics." *Journal of the Optical Society of America and Review of Scientific Instruments* 16 (February 1928): 71–87.

8 PHYSICISTS READING PHYSICS

SCHEMA-LADEN PURPOSES AND

PURPOSE-LADEN SCHEMA

Just as a scientist writes as part of an active life within a research community, the scientist reads as part of the continuing activity of research. If texts are not—cannot be—produced by the simple transcription of natural fact, no more can they be read as a direct apprehension of contextless meaning. Readers make their readings, each for their own purposes and by their own lights.

Yet, although each reading is a personally constructed event, the individual reading is embedded in communally regularized forms, institutions, practices, and goals. The reading is part of the historical realization of a communal project. In the same way that each scientific article, although a totally new document, bears significant similarities and relations to prior and future texts, each reconstruction of meaning through reading coheres with other readings as well as other structured elements of the scientific endeavor. Twentieth-century physicists read articles in physics within the activity and structure of twentieth-century physics. Their reading is motivated and shaped by their participation in that communal endeavor.

Although reading consumes a substantial part of a research scientist's working life, science studies have not looked very far into exactly what happens when a scientist reads and how this reading is precisely related to scientific activity. Macroscopic surveys have documented the amount of time scientists in different specialties read, what kinds of documents they read, and from which source they identify documents they might read. But in these studies, largely driven by information science interests in improving accessibility to information, the process of reading itself has not been considered problematic. The only substantial research into the processes by which the scientific literature is read has been through examination of citations to articles in subsequent literature. These studies of citation use and transformation (most notably the work of Cozzens and Small) have indicated some of the patterns by which interpretations and evaluations of read texts become meaning-carrying elements in new writing. These citation studies suggest strongly how intimately reading and writing are tied together in an intertextual system of knowledge creation. Yet these studies still have only looked at the reader after he or she has written a new text. They have not yet looked at the reader reading, or even at the reader in the process of writing, relying on earlier texts.

Literary studies and cognitive psychology have turned more thoroughgoing attention to the problem of constructing meaning from reading. Literary studies, concerned with poetic meaning, have turned from both the intentions of the author and the text itself to the reader's construction of meaning from the fixed set of words of a text. Iser and Eco, for example, have been concerned with how texts guide those constructive processes to varying degrees, whereas others, such as Holland and Bleich, see the construction of meaning as almost wholly guided by the reader, so much so that the text has little role in determining meaning. Extensive annotated bibliographies appear in Tompkins, Suleiman and Crosman, and Holub. At the same time as the meaning of the text is seen to reside within the reader, that meaning is also seen to develop out of a web of relations with other texts. The reader reads not a single text, but an intertext which creates both the traces of language familiar and meaningful for the reader and the presuppositions on which the reading rests. Kristeva first developed the concept of intertextuality; a recent survey appears in Orr.

In cognitive psychology, studies of children learning to read have considered comprehension as a product of a reader's interaction with a text. Readers actively employ their structured background knowledge (or schemata) in order to understand a text (Rumelhart and Orotony; Spiro; Reynolds et al; Steffensen; Bruce). Furthermore, the reader's purpose in reading helps the reader define a reading strategy and select what information to glean from the text (McConkie, Rayner, and Wilson; Reynolds and Anderson). Differences in schema or purpose that the reader brings affect both the process of comprehension and the meaning constructed from the text. Johnston reviews much of this work.

In making the meaning of a text a socially active phenomenon, these constructivist approaches to reading problematize scientific knowledge by calling into question the concept of a fixed text. On the other hand, the study of reading processes can also illuminate how reading is placed against experience and how shared meanings form. Meaning construction has empirical and sociological elements as well as psychological. The reader is not an isolated mind, devoid of experience and community.

Physicists Reading Physics

In this chapter I report on the reading processes of seven research physicists, based on data gathered from a series of interviews and observations. Throughout the interviews and observations, two themes from contemporary reading research proved indispensable in understanding how these professionals manage the literature in their fields: the reader's purpose and schema of background knowledge. The researcher's own need to carry on research and his or her own understanding of the field clearly shape the reading process and the meaning carried away from the professional literature. Moreover, purpose and schema are intertwined, so that the reader's schema incorporates active purpose and purpose is framed by the schema. In this dynamic interplay any article has the potential for reshaping the reader's schema and purpose.

Since the purposes for reading derive from the reader's own active research program and the schema are constructed around that program, interpretation and evaluation of read texts are intimately bound up with the empirical experiences and emerging empirical projects of both the individual and the discipline. Scientific reading is drawn into that same structured web of doing and formulating that constrains and occasions scientific writing. Texts are read against a continuing disciplinary activity in the world and judgments about how that activity might be most successfully continued. With readers already in motion, mentally and physically, texts are drawn into constant and consequential contact with the natural world.

The Interviews

The seven physicists I interviewed and observed represent a variety of specialties: three (T1, T2, and T3) are small particle theorists; two (BP1 and BP2) are experimentalists in biophysics; and two (RS1 and RS2) work with applied theory in the area of remote sensing. Five are from the same middle-sized private technological university (RS1, BP1, T1, T2, T3). RS2 is from a nontechnological branch of a large public university. The last, BP2, is the head of a lab at a major research university. He is the only one who regularly works as part of a consistent lab team. The rest either work individually or collaborate intermittently.

Single interviews with each lasted from 90 to 120 minutes (except for T3 whom I interviewed for about 250 minutes over three sessions) and were tape recorded. Each interview included a discussion of the subject's reading practices; with four subjects I was able to observe reading activities—library search for materials, scanning tables of contents, reading of articles—and then discuss what happened. The three interview sessions with T3 consisted of (1) a general interview; (2) an observation and focused interview of a library search for materials and quick first reading of those materials in the library; and (3) observations and focused interviews on careful readings of two articles.

With all the subjects who were observed and interviewed, I noticed no obvious differences between their accounts of their practices and the observed practices. They seem to do what they say they do. The observed activity, however, did lead to a more detailed discussion in the consequent focused interview.

Purposeful Choices

All through the reading process the physicists interviewed carefully select what they pay attention to and retain based on the needs of their own research. The continuation of their own research projects forms the purpose for the reading and, thus, determines what they want to get from reading.

The range of these physicists' serious reading is defined by what they feel necessary for current or anticipated work. If their work is on well-known puzzles with a substantial literature, they read mostly work similar to their own. If they perceive their current work touching on many fields, they search more widely for relevant work. Furthermore, they all accept the distinction between core reading close to their own work and peripheral reading. Finally, some read for prospective work: to tutor themselves, to gather information, or to window shop for potential problems to work on.

In terms of amount of reading, all define their "must" reading by the amount available and relevant to their issues, whether the amount is large (BP2) or small (RS2). Where time pressure interferes, it affects the more peripheral reading, which gets a more cursory scan.

In order to find the articles necessary for the continuation of their work, almost all these physicists periodically scan the tables of contents of selected journals—whether through *Current Contents* or in the actual journals. They sometimes supplement these scans by computer searches, reviews of the literature, abstract publications, and the *Science Citation Index*.

Schema for Making Choices

In making these early choices of what articles to read, each reader calls on personally organized knowledge. This schema extends beyond textbook knowledge of accepted facts and theories to include dynamic knowledge about the discipline's current practices and projections of its future development. The schema even includes judgments about the work of colleagues.

In selecting the range of reading the physicists must, of course, have a sense of the various fields of current work. Moreover, in deciding the urgency of reading the physicists must rely on an image of how rapidly work moves in their fields. All the pure theoreticians and experimental biophysicists go to the library at least once a week to search the tables of contents of newly arrived journals because they perceive their fields as moving rapidly and they must keep current to do adequate work. Both physicists in remote sensing, however, choose less timely methods of search—one using *Current Contents* and the other using abstract indexes. When questioned about the slowness of their search techniques, both said that their field did not move fast enough for that to matter.

The scanning processes of these physicists give evidence about how deeply these schema are impressed in the subconscious. The subjects scan so rapidly over tables of contents that they cannot give conscious thought to each title. Rather, certain words seem to trigger the attention and make the scanner question a particular title more actively. Indeed, both BP1 and T3 described how certain words seem to pop out of the page in some form of rapid unconscious processing. When I asked the subjects about particular titles they chose to look at further, they always attributed their interest to particular words.

These words are of three kinds, indicating domains of organized knowledge within which the word is immediately and unconsciously placed, then give value:

Names of objects or phenomena. These are the same as or closely related to objects or phenomena being studied by the researcher. Typically T3 reported an interest in an article by the term "atom–diatom collisions" and in another by "spin polarized hydrogen" because in each case that was just the thing with which he was working. BP2 reported that he had "quite a large number of such names. . . . I have a fairly organized view of this field, so I immediately categorize these nouns . . . into a context and make a judgment as to their value."

Names of approaches or techniques. These are not objects themselves, but ways of knowing those objects. RS2 in searching articles and indexes always looks for "remote sensing."

Three: Typified Activities in Twentieth-Century Physics

Names of individuals or research groups. All interviewees expressed awareness of who was doing good work in their field, with the three theoreticians being certain about comprehensive knowledge of all the significant actors in the field. Each of the scanners indicated that they were frequently attracted to an article by the name or research group of an author, even if nothing in the article title attracted them.

The importance of knowledge of the important actors in the field is furthered by the role of preprints and recommendations in determining reading. All the interviewees mentioned that many of the most important articles came through the mail as either preprints or reprints, and they paid at least some attention to all articles that arrived in this way. On a few occasions, as the subjects scanned journals, they commented that they would read a particular article, except that they had already seen it in preprint. And most of the interviewees mentioned recommendations by colleagues as an important source of articles.

Complex Choices: Complex Schema

The way that kinds of knowledge fit together in article selection decisions reveal the complexity of the reader's overall schema. Once the scanner's attention has been grabbed by a single term, he or she then will look at the other words in the title. In the observations I made, only about one quarter of the titles that triggered attention on the basis of a single term were actually looked at. All others were deleted on the basis of the other information of the title and author.

In the simplest deletion cases, further words in the title defined the phenomenon or the technique more precisely so as to place the article outside of the researcher's interest. For example, RS2 would regularly eliminate titles signaled by the keyword "remote sensing" if the title indicated any wavelength region other than infrared, for not only was the specific information different, the problems of measurement were also different in the other wavelength regions.

Similar, but more interesting, were the cases in which the technique that triggered attention was, on closer inspection, discovered to be applied to a different phenomenon. T3, for example, was attracted to the acronym DWIA (meaning Distorted Wave Impulse Approximation) in a title because he had used that method before and had referred to the acronym in his published work. He eliminated the article, however, when he saw that the research site was a molecule much more complex than the one with which he was working; he anticipated that the calculations would look entirely different. With respect to a similar example, he commented that the elaboration of a technique would be totally different in a new domain.

Conversely, the phenomenon may be right and trigger attention, but then a glance at the technique term of the title will eliminate the article because the reader feels that the method or technique cited is either unpromising or unlikely to produce calculations or results interesting to the researcher. T2 does not find work produced by the shell model currently interesting; although he was attracted by a title mentioning a phenomenon directly related to his problem of nuclear shape, he bypassed the article because it used this model.

The intersection between names of authors and the substance of their titles allowed readers to predict how a piece of work might go and thus how useful it might be. T2, for example, although attracted by the substantive terms of a title, passed over an article on the basis of the authors because he felt that they were only redoing what they had been doing for the past five years, only calculating higher-order terms. He called this work "too messy . . . extremely long and complicated. . . . I am sure the calculations are right, but it is the wrong approach." On the other hand, he also expressed some interest in an article, despite a title indicating work totally outside his area, because he knew the author to be doing interesting work that might be of importance for the whole field.

When the title and author provide inadequate, ambiguous, or misleading information the reader will turn to the abstract to decide whether the article is worth reading. Because the abstract usually contains more information, it allows a more precise placement of the article within the schema, and the process of placement reveals the complex multidimensionality of the schema. In one particularly revealing example, T2 was first attracted to the coauthored article both by the name of one author and by the title of the article. However, as he read the abstract he became confused, saying this "went beyond the previous work." Then he seemed unsure about what the abstract was saying. Finally, he realized that the article was based on the work of the research group of the other coauthor. The meaning of the abstract came clearly into focus, but as T2 did not find the current work of the other research group nearly so interesting, he dropped the article at this point. Thus, the same topic, in part from a respected author, because it came from a different research program, suddenly became judged less interesting or less consequential to the reader's work. This example bespeaks the reader's highly articulated and purposeful sense of the work going on in the field.

Three: Typified Activities in Twentieth-Century Physics

Purpose-Laden Schema

Through these examples we can see that in deciding whether to look further into an article, the reader is actually placing the article within his or her personal map or schema of the field. As in Steinberg's famous drawing of a New Yorker's map of the world, the items are given various importance or size based on the observer's perspective—in this case the reader's own work. Some items loom large and must be investigated in detail, whereas others seem to fall off the end of the known personal universe. The map is so well developed that just from the clues of the title, author, and perhaps the abstract, the reader can make strong predictions about what an article in a significant area in the map is likely to contain. T2 was able to predict correctly that an article would use techniques twenty-five years old in familiar expansions, because new techniques under the same name had not yet diffused to the geographic locale of the author and the applications indicated in the article title.

Unlike Steinberg's terrain of fixed landmarks (analogous to a codified picture of nature), however, the working physicist's map applied to his or her reading is a dynamic exploratory one built on the problems on which the field is working, the way the problems are being worked, and which individuals are working on what. The map embodies the physicist's personal perceptions of the forward motion of the discipline of which the researcher considers himself or herself a part. The personal map changes to reflect changing events—new problems being opened up, new actors appearing on the scene, and old problems and actors vanishing. A recent workshop at their university, for example, introduced T2 and T3 to the work of the workshop leader and, consequently, both picked up an article of his in a current journal.

This map, moreover, is seen through the perspective of the reader's own set of problems and estimate of the best ways to solve these problems, so that the map changes as the reader's own problems and guesses about the best approach or technique change. BP2, for example, was once interested in an approach to his subject through the study of divalent cations, but experiments in his lab as well as the inconclusiveness of the large number of articles with this approach convinced him that this was a dead end. Now he does not even look at an article with "divalent cations" in its title.

Purpose-Laden Schema in Understanding the Article

This doubly dynamic schema (a vision of a field in the process of trying to solve problems as seen through the individual's own research interests) provides the framework against which the reader comes to understand an article. The reader will process information that has significance for the existing schema and will view that information from the perspective of the schema. Thus, the way one reads is a strategic consequence of what one is trying to accomplish. How to read turns out to be as fundamental a decision as what to read.

The majority of interviewees read the larger part of articles selectively, seeking what they consider the news—that is, what will fill out or modify their schema or picture of subject and field. But what the news is depends on individual interests and purposes. Theoreticians, for example, may go right to the results of experimental articles to see what kind of data is obtained and must be accounted for by their theory; they are likely to skip over methodological sections as uninteresting and theoretical sections as familiar. Even problem formulations and conclusions may not contain much that is helpful to them.

In work very close to the reader's own, the reader often skips past the largest part of the article as thoroughly familiar, only to stop at the new equation or technique or trick. BP2 reports that a main activity of his reading is to notice things that don't fit his expectations. "There are some things that go against what you expect, that trigger the attention: 'Is this right?' If so, then something is missing [from our knowledge]. . . . From our theoretical knowledge and our basic understanding we know a great deal how things are supposed to go. . . . Some other things are a little surprising. . . . Somebody should check that."

Frequently the interviewees read backwards, or jump back and forth, depending on their interests or as one section raises questions about earlier ones. They generally do not read articles sequentially. In quite a number of cases, both reported and observed, the readers looked at the introduction and conclusions, perhaps scanning figures, to get a general idea of what the writer was trying to do. Then they simply filed the article for possible later reference. They only gave the article more careful reading at that time if the article seemed important to their work.

Even when articles are read sequentially, to reconstruct the author's argument, frequently the detailed mathematics are skipped over, with only a look at the kind of equations that result. The derivations are simply assumed to be correct.

Purposeless Information and Hazy Schema: Confusions and Black Boxes

Because readers gain the meaning of articles through their schema, parts of articles that do not readily fit against the comprehension schema create difficulties. Some parts of articles appear irrelevant and thus fall off the edge of the map; others are terra incognita—part of the relevant world but not sufficiently well-known. Some are not drawn clearly enough to clarify one's existing picture; and some do not fit well against existing schema and thus seem confusing in meaning. How readers deal with these lapses in comprehension depends on their perception of how potentially significant the passage is.

Where articles contain unfamiliar or difficult material, the reader weighs the cost of working through the difficulty against the potential gain. Such situations occur when the article requires technical knowledge outside the range of the reader or contains detailed calculations or derivations. All of the interviewees at times have to look up background material in reference works and textbooks. On the other hand, RS1 and RS2 both find their field so interdisciplinary that they inevitably must live with wide ranges of relevant ignorance. T1, T3, BP2, and RS2 frequently skip across complex mathematics, only identifying the techniques, the general gist of the derivation, or the results, unless they feel they have to know the derivation for their own work. A significant subcategory of this is the computer program used to generate results. Only in exceptionally significant situations will the reader request a printout of the program for detailed analysis.

Sometimes the articles are so poorly written that the reader cannot follow the argument or its meaning. Here, one must calculate not only the effort, but the possibility of adequate reconstruction. Enigmatic conciseness or disorderly presentation of the key steps of the derivation lead to troubling obscurity for all three theoreticians. Furthermore, bad writing signals a poorly framed problem, inadequately defined assumptions, fuzziness of method, or unclear results. T3 reports that such difficulties lead to "a false sense of the connection between that work and yours." BP1 comments that when he finds a model fuzzy, it "may be because I don't understand the model or the author does not understand the model."

Another form of haziness occurs when, despite clear presentation, the data are not clearly significant. This is of particular concern to BP2, who works in an area with many experimental results being published for many of which, BP2 feels, the problem addressed is inadequately defined or the techniques are not appropriate.

When articles project representations of nature that do not correspond with either accepted data or related accepted theory, the reader can have trouble figuring out what the author has in mind. Coherence with contextual knowledge is important in enabling the reader to interpret a set of claims. BP1 reported being baffled by an article and bringing it to an expert in the area who said the article "was just wrong. You know, wrong. It should not have been published." Meaning seems to come from being able to fit the article in with what you know.

If the new message cannot be meaningfully associated with what the reader knows, the reader finds it difficult to obtain a meaning from it. Moreover, he or she has difficulty reading it like a fiction—the presentation of a hypothetical world. In reading, as in the rest of their work, these physicists are guided by the purpose of building up a picture of the actual world. If a statement does not fit in with the endeavor, it does not convey a significant meaning.

At times articles may be only temporarily confusing, for upon consideration the reader readjusts the schema to incorporate the puzzling material. After reading a particularly profound article, RS2 thought about it for a number of days before she felt she understood it fully in all its consequences. BP1 reports a more subconscious version of the process of schema reshaping or refinement: "I may say, 'gee, I don't understand it,' and put it in a drawer for a week or two . . . then I look at it again and the penny drops." The temporarily confusing statement requires one to think differently, and is confusing only as long as it takes to change one's way of thinking. The statement must be of such apparent promise and importance that the reader will reshape the schema for it.

Opening Up Black Boxes

Two reasons motivate the interviewed physicists to work through comprehension difficulties: either to add to their background knowledge or to mobilize aspects of the article in their immediate work. Each reason leads to a different reading strategy.

In filling in one's ignorance, one is likely to read trustingly and uncritically. One adds new information to one's schema, familiarizing oneself with new concepts and techniques. RS1 describes his method of using the article as a tutorial: "I will read it in various stages. After I have read it once I will go through it again. I will look at some of the basic crucial references. . . . Then, I will try to verify some of the equations . . . and chances will be I won't know where they got them. . . . In order to verify the equations I would have to spend some time . . . look on some other papers, on the other references, occasionally they may come from a textbook. . . . Then I would consider it as part of the background I would understand."

On the other hand, a second reading in anticipation of immediately using the results is likely to be more critical, concerned with placing the article in and against one's existing schema, deciding carefully just what role the new material ought to take. Because one will be building one's own actions and statements on the material, one considers the argument, methods, evidence, and claims cautiously. Deciding to integrate another's work into one's own is the core of the communal endeavor of science. But it is a wary communal endeavor.

The following extended example reveals how detailed reading involves detailed schema matching and judgments as to the value of integrating the material more deeply into one's schema. T3 read twice through an article about a mathematical technique that he was interested in applying in his own work. After selecting the title on the basis of the name of the technique, he immediately "knew roughly what the article was trying to do." The issue now was whether the technique was worth the effort to acquire and employ.

On the first quick five-minute reading, T3 skipped through the article, looking at the equations and a results table to note the difficulty of the equations and the accuracy of the results in comparison to experimental figures. At this point he noted that the method would get accurate results, but only after a fifteen-term expansion. He would have been happier with accuracy after a five-term calculation, but he still considered this method worth a further look, particularly after skimming the conclusion that said that the method was "practical and numerically stable." As T3 knew the authors and respected their judgment, he gave the article another, more careful reading from the beginning, for an additional half hour.

In this slower reading he followed the mathematical reasoning more closely, although he still did not derive or work through all the equations. He noted the expansions of the equations used, but could not find any reason for the choice of these particular expansions. Also, he noticed many subproblems involved in completing the expansions. The error/accuracy estimates and the method of generating certain functions required more computer capability than he had available. As he read two textbook-type examples, he felt the desire for a more complex example. The method began to seem less attractive to him, requiring great efforts to solve insufficiently complex problems. He consequently reinterpreted the author's judgment of "practical and numerically stable" as a rather lukewarm evaluation. At this point he decided not to work through the equations, which would have constituted his third reading if he had found the method more attractive. Through the comparison between the article's proposed method and methods already familiar to T3, the article, which at first seemed a potentially major contribution to T3's schema, shrank to inconsequentiality.

Evaluating Articles: Criteria for Modifying Schema

Detailed reading motivated by anticipation of using the results, as we have just seen, merges into evaluation. For fitting new material into an already highly articulated schema is a judgment-laden process, affecting each reader's future work. The accumulation of such individual evaluations of reading influences the course of the whole community's knowledge and work.

All articles in the process of comprehension undergo evaluations of usefulness and importance. The article that remains unread, unused, and uncited suffers a harsh judgment. But even the articles that are read undergo evaluations of apparent importance. The general criterion reported for importance is the amount of news contained in the reading—that is, how significantly the article adds to or shakes up the current schema of what is known and how the field should go about knowing more. This criterion cannot be separated from the individual researcher's basic purpose in reading—finding out what one needs to know to pursue one's work. BP1 finds that an important paper "redefines an area . . . gives you hard information as to where you should be restricting your search." All interviewees associated news with future action as well as a current picture of nature.

Although all articles go through at least an implicit judgment of importance, only some articles undergo significant immediate judgments of their truth or quality. Most articles are considered reliable, on the face, because most of the interviewees read most articles for self-instruction or information in areas beyond their intimate knowledge. Only where prior knowledge is highly focused and articulated is the reading likely to conflict in substantial ways with the reader's schema. As BP1 commented of one article, "From then on, I am not competent to judge whether he is right, so I will be learning." Experimentalists generally do not question theoretical articles, and theorists generally do not question experimental papers.

Only BP2, with a comprehensive field-wide schema, tends to be critical of most of his reading. Whereas, for example, other interviewees report using a pencil and paper while they are reading for self-instructive functions (working through derivations, making notes and outlines), BP2 always reads with a pencil in his hand, making critical evaluative comments: "I scribble something awful."

The judgments, when made, often reflect a vision of how such works should go, rather than a sense of the substance of the statements. That is, readers compare the articles with the parts of their schemata that suggest how work should proceed rather than state what results should be. Internal evidence and stylistic features give the readers clues to the article's reliability. BP1 relies on the wording as an indicator: "The way a paper is phrased tells you if he is of this epoch and knows what he is talking about. Often you will get papers whose wording is wrong. . . . Sometimes it is really so strange you know something is odd." Both BP1 and BP2 are positively impressed when the author admits experimental or methodological difficulties, particularly if they are aware of the difficulties from their own experience. BP1 said, "only a careful guy does these things." BP2 commented: "Some . . . experimental sections are crisply clear and little goodies are buried in it, like 'it turns out that one cannot do it this way because' . . . or 'there is a little artifact in these results' and the guy spells out how he avoided it. Very good. This kind of paper you can believe because the guy clearly knows what he is doing."

The clarity of the model being presented also concerns the readers. T2, for example, finds an article suspect if the assumptions, methods, or results are not laid out clearly, for such fuzziness of presentation may indicate fuzziness in the work.

To evaluate the substance of statements, the interviewees generally rely on their own methodological experiences. The experimentalists interviewed examine experimental technique to see if it accords with their own experience of how such experiments should be run. BP1 asks, "What techniques, what kinds of techniques did they use? Did they follow the necessary protocols?" The theorists who create simplified models of complex systems question the simplifying assumptions of articles being read based on their own experiences in working with various assumptions. T2 calls the evaluation of assumptions the most critical evaluation he makes, for given the article's assumptions, the consequent calculations are rarely wrong. The whole problem of his field is to choose the right simplifying assumptions.

The existing body of published experimental results also plays an important role in the evaluation of both theoretical and experimental articles. In evaluating theoretical results, T1, T2, and T3 all look to see how well the calculated values compare with experimental results. This is then balanced against the simplicity or cumbersomeness of the method of calculation. When looking at experimental results, the experimentalists BP1 and BP2 note whether the results fit their expectations. BP1 comments: "Are the effects that should be there, there, and the effects that shouldn't be there, not there?"

In some cases, for some purposes, an article may not accord with the reader's perception of the problems, the significance of previous literature, or the meaning of the current results, but the reader will ignore those differences to take from the paper what appears novel or important. This selective evaluation is strong evidence for the priority of one's individual schema in evaluating results over an absolute, textually based standard. That is, arguments are generally evaluated not with respect to the correctness of the entire argument, but to how the reader can assimilate pieces into ongoing work.

Evaluation Changes over Time: Changing Schema and Changing Field Purposes

The judgments made upon reading articles are not necessarily final. BP1, for example, notes, "Sometimes I miss things. . . . I think things are not particularly interesting, and then I kick myself later for having missed it." Later work may show an error in a piece of work, but more often evaluations change because the field in some way leaves the work behind (or in a few cases, catches up): either new methods and experiments prove to be stronger or the general thinking of the field has changed so as to alter the schema against which the article is placed. In BP1's words: "My model of the universe would change . . . along with a majority of the people in the field. . . . There is sort of a drift."

T1 shows a similar awareness of the evolutionary nature of the field and how one's changing schema is tied to that evolution. When he evaluates the quality of the results from a method of approximation, he allows a greater margin of error for the first attempt at a theoretical calculation than he does after a number of people have proposed solutions.

Schema-Laden Purposes

Articles, in their challenge to existing statements, foment new work. Plausible new methods, evidence, claims, and interpretations change the landscape against which the researcher plans and

Three: Typified Activities in Twentieth-Century Physics

realizes research purposes. Just as schema embed the purposes of the individual researchers, purposes embed the researcher's schema.

In an immediate way, both experimentalists and theoreticians report doing more work to confirm striking results in their field. RS1 said: "If I am working on the problem, then of course I would do a series of things to verify and test" the novel results. BP2 similarly said he would carry out or assign one of his subordinates in the lab to carry out further experiments to explore and test novel results, as when one of his graduate students showed that some published results were artifacts.

Over the long term the body of claims from the corporate literature that are integrated into the individual's schema will close off certain problems and methods and open up others. A changing picture of nature and the dynamics of investigation, all garnered from reading, will modify research purposes. The researcher acts on what he or she knows, and much of what the researcher knows comes from reading.

Constructing a Literature

Given this evolutionary understanding of their work and their colleagues' work, the interviewees recognize that their thinking and knowledge reflect the joint endeavor of constructing a literature. Their view of nature is directed toward making more statements about nature and their statement-generating actions are based on schema arising from previous statements.

The interviewees express a variety of opinions about their vision of nature, but none claim an unmediated, clear, and certain access to nature. T1 most directly states that he does not believe in such a thing as a truth about nature, but only greater or lesser solidity in the statements we make about nature. T2 and T3 admit having only an impression of the phenomena they theorize about through what is reported in the literature. Although T2 does admire some of the experimentalists he works with who seem to have a concrete feel for the actual phenomena, he has learned never to say "nature is not like that," rather, only "nature could not be that complicated." The experimentalists interviewed, indeed, seem to have more of a feel for concrete nature, but they still find it hard to disentangle nature from the impression created by the literature.

If the literature is then understood, criticized, and evaluated against an image gleaned from the literature rather than against nature itself, we are confronted once again with the epistemological problem of the socially constructed nature of science and scientific knowledge. In this study we find texts being read piecemeal for specific pieces of informa-

Physicists Reading Physics

tion. We see the information being placed within and against personal frameworks of knowledge. We see individual purposes and uses driving and shaping the reading. We see new statements being accepted based on how well they integrate with existing schema of how work should go. We see much reading accepted noncritically, from lack of experience with the work being discussed.

Evaluations, moreover, seem to be deeply enmeshed with ad hominem judgments. BP1, for example, does not necessarily look too closely at the experimental section of a paper if he knows the colleagues and their work well—he is personally familiar with their experiments. Even personal factors enter into the process of criticism. RS1 notes: "If you are stepping on someone's toes, it may be very difficult." When you step on toes demands of proof are higher. Proof criteria similarly go up when results are startling, as RS1 points out, for then, in a sense, you are stepping on everybody's toes, making them all reevaluate their schema. Thus, even standards for public argument are situational, depending on the degree of competition and conflict.

Furthermore, reading habits and procedures seem affected by psychological and sociological variables. BP2, for example, as head of a laboratory, has wide reading responsibilities and a critical function, but he also reports that ever since childhood he has read broadly and critically. Whether he became a lab head because of these habits or developed these habits as part of his rise and then reinterpreted his childhood to fit his new self-conception, there is role-appropriate behavior.

Within this welter of individual mind and circumstances, various purposes, limited criticism, and evanescent texts, we begin to wonder how such a thing as shared understanding of a field is possible, how ideas gradually become accepted or validated, how consistent criteria are possible, or how a coherent canon of knowledge can develop.

Yet, on another level these findings suggest merely that texts communicate from one mind to another, and each mind is organized and purposeful in its own way. In a social system relying on originality and individuality of judgment, each person will take and judge differently. Where they know more they can question more deeply. Where they have questions they question harder.

Communication is a social process. In the comparison of schema across the printed page some shared understandings are reached. These shared understandings are based on many individuals each being individually satisfied that claims accord with experiences and best judgments about how the world should be conceived and science conducted. Moreover, those whose work is closest to one another most often have to judge one another's work to carry on their own. What emerges from the

Three: Typified Activities in Twentieth-Century Physics

conflict and integration of the schema of those closest to the material does, then, represent a consensus to be taken seriously. As BP1 remarks, through phrases such as "the current mythology" and "our faith is," colleagues at conferences recognize agreement even on matters for which there is little solid evidence. All of the interviewees assumed a wide range of shared perceptions with their colleagues except in specified areas of difference or well-known open problems.

The long-term process of scientists building on one another's results, moreover, seems a powerful corrective to the idiosyncracies of individual work and short-term misunderstandings and misevaluations. Although individual experiments and calculations may be plausible or implausible, correctly or incorrectly understood and evaluated, replicated or not replicated, in the long run they must accord with the continuing experience of a range of researchers in order to maintain current acceptance. The statements that will have a continuing life in the literature will be those that readers will consistently integrate into their work.

Within such a social understanding of the construction of a scientific literature even such potentially disillusioning behaviors as the necessity of publicizing one's own work at conferences take on important functions. If one can get other people to see how one's work might bear on theirs, they may use it, develop it, refine it, add significant related results to it. T2 comments on the importance of "salesmanship": "A lot of people in nuclear physics . . . have had great success because of a very interesting model to start with, but also in the sense of having done a good sell. This is very important. People may be able to feed back ideas or information in the model. . . . If they see any kind of connection at all [with their own work] they will become interested in it. . . . The more people working in an area, the more ideas will be generated; some of them will be good ideas. . . . In the long run that will help the reputation of the model." Idea development is a communal development.

The short-term reading processes examined in this article fit into this longer-term emergence of scientific knowledge. Each scientist forms a personal view of the field, yet remains willingly accountable to experimental results and reasonably open to any powerful suggestion that comes along in the literature that might affect a work-directed schema upon which individual plans ride. Within this framework what turns out to be most useful to the most workers in the field over a long period has more than faddish significance. *Usefulness, if it is constantly tested from many angles against an uncooperative nature, is in the long run much more than a pragmatic criterion.*

These working research scientists have an extraordinary commitment

Physicists Reading Physics

to the literatures of their fields. They work hard to keep up with the literature and are willing to change not just their minds, but their plans and work on the basis of what they read. As a number of commentators have noted, the literature for scientists seems similar to scriptures for fundamental believers of the divine word. Yet the differences are major, for scientific reading does not attempt to return to a primary vision. The constant attempt is to add to the scripture, to move on to a better understanding, a new vision. Old parts of the canon are willingly scrapped, despite the resistance that sometimes attends new findings. Most of all, the literature constantly is being held accountable to an outside measure, whereas scripture is usually held to be hermetically true, no matter what the world tells you. Although each scientist is moved to do his or her own good works through individual conscience and reading of the shared texts, ultimately the individual must bend to the world, for that is where the researcher believes good works are to be found.

PART FOUR

THE REINTERPRETATION OF

FORMS IN THE

SOCIAL SCIENCES

9 CODIFYING THE SOCIAL SCIENTIFIC STYLE THE APA *PUBLICATION MANUAL* AS A BEHAVIORIST RHETORIC

The intellectual, practical, and social successes of the natural sciences have made their ways of going about their business highly attractive to other communities that create knowledge. Not only have the natural sciences seemed to have found a way of producing statements of great detail and reliability, expanding our powers of prediction and control over nature, they have also been able to develop wide agreement on a large number of statements within their communities and have gained the respect and support of the broader society. Thus the natural sciences have generated wide social, political, and economic power as well as power over nature.

In particular, those communities concerned with issues of human mind, society, and culture have been moved to adopt (and adapt) what they perceive to be the methods of the physical and biological sciences. Just as natural philosophy gradually was reorganized as the natural sciences over the seventeenth and eighteenth centuries, many other parts of philosophy since the late nineteenth century have been in the process of being reorganized into what are called variously the *social sciences, behavioral sciences, cognitive sciences,* or *human sciences.*

Central to the reorganization of these knowledge-creating communities has been an imitation of the forms of argument developed within the natural sciences. The compelling force of these arguments, the consensus developed over the aggregate results of these statements, and the power over natural forces achieved through the understanding constructed from these texts, seem to remove them from the traditional realm of rhetoric—those things about which we are uncertain, as Aristotle remarks at the opening of his *Rhetoric.* By arguing without seeming to argue and compelling without apparently urging, the scientific manner

Four: The Reinterpretation of Forms in the Social Sciences

of formulating knowledge seems to offer a way out of the deep divisions of belief and imponderable conundrums that seemed to pervade psychological, social, moral and cultural questions.

However, as we have seen in the previous chapters, the literary forms of scientific contribution have developed out of active argumentative situations in particular forums. Scientific discourse emerged as a way to win arguments rather than as a way to avoid them. They remain in the realm of rhetoric because there is no certainty in science, no absoluteness of statement. Problems of induction, reference, skepticism, and intersubjectivity haunt the lowest strata of our empirical knowledge and scientific representations. Scientific modes of communication developed as a series of solutions to the problems of persuasion. These solutions emerged within developing communities, and were embedded within emerging empirical, social, and rhetorical practices.

Scientific writing is no unitary and unchanging thing, defined by a timeless idea. Varieties of scientific writing have developed historically in response to different and evolving rhetorical situations, aiming at different rhetorical goals, and embodying different assumptions about knowledge, nature, and communication. The form of the experimental report, in particular, solves a changing rhetorical problem: given what we currently believe about science, scientists, the scientific community, the scientific literature, and nature, what kind of statement about natural events can and should we make? To treat scientific style as fixed, epistemologically neutral, and transcending social situation is rhetorically naive and historically wrong.

In attempting to mobilize the powerful forms of argument developed within the natural sciences, the human sciences neither escape rhetoric nor eliminate rhetorical choice. Though some practicing social scientists might wish to escape the uncertainties of human discourse by embracing a single, correct, and absolute way of writing science, any model of scientific writing embeds rhetorical assumptions. Recognizing and examining these assumptions reasserts our control of choices that may otherwise be determined by unconsidered tradition, stereotype, and ideology. The forging of a scientific language is a remarkable achievement; but since it is a human accomplishment, it must be constantly reevaluated and remade as the human world changes.

This reevaluation is all the more important because the assumptions of forms of scientific communication involve the fundamental practices and organization of the disciplinary community. Attempts to transplant rhetorical forms from one community to another engage basic issues of what these communities are doing and how they go about it. The form will either be changed by the soil and climate of the new disciplinary community or it will struggle with maladaptation. This chapter and the next discuss two cases of the transplantation of the experimental report into the social sciences. In the first case, the development of experimental psychology gives a particular interpretation to the experimental report that achieves a highly codified, institutionalized form. This codification stabilizes particular intellectual beliefs, empirical practices, and social relations around assumptions of a particular kind of research program. In the second case, political science seems to have had greater difficulties in defining a consistent, stable interpretation of the experimental report despite energetic attempts to do so. The task, concerns, methods, and organization of political science seem to bring to bear many pressures on the language, which have not yet seemed to crystallize around a satisfactory form.

A Scientific Style for the Social Sciences

To understand the scientific style that emerged in the human sciences over the last century we need to look closely at experimental psychology. Experimental psychology was the first human science to establish a specialized discourse, distinguished from traditional philosophic discourse. Experimental psychology became the model and set the standards for all the psychological specialties that aspired to the status of science. In time, it played the same role for sociology, which did not start to develop a predominatly scientific style until the 1920s, and political science, which followed suit in the 1950s. Today the American Psychological Association *Publication Manual* symbolizes and instrumentally realizes the influence and power of the official style.

The official APA style emerged historically at the same time as the behaviorist program began to dominate experimental psychology. Not surprisingly, the style embodies behaviorist assumptions about authors, readers, the subjects investigated, and knowledge itself. The prescribed style grants all the participants exactly the role they should have in a behaviorist universe. To use the rhetoric is to mobilize behaviorist assumptions.

Recent versions of the *Publication Manual*, filled with detailed prescriptions, convey the impression that writing is primarily a matter of applying established rules. The third edition, published in 1983, offers approximately two hundred oversized pages of rules, ranging from such mechanics as spelling and punctuation through substantive issues of content and organization. The important section on "Content and Organization of the Manuscript" focuses almost exclusively on experi-

Four: The Reinterpretation of Forms in the Social Sciences

mental reports, for although it recognizes genres such as review articles and theoretical articles, it comments that "most journal articles published in psychology are reports of empirical studies."

The experimental report is to have the specified sections: title, abstract, introduction, method, results, and discussion. Each of the last three sections is to be so titled. Each section must conform to detailed instructions, at times resembling a questionnaire in specificity. In the methods section, for example, one must include separately labelled subsections (usually *subjects, apparatus,* and *procedure*), each reporting specified content. The instructions for describing the experimental subjects indicate the level of prescribed detail:

> Subjects. The subsection on subjects answers three questions: Who participated in the study? How many participants were there? How were they selected? Give the total number of participants and the number assigned to each experimental condition. If any participant did not complete the experiment, give the number of participants and the reasons they did not continue.
>
> When humans are the participants, report the procedures for selecting and assigning subjects and the agreements and payments made. Give major demographic characteristics such as general geographic location, type of institutional affiliation and sex and age. . . . (26)

And so on for another two and a half paragraphs.

Few could question, given the collective experience of the discipline, that such information is often important for understanding and evaluating the experimental results. But the assignment of the information to a fixed placed in a fixed format lessens the likelihood that researchers will consciously consider the exact significance of such information, whether it and other possible information should be included, and exactly how this information should be placed in the structure of the whole article. The prescribed form of fixed sections with fixed titles creates disjunctions between mandatory sections: the author does not have to establish overt transitions and continuity among the parts. The method section is a totally separate entity from the introduction or results. Although problem, method, and results must correlate at some level, the author escapes the need for transitions to demonstrate the coherence of the enterprise.

The foreword of the *Publication Manual*, well removed from the substantive prescriptions, does contain several disclaimers about linguistic evolution and flexibility. It notes, for example, that

Codifying the Social Scientific Style

> Although [the manual's] style requirements are explicit, it recognizes alternatives to traditional forms and asks authors to balance the use of rules with good judgment. . . . It is a transitional document. It looks at the literature itself to determine forms rather than employing style to contain language. (10)

Yet the introduction to the actual organizational prescriptions takes a hard line:

> Consistency of presentation and format within and across journal articles is an aspect of the scientific publishing tradition that enables authors to present material systematically and enables readers to locate material easily. Finally . . . the traditional structure of the manuscript allows writers to judge the thoroughness, originality, and clarity of their work and to communicate more easily with other individuals within the same tradition. (18)

In addition to the appeal to tradition—a tradition we will find shorter and more varied than one might guess—this passage urges uniformity on three other grounds: efficiency of reference, evaluative usefulness, and ease of communication. The second reason presupposes one right way to present an experimental report and that wandering from the form is bad science, or at least keeps bad science from being evident. The other two reasons suggest an encyclopedic function for an incremental literature; the concept of incremental encyclopedism will be examined later in this chapter.

History of the APA *Publication Manual*

The prescriptiveness evident in the current publication manual has only gradually developed since the first "Instructions in Regard to Preparation of Manuscript" appeared in the February 1929 *Psychological Bulletin*. This original stylesheet was only six and a half pages long. About a page discussed "Subdivision and Articulation of Topics," a third of which was explicitly devoted to experimental articles. Despite a "natural order" for the presentation of experiments, internal titles are discouraged: "Necessary Headings only should be inserted" (58). Advice was of a general kind; for example, to include sufficient detail to allow the reader "to reconstruct and to criticize the experimentation and to compare it with other procedures and results" (59). The committee

Four: The Reinterpretation of Forms in the Social Sciences

preparing this set of instructions avoided an authoritative stance, presenting these suggestions for "general guidance" only.

The 1944 stylesheet, "The Preparation of Articles for Publication in the Journals of the American Psychological Association," grew to 32 pages. Guidelines for bibliographical reference and the use of tables and graphs correspondingly increased in length, as did the explanation of the editorial policies of the APA journals. On the structure of the experimental article, however, the stylesheet says little more than the previous edition, although now conceding that the form "has now become structured into a fairly developed pattern" (350). Moreover, the stylesheet encourages use of headings to indicate "the main features of [the article's] framework" (351). The authors offer their advice for the "younger members of the profession, many of whom are writing for publication for the first time" (345). Thus pedagogy allowed prescriptions without committed prescriptiveness.

The 1952 *Publication Manual,* now a 61-page separately bound supplement to the *Psychological Bulletin,* no longer hedges its prescriptive intent: "The purpose of the publication manual is to improve the quality of the psychological literature in the interest of the entire profession" (389). The manual is the standard. And as a standard it lays out explicitly just what is demanded. The section on organization lists the familiar parts of the experimental study, but suggests that headings reflect "the particular requirements of a study," rather than the standard part titles. Nonetheless, the manual prescribes what should be included within each. For example, the method section "should describe the design of the research, the logic of relating empirical data to the theoretical propositions, the subjects, the sampling and control devices, the techniques of measurement, and any apparatus used" (397).

The 1957 and 1967 revisions, although differing in some specifics, retain the general length and detail of the 1952 manual. The 1974 edition doubles the length and detail of prescription again, devoting 12 of the total 132 pages to content and organization. The 1983 edition "clarifies" and "amplifies and refines" this second edition, but does adhere to much of its wording. Notably, to ensure that standards are met on all levels, this last edition adds a section on grammar.

Two further style changes concerning the summary and reference formats are worth noting here. In the 1927 stylesheet, the last section of a paper was defined as a summary entirely separate from the abstract to be submitted to *Psychological Abstracts.* The 1944 stylesheet clarifies that the summary should be a serially numbered list of conclusions. In 1952, the summary, no longer a list, becomes a description of the entire argument, covering "the problem, the results, and the conclusions." This formal summary could also be used for *Psychological Abstracts*. Beginning in 1967, however, the abstract appears at the front of the published article, eliminating the final summary.

The reference format changes from traditional footnotes in 1927, to cross references, to a numbered bibliography in 1944, to the current system of author and date amplified in a reference list at the end, first prescribed in 1967. These changes help bring the references into the flow of the discussion as items for conscious attention. Both the dates and the names of authors now serve as kinds of facts in the argument.

Early Articles in Experimental Psychology

The evolution of the published articles in experimental psychology reveals the nature of the rhetoric embedded in the *Publication Manual*, for the history of the articles shows the rhetoric in action. The characterizations that follow are based on analyses of over 100 articles and examination of several times that number from the chief journals of experimental psychology, clustered in the early period (last decades of the nineteenth century), the periods of behaviorism's rise (1916 to 1930) and dominance of behaviorism (1950 and 1965, taken as sample years), and the current period (1980 as a sample year). The selection of articles analyzed and examined is large enough to reveal the major trends, but the dates attributed to the first emergence or dominance of any particular feature are necessarily approximate. Further, any characterizations of large numbers of texts will inevitably obscure differences among texts and may not be accurate for specific features of individual texts; however, as the official behaviorist style emerges, texts become much more uniform. That movement toward prescriptive uniformity forms a central part of the story.

The founding journals of the discipline defined the acceptable range of writing for the field by the articles they published: *Philosophische Studien* (hereafter *PS*), founded by Wilhelm Wundt in 1883; the *American Journal of Psychology* (*AJP*), founded by G. Stanley Hall in 1887; and the *Psychological Review* (*PR*), founded by J. M. Cattell and J. M. Baldwin in 1894. Each began the first issue with an editorial or article discussing the emergence of a new scientific psychology based on experimental results.

Despite these rigorous programmatic statements, the early issues of these journals, particularly the two American ones, contain a wide variety of articles, only some of which could be labelled experimental. The first two volumes of the *AJP* do contain such narrowly experimental studies as "Dermal Sensitiveness to Gradual Pressure Changes," but also contain "A Study of Dreams," "Winter Roosting Colonies of Crows," "Extracts from the Autobiography of a Paranoiac," "The Place for the Study of Language in a Curriculum of Education," "Folk-Lore of the Bahama Negroes," and "On Some Characteristics of Symbolic Logic." Many articles sought to bring empirical data to the philosophic inquiry into the mind. Indeed, the editor's manifesto in the first issue claims a broad audience for the *AJP:* teachers of psychology, anthropologists interested in primitive manifestations of psychological laws, physicians interested in mental and nervous diseases, biologists and physiologists, and anyone else interested in the advances in scientific psychology.

Early experimental work measured such quantifiables as perceptual sensitivity and reaction times, but these measurements served only as empirical entry ways into the mysteries of the mind. Although they followed the general structural pattern of experimental reports already established in the natural sciences, the early articles had more the character of philosophic essays. For example, an article in the first issue of the *AJP* by Hall and Motora begins with a Greek epigraph from Plato (72).

The two American journals did not use any internal headings in the articles; consequently, words had to bridge the parts, explaining how the whole inquiry fit together. In the first volume of *PR*, for example, Hugo Munsterberg presents a series of five "Studies from the Harvard Psychological Laboratory." These studies have no internal divisions, although they clearly follow standard experimental order. Each part grows out of the previous one. The third study, "A Psychometric Investigation of the Psycho-Physic Law," demonstrates this strikingly. The opening theoretical discussion of the psycho-physic law argues that a new kind of measurement is needed. The experimental design then provides the desired measurements. Moreover, each aspect of the experimental method is justified and explained in terms of current knowledge about the psycho-physic law. The specific parameters for measurement refer back to the theoretical problem, and the actual results follow immediately as a response to the specific parameters. Discussion of the consequences of the results for the psycho-physic law follow naturally as part of the thematic continuity of the whole essay.

Articles in the German *PS*, although they frequently use standard section headings, provide heavy continuity among the parts. Often the first paragraph or two of a labelled section considers either general thematic material or the issues raised in the previous section, so that the substance of the section is not directly discussed until it is firmly tied to the total structure of the article.

In these experimental essays, the authors reveal themselves as problem-solving reasoners, figuring out how quantitative experiments might aid understanding of philosophical issues. The discussion of methods plays a crucial role, raising and answering the problem of how one can translate the theoretical problem into concrete empirical results. For example, Munsterberg, in the series mentioned above, repeatedly proposes his methods as correcting the failure of previous methods to make proper distinctions. The effort devoted to the presentation of the methods shows clearly that they are a significant part of the intellectual achievement of the work presented in the article. Similarly, the first experimental article in the premiere issue of *PS* devotes an eight-and-a-half-page methods section to deriving the methods from the nature of the phenomena to be investigated and to evaluating alternative methods (Friedrich).

The early authors believed that psychological phenomena were internal, subjective events and that the measured data were only external indicators of what was going on inside. Trained introspection provided evidence in conjunction with more external quantitative measures. Thus the subjects of the experiments emerge as active characters in the experimental report. Individual experimental subjects, which included trained psychologists, were often identified by name. (Wundt himself was a subject in many experiments performed in his laboratory.) In the experimental report the identification of subjects shows their training and credentials for making accurate introspective judgments. The author of "Experiments in Space Perception," James Hyslop, is himself the experimental subject. Combining psychological knowledge and an unusual ability to use his eyes independently, he devised certain tricks or exercises for himself that help to elucidate principles of perception. The two-part article is imbued with first-person accounts of what he did and what he perceived.

The readers were sometimes treated as being quite knowledgeable about current work, so much so that much technical background was left understood, as, for example, in Hall and Motora's article on dermal sensitivity in the opening issue of *AJP*. Nonetheless, the audience was generally treated as concerned with broad issues of psychological understanding. The early articles almost always begin at some issue of general psychological interest and connect the specific study to that issue. In fact, that technical article by Hall and Motora is the one prefaced by the Greek quotation and appears in the same issue as Hall's editorial anticipation of broad readership for the journal.

These articles review the literature only sporadically. At most, short summaries present assorted experimental results, without establishing definitive findings that lay a stable groundwork for current studies. Fre-

Four: The Reinterpretation of Forms in the Social Sciences

quently articles begin without any specific mention of previous work. In short, the articles give the general impression of a new beginning, to be grounded thoroughly on empirical results, as opposed to the implicitly rejected nonempirical earlier work. This is consistent with a philosophic tradition that treats each new approach as a fresh attempt to ground philosophy on its true footing.

Wundt's role in his journal, which largely published the results of his own laboratory, best reveals the philosophic nature of the endeavor. Wundt, although the founder of the first regular lab and frequently called the father of experimental psychology, did not publish any experimental reports in *PS* (the experimental reports were written by his subordinates). Nonetheless, articles by Wundt appeared in the journal at least two or three times a year, and as often as eight, discussing ideas, methods, and large philosophic issues well removed from psychology. These discussions often appeared as reviews or critiques of the work of others, but always with the purpose of explicating fundamental issues. Wundt kept the empirical work of the new discipline firmly in philosophic, reasoning focus. Although his students and other followers stayed much closer to the data—and no one seemed to be granted his same right to philosophize at length in the pages of the journal—he helped maintain the philosophic thrust of the discourse.

Despite the desire to subordinate the experiments to philosophic inquiry, the experimental data proved too complex and too removed from philosophic issues to resolve the problems posed. Typically in the early articles, the continuity of rational discussion breaks down when the results section is reached. The argument bogs down in extensive tables, reporting massive amounts of data—much of it raw or subject only to simple aggregating calculations. As in an 1894 study by Jastrow in *PR*, the discussion often no more than repeats the tabular data with a few, low-order statistical generalizations. Characteristically, no conclusions relative to a substantive problem are drawn, and the ultimate meaning of the data remains murky. Authors often caution against generalizing too quickly on the basis of uncertain results in situations that remain too multifactored to analyze fully. Future, more decisive results are promised. When substantive conclusions are drawn, the intermediate analysis of the data may be missing, such as in one of Munsterberg's studies which bypasses specific explanations through phrases like, "it is evident," "of course," and "the reason lies evidently in the fact that."

The inability of this massive data to resolve philosophic issues, such as the natures of memory and perception, soon led to a divorce between philosophic and empirical work.[1] Articles turned to establishing low-

---

1. Indicative of the early divorce between philosophy and psychology is the changing

level generalizations descriptive of the results. Literature reviews grew longer as the literature grew, and there was some attempt to find common denominators or clear patterns of disagreement among the prior results and set up the current experiment as a resolution. Methods became standardized and were frequently referred to by eponyms or citations. But the results generally did not resolve substantive issues. Conclusions were often a series of numbered statements, repeating the data. Even where the numbered conclusory statements addressed the originating question, as in the 1916 article in the *Journal of Experimental Psychology* (hereafter *JEP*) "A Preliminary Study of Tonal Volume" by G. J. Rich, only minimal substantive discussion related results to the problem. The complex data, both psychophysical and introspective, were left largely to speak for themselves.

Since the true object of inquiry remained internal phenomena, the subject of the experiment remained an important independent actor in the story. Subjects were described to show expertise or particular qualifications for accurate observation. In Dallenbach's articles throughout the period, for example, subjects are characterized as trained in psychology and familiar with the purposes and methods of the particular experiment. Introspective accounts provide data and, importantly, possible interpretations of the measured data. As late as 1930, in a study by Ferral and Dallenbach, the introspective accounts of the subjects (which include Dallenbach) are used to guide the analysis of the other results. Another striking example, "An Experimental Study of Fear" by V. Conklin and F. Dimmick, is based entirely on introspective accounts of emotional responses to the experimental situation.

Other methods of gaining evidence about the internal processes of humans were also still acceptable. A study of the foster-child fantasy is based on a survey of adolescents rather than on an experiment (E. Conklin). Another study was an anthropological observation of "The Gesture of Affirmation Among the Arabs," to clear up some incorrect and misinterpreted facts used by Wundt (George). Studies of literary fig-

---

character of the articles in the English journal *Mind*, founded in 1876 with the stated intention of being the first journal of the new psychology. The philosophic climate in England, however, did not prove conducive to the flowering of experimental psychology. Although early volumes contain glowing reports of the experimental work in Germany (for example, J. Sully, "Physiological Psychology in Germany," *Mind* 1 [1876]: 20-43), reviews of experimental work became increasingly critical (for example, G. C. Robertson, "The Physical Basis of Mind," *Mind* 3 [1878]: 23-43; and E. W. Scripture, "The Problem of Psychology," *Mind* 16 [1891]: 305-26. The general complaint against experimental work was grounded in the mind/body dichotomy; these philosophers found physical data of no value for understanding issues of mind. By the turn of the century discussion of experimental psychology ended altogether, leaving the journal as a purely philosophic one.

ures based on their works still appeared in *AJP* as late as 1920, when analyses of Charlotte Brönte and Edgar Allen Poe were published (Dooley; Pruette).

The author thus remained a problem solver, trying to gain some understanding of mental processes using empirical data, even though the discussion had now switched from a general philosophic to a more particular descriptive mode. Articles through 1920 still read as continuously reasoned arguments, with internal headings used sporadically and flexibly. Headings, when used, often reflected the specific content of the article and were not typographically prominent.

The implied audience as well remains varied—interested in the problems, but not necessarily involved in research. Through the 1920s articles still frequently start with familiar problems of everyday experience (such as fear, fantasy, and the sensation of burning heat), and they take a variety of approaches to study the problems. The articles are aimed at a wide range of people interested in the workings of the mind.

Behaviorism Finds Its Voice

As behaviorism in its many forms came to dominate psychology between the two world wars, a rhetoric consistent with behaviorist assumptions narrowed rhetorical possibilities and became the basis for the official style reflected in the *Publication Manual*. By behaviorism and behaviorists, I mean the general turn toward behavior and away from mind as the proper subject and data for psychological investigation. Many varieties of explicit behaviorism developed, not just the classic versions of Watson and Skinner. Additionally, many other schools of experimental psychology followed behaviorist procedures, although they did not explicitly espouse behaviorism.

Toulmin and Leary associate the dominance of behaviorism and neobehaviorism with a "cult of empiricism" fostered by an alliance with logical positivism, popular during the same period between the wars. The positivist principles of "physicalism" and "operationalism" legitimated the behaviorist limitations of allowable questions, method, and data. The behaviorist method then could be considered identical to scientific method, excluding other forms of psychological investigation as unscientific.[2] And the behaviorist rhetoric could be identified as the only proper way to write science.

2. Lawrence D. Smith makes a similar point in "Psychology and Philosophy: Toward a Realignment, 1905–1935."

The proper way in which to write positivist, behaviorist science did not, however, appear immediately on the scene, invented in a burst of self-conscious rhetorical creativity. Instead, the style emerged over a number of years as many individuals gradually discovered the form most congenial to their ideas and work. Early works appeared in a variety of styles consistent with the patterns of the past.

John B. Watson, although often credited as the founder of behaviorism, published little behavioristic experimental work. Rather, what is taken as his seminal work, "Psychology as a Behaviorist Views It," is a polemic. It is continuous, persuasive, and aimed at a general audience; it considers a general problem and presents the author and audience as reasoners capable of making intelligent judgments. Furthermore, as editor of *Psychological Review* from 1910 to 1916 and then of the newly founded *Journal of Experimental Psychology* for another ten years, Watson presided over the kinds of articles described in the previous section.[3]

The famous article "Conditioned Emotional Reactions" (1920), which Watson coauthored with Rosalie Raynor, reports one of his few published experimental studies. This unusual article, although different in many respects from both articles that came before and those to come after, still bears more resemblances to the earlier rhetoric than to the later. The study, which describes the conditioning of an infant to fear rats, is told as a coherent story with no real headings or strong divisions to interrupt the flow of argument. The only marked divisions are four questions identified by Roman numerals and passages from the laboratory notes, identified chronologically. The typical structure of introduction, method, results, discussion is not even maintained. Rather the theory to be demonstrated dominates the organizational pattern, with aspects of the method and results separated and subordinated to the different questions to be answered.

Thus the authors emerge as reasoners and persuaders, constructing an argument using experimental results to persuade the readers of the truth of a general theory. The authors use the first person throughout in order to present themselves in a number of roles: as doers of the experiment, as holders of certain expectations, as investigators desiring tests of certain questions, as makers of observations, as provers of certain propositions, and as interpreters of results. Furthermore, they present the experimental results in the rather personal form of the lab notes, replete with disjointed phrases and sentence fragments. Even though the notes present the events without reference to internal processes or

---

3. The *Journal of Experimental Psychology* was founded as an offshoot of the *Psychological Review*, and the two journals shared editorial boards.

imputations, rhetorically they serve to show the events through the eyes of the narrator.

The authors also stand well back from the literature, which is presented largely as speculative and unfounded, even including Watson's own writing on the subject. This article is, in short, another attempt to begin inquiry into basic matters *de novo*. Here again we see the independent philosopher, impatient with earlier false starts and misguided work. The tone of the opening paragraph reviewing the state of the problem is brusque and mildly contemptuous; that of the next to last paragraph comparing the authors' conclusions with Freud's is gratuitously and gleefully nasty, reminiscent of the delightfully vitriolic exchanges of nineteenth-century German philosophers.

Thus the audience is witness to a knock-down intellectual argument and is invited to choose sides, not just between ideas, but between persons: Watson and Freud. The choice rests on the audience's response to a first-person account of a single incident: in essence, a short story. In its narrative simplicity, clarity of argument, and broadness of issue, the article clearly aims at a wide audience. Its vigor of argument assumes that readers can and will make a choice—in favor of Watson.

The subject of Watson's experiment, the infant Albert B., has an immediate presence in the drama of the piece. The detailed description shows how, by virtue of his stability and lack of fear, he is mentally fit for the test to which he will be submitted. He emerges as an individual character in an engaging narrative account of his induced phobia, very much in the tradition of the clinical accounts of the mentally ill that had until recently shared the pages of the journals with experimental reports.

However, two differences set the treatment of Albert as a subject apart from the treatment of subjects in previous articles. First, the details of his background establish that his mind is a clean slate, unaffected by special quirks, foreknowledge, or other hindering factors. The subject's identity, in other words, stands as a sign of the experimenter's control of variables, rather than as a sign of the subject's special capacity to observe his own reactions. Second, the authors exclude introspection or any other attempt to gain knowledge of the subject's internal processes or sensations. This is the obvious mark of behaviorism. Yet, despite the attempt to turn Albert into an impersonal object of study, the fullness of narrative reveals a poignancy to the story. As Albert's phobia grows, the reader sees him become a victim, moved by the manipulations of the experimenter.

Stabilizing an Objectified Rhetorical Universe

In the period following the publication of this article, the objectification of the subject increases. Author, audience, and literature as well become more objectlike. All the aspects of the drama of experimental article move into a behaviorist universe. The rhetorical decisions made in the 1920s are elaborated, rigidified, and standardized in subsequent decades. The first APA stylesheet appeared in 1929; the increasing certainty and detail of prescriptions in the successive stylesheets follow and confirm the growing influence of this behaviorist style in the journals. Articles begin looking like one another, so that we can clearly identify an official style that lies behind the prescriptions of the publication manual.

Only when a community decides there is one right way, can it gain the confidence and narrowness of detailed prescriptions. In rhetoric, "one right way" implies not only a stability of text, but a stability of rhetorical situation, roles, relations, and actions, so that there is little room or motive for improvisatory argument. Within a stabilized rhetorical universe, people will want to say similar things to each other under similar conditions for similar purposes. In this context, prescribed forms allow easy and efficient communication without unduly constraining needed flexibility. The behaviorist picture of the world allows that stability and lack of free invention.

As we have seen in the article by Watson and Raynor, the behaviorist world view first made itself felt in characterizations of the experimental subject and the phenomena investigated. Not only do behaviorists categorically eliminate imputations of internal processes and introspective accounts, they no longer consider the external data as indicators of some mental process. The experimental problem switches from one of indicators to one of controls, from getting some hard data on complex individual internal processes to keeping the history of the subject and the environment sufficiently clean. The kind of narrative that Watson provides of Albert B. soon vanishes, for such a narrative grants too much personality to the subject, who is to be reported more as a type exhibiting very specific behaviors in highly controlled circumstances.

The previous tendency toward low-level conclusions that give only aggregate descriptions of the behavior observed no longer is a difficulty—it is the whole extent of the enterprise. One looks only for patterns of behavior, not underlying principles or mental operations. The increasing statistical sophistication of experimental articles serves to

Four: The Reinterpretation of Forms in the Social Sciences

exhibit and validate patterns of behavior across large numbers of subjects. The results themselves appear in increasingly calculated and patterned ways. Individual behavior disappears in a pattern, displayed in a graph or a table of secondary calculated values, rather than as a raw number. The results sections increasingly begin by describing the display tables and figures. By 1950, statistical talk, describing the statistical methods used and the limits of statistical reliability, becomes a standard part of the results section, usually immediately following the presentation of the numerical display.

Instead of a reasoner about the mind, the author is a doer of experiments, maker of calculations, and presenter of results. The author does not need to reason through an intellectual or theoretical problem to justify or design an experiment, nor in most cases does he or she need to identify and take positions on arguments in the literature. To produce new results, the author must identify behavior inadequately described and design an experiment to exhibit the behavior in question. With the methodological problem reduced to obtaining uncontaminated results, carefulness rather than good reasoning becomes the main characteristic to be displayed in the methods section. The methods section becomes less substantively interesting. Starting about 1930, the section is demoted to small print, where it remains today. Nor are methods customarily covered in summaries or abstracts.

This rhetorical diminution of methods in a science devoted to obtaining experimental results only makes sense once we see that the main rhetorical function of the methods section is not to present news or innovation, or even to help the reader conceptualize the event that produces the results. Its main function is rather to protect the researcher's results by showing that the experiment was done cleanly and correctly. In the articles from sample year 1950 that I examined, this desire to protect results by constantly demonstrating that one has done things correctly on all counts, from examining the prior literature to using proper statistical methods, becomes obtrusive and accounts for much of the length of the articles. As the conventions for demonstrating proper work become stabilized, by the growing prescriptiveness of the stylesheets and by repeated practice, this competence display is done more rapidly, so that by 1965 these preliminaries take much less space.

Because the methods section no longer serves as an intellectual transition between the problem and results, the article tends to break into disjointed parts, increasingly labelled by standard headings, as reflected in the successive stylesheets. The results become the core of the article. Discussion often merely sums up the data and is sometimes relegated to small print. Conclusions do little more than repeat confirmation of the descriptive hypotheses.

Codifying the Social Scientific Style

With the article primarily presenting results, constrained and formatted by prescription, the author becomes a follower of rules to gain the reward of acceptance of his results and to avoid the punishment of nonpublication. Accepting this role, he subordinates himself to the group endeavor of gathering more facts toward an ultimately complete description of behavior—a project of incremental encyclopedism. As behaviorism gradually gained influence, authors began presenting results as ends in themselves, to fill out gaps in other results, rather than as potential answers to theoretical questions. In the mid-1920s, introductions rapidly changed from raising a problem to giving a codified review of literature, with each item associated with a specific contribution. The experiment to be reported in the article was then presented simply as some form of continuation of the prior work. After a brief period when close analysis of the literature was allowed in small print, disagreements over theory, results, or formulations in the previous literature tended no longer to be discussd. Articles were treated as accumulated facts; literature reviews in the articles lacked synthesis, problem-orientation, or interpretation. In 1930, Edwin Boring, then an editor of the *American Journal of Psychology*, in a note in that journal attempting to domesticate the Gestalt movement, articulated the principle: "The progress of thought is gradual, and the enunciation of a new crucial principle in science is never more than an event that follows naturally upon its antecedents and leads presently to unforeseen consequents" (309).[4] This communal vision—much narrower than the traditional "shoulders of giants" formulation—diminishes the role of any individual as a thinker.

Several other rhetorical consequences flow from this incrementalism. First, since the function of the article is now to add a descriptive statement to an existing body of such statements, and since the new statement would achieve this goal only if it passes certain tests, strong rhetorical pressure pushes the candidate statement (the hypothesis) near the front of the article. Only then can the reader, in reading the body of the article, judge whether the claim passes the criteria. Thus the descriptive generalization moves from a conclusion to an opening hypothesis that takes on an increasingly central role in the presentation of the experiment.[5] As the main unifying element in the article, the hypothesis often comes to be repeated four or more times in a single article. Similarly, as the abstract switches from a summary of results to the presenta-

---

4. Boring had earlier formulated this principle in "The Problem of Originality in Science," *American Journal of Psychology* 39 (1927): 70–90.
5. The common methodological belief that the formulation of a hypothesis must precede the design of an experiment in the actual research process may in part derive from this rhetorical order.

Four: The Reinterpretation of Forms in the Social Sciences

tion of problem, results, and discussion, the "problem" comes to mean the test of the hypothesis and the "discussion" the confirmation of the hypothesis.

Second, since they are adding only bits to a larger descriptive project, articles decrease in scope and length. The single experiment replaces the series of experiments with minor variations in conditions or procedures. The confirmation of a single descriptive statement replaces the examination of a large phenomenon from a number of angles.

Articles also become shorter with the codification of format and of surrounding knowledge. With a fixed framework of knowledge and communication, one can add one's single additional bit more rapidly. In the selection of articles I examined, the low point of article size was in the mid-1960s. Articles from the same period also show significant increase in the technical vocabulary, indicating a dense specialized knowledge. Earlier most of the technical terms (except for statistical terms) were ordinary language terms, only given more precise definition; for example, *stimulus, condition, fatigue*. Even such unusual coinages as *retroactive inhibition* are not far removed from ordinary usage. But in the 1965 articles, terms, although originating in common-use vocabulary, take on such narrow concrete meanings that they diverge from normal meaning. The terms then get used in tight combination with other such terms. As well, key terms start being replaced by acronyms or abbreviations. Only those familiar with the technical background can be sure that they know exactly what is being discussed in a phrase such as, "the effects upon verbal mediation of the delay intervals interpolated between the two acquisition stages of a mediation paradigm or between the second acquisition stage and the test trial" (Peterson, 60).

Third, the *Publication Manual* adopted the new reference style, wherein the author and date of an article appear as facts or landmarks in the course of the article, visibly demonstrating the incrementalism of the literature. As anyone who has worked with this reference system can attest, it is very convenient for listing and summarizing a series of related findings, but it is awkward for extensive quotation or discussion of another text, and even more awkward for contrasting several texts in detail. The format is not designed for the close consideration of competing ideas and subtle formulations.

Finally, readers are no longer cast in the role of people trying to understand or solve some problem. Rather they are presumed to be looking for additional bits of knowledge to fit in with their previous bits. They are assumed to be looking for faults, because such faults would disqualify the experimental report as a valid increment to the descriptive encyclopedia. The author must display competence to the audience, rather than persuade readers of the truth of an idea. If properly demonstrated by a proper experiment, the hypothesis must be accepted by the audience. In an intellectual sense, the audience has little to say about the meaning of an experiment or even about the truth of a hypothesis. Its role, rather, is to judge the propriety of the experimental proof.

Within this rhetorical world, the chaos of intellectual differences is eliminated. Individuals assumulate bits, follow rules, check each other out, and add their bits to an encyclopedia of behavior of subjects without subjectivity. There is not much room for thinking or venturing here, but much for behaving and adhering to prescriptions. Thus we get to the ever-expanding *Publication Manual*.

Over the last twenty years, a major style change in the psychological journals has again started to take place, the result of the rising influence of a cognitive psychology based on the computer model. This new approach brings with it a new epistemological and rhetorical universe. It is too soon to give a full account of this new style, nor is it clear how pervasive it will become in the face of the continuing behaviorist rhetoric. One thing is clear: this new style has not yet affected the *Publication Manual* in any significant way. The APA manual still serves basically as a codification of behaviorist rhetoric.

For those social scientists who believe that the behaviorist, positivist program creates an accurate picture of the human world and provides the surest (if not only) path to knowledge, the prescriptive rhetoric of the *Publication Manual* is precisely the right one. It offers a programmatically correct way to discuss the phenomena under study; moreover, it stabilizes the roles, relationships, goals, and activity of individuals within the research community in ways consistent with the community's beliefs about human behavior. The invention of a way to communicate consonant with beliefs constitutes a major accomplishment. Nonetheless, the realization that behaviorism has not escaped rhetoric, but has merely chosen one rhetoric and excluded alternatives, may temper adherents' certainty about their mode of communication.

For those who have received any rhetoric as a given, the recognition of the implications of an official style reopens the question of how to write. Rhetoric is always sensitive to beliefs about the world. The human sciences undergo a particularly immediate form of this rhetorical sensitivity, for these sciences create and argue for beliefs about human beings, the inevitable main actors in the drama of communication. If a social science changes our view about the nature of ourselves, we need to change our way of talking to each other consonant with our changing self-image. To neglect the implications of our rhetoric is to lose control of what we say.

Four: The Reinterpretation of Forms in the Social Sciences

Versions of the *Publication Manual*

Bentley, Madison, et al. "Instructions in Regard to Preparation of Manuscript." *Psychological Bulletin* 26:2 (February 1929): 57–63.
Anderson, John, and Willard Valentine. "The Preparation of Articles for Publication in the Journals of the American Psychological Association." *Psychological Bulletin* 41; 6 (June 1944): 345–76.
Council of Editors. *Publication Manual of the American Psychological Association, Psychological Bulletin* 49; 4, pt. 2 (July 1952 supplement): 389–449.
American Psychological Association. *Publication Manual.* Rev. ed. Washington, D. C.: American Psychological Association. 1957, 1967.
American Psychological Association. *Publication Manual.* 2d ed. Washington, D. C.: American Psychological Association, 1974.
American Psychological Association. *Publication Manual.* 3d ed. Washington, D. C.: American Psychological Association, 1983.

Primary Articles Discussed in This Chapter

Boring, Edwin G. "The Gestalt Psychology and the Gestalt Movement." *American Journal of Psychology* 42 (1930): 308–315.
Boring, Edwin G. "The Problem of Originality in Science." *American Journal of Psychology* 39 (1927): 70–90.
Conklin, Edmund S. "The Foster Child Fantasy." *American Journal of Psychology* 31 (1920): 59–76.
Conklin, Virginia, and Forrest L. Dimmick. "An Experimental Study of Fear." *American Journal of Psychology* 36 (1925): 96–101.
Dallenbach, K. M. "The Measurement of Attention." *American Journal of Psychology* 24 (1913): 465–507.
Dallenbach, K. M. "Attributive vs. Cognitive Clearness." *Journal of Experimental Psychology* 3 (1920): 183–230.
Dallenbach, K. M. "The Measurement of Attention in the Field of Cutaneous Sensation." *American Journal of Psychology* 27 (1916): 445–60.
Dooley, Lucille. "Psychoanalysis of Charlotte Brontë as a Type of the Woman of Genius." *American Journal of Psychology* 31 (1920): 221–72.
Ferrall, S. C., and K. M. Dallenbach. "The Analysis and Synthesis of Burning Heat." *American Journal of Psychology* 42 (1930): 72–82.
Friedrich, Max. "Über die Apperceptionsdauer bei einfachen und zusammengesetzten Vorstellungen." *Philosophische Studien* 1:1 (1883): 40–48.
George, S. S. "The Gesture of Affirmation Among the Arabs." *American Journal of Psychology* 26 (1916): 320–24.
Hall, G. Stanley. "Editorial Note." *American Journal of Psychology* 1, 1 (1887): 3.
Hall, G. Stanley, and Yuzero Motora. "Dermal Sensitiveness to Gradual Pressure Changes." *American Journal of Psychology* 1, 1 (1887): 72–98.

Hyslop, James. "Experiments in Space Perception." *Psychological Review* 1 (1894): 257–73, 581–601.
Jastrow, Joseph. "Community and Association of Ideas: A Statistical Study." *Psychological Review* 1 (1894): 152–58.
Munsterberg, Hugo. "Studies from the Harvard Psychological Laboratory." *Psychological Review* 1, 1 (1894): 32–60.
Munsterberg, Hugo, with the assistance of W. T. Bush. "III. A Psychometric Investigation of the Psycho-physic Law." *Psychological Review* 1, 1 (1894): 45–51.
Peterson, Margaret Jean. "Effects of Delay Intervals and Meaningfulness on Verbal Mediating Responses." *Journal of Experimental Psychology* 69 (1965): 60–66.
Pruette, Lorine. "A Psychoanalytic Study of Edgar Allan Poe." *American Journal of Psychology* 31 (1920): 370–402.
Rich, G. J. "A Preliminary Study of Tonal Volume." *Journal of Experimental Psychology* 1 (1916): 13–22.
Watson, J. B. "Psychology as a Behaviorist Views It." *Psychological Review* 20 (1913): 158–77.
Watson, John B., and Rosalie Raynor. "Conditioned Emotional Reactions." *Journal of Experimental Psychology* 3 (1920): 1–14.

10 STRAINS AND STRATEGIES IN WRITING A SCIENCE OF POLITICS

THE UNSETTLED RHETORIC OF THE *AMERICAN POLITICAL SCIENCE REVIEW*, 1979

Psychology, by treating the individual as a separate biological behavioral unit can create a disengaged, objectified discourse that seems to separate both the experimental object and the experimenter from the historically evolved forms of culture in which humans act. Indeed, as we have seen, one of the important themes in the rhetoric of experimental psychology is to represent one's experimental subjects as sufficiently clean *tabulae rasae* and the conditions of one's experiments far enough removed from daily life so as not to be contaminated by the uncontrolled complexities that move our lives. But other social sciences, such as economics, sociology, anthropology, and political science, must deal more immediately with the complexes of human-made culture, for these human-made complexes are exactly their subject. As a result, when they come to try to represent any particular case, they must contend with many forces that cannot be contained within the laboratory walls. Culturally embedded studies must overcome many obstacles in arguing from the particular to the general, for the complex of details and local variables of each case can generate unending alternative descriptions and generalized accounts of the processes involved. To move from plausible conjecture to forceful persuasion to compelling argument, the researcher of human sciences must develop a rhetorical tool kit, different and perhaps more subtle than that developed in the natural sciences. And that rhetorical tool kit will also have likely consequences for the interaction with the object of study, the structure of communication, the social system of the discipline, and the discipline's goals and activities.

Strains and Strategies in Writing a Science of Politics

Despite the rhetorical problems posed by the social sciences, many social scientists have attempted rather direct importation of what they perceive to be the methods and communication styles of the natural sciences. As in the case of experimental psychology, the model of scientific communication adopted is likely to be a simplified abstraction (often supported by a prescriptive philosophic position), that ignores the complex rhetorical dynamics and historical fluidity of actual communication in the social sciences. In some respects the models of scientific communication transplanted wholesale into the social sciences more resemble that of high school laboratory courses. The high school science laboratory is an orderly and predictable place filled with well-defined objects, well-established formulations from textbooks, fixed expert-amateur social relationships, and predetermined discoveries. The social, intellectual, natural, and creative worlds are held constant so that students can rehearse set operations to be reported in set formulations. In such stabilized conditions, language can appear an unproblematic representation of a stable reality.

There are gains and costs in such stabilizing simplifications. In experimental psychology, both the gain and cost have been a thoroughgoing commitment to a particular kind of research program that has seemed appropriate to large and influential parts of the research community. In economics, as Donald McCloskey argues persuasively in *The Rhetoric of Economics*, the gain has been clarity about the mathematical realization and relationship of economic forces, but the cost has been a kind of hypocrisy of the discourse that leads important issues and forms of argument to appear in only covert ways. The official style of contemporary economics seems to exclude a wide range of nonmathematical disciplinary reasoning, individual and cultural dynamics in economic participation, and traditional moral, social, and policy questions about economic choices. However, as McCloskey argues, these excluded forms of discourse have not vanished; they have just become hidden, making their discussion fragmentary and insufficient. He believes explicit recognition and acceptance of these topics will lead to a more satisfactory and productive discussion among economists without losing the clarity gained by the current official style.

A similar debate has been going on in anthropology concerning the status of ethnographies. Under the banner of scientific objectivity, ethnographies had been represented as impartial, disengaged observations of stable social realities, recorded in a socially inert, acontextual manner. Recently, however, issues of the social and literary participation of both ethnographers and informants have been raised. The ethnographic text reflects the interactions of ethnographer and informant

Four: The Reinterpretation of Forms in the Social Sciences

with each other, with the tribal community, with western society, and with the professional community of anthropologists. The text also serves as a form of social action within all these collectivities. Through critique and practice, anthropologists such as Clifford, Fabian, Geertz, Marcus and Cushman, Rosaldo, and Tyler have been attempting to reformulate ethnographic writing to consciously address the rhetorical complexity of the documents.

Political science, as well, in adopting what it considers a scientific style of communiction has relied on simplifications both of scientific discourse and of the rhetorical problem the discipline faces. Unlike anthropology, however, political science has not developed a significant reflexive literature to consider the true complexity of its rhetorical task. The discourse of political science suffers from a number of unrecognized strains, which I hope to begin to uncover in this preliminary study.

Political Science's Version of Scientific Writing

Since the middle of this century, the study of politics has been developing a form of scientific presentation relying heavily on mathematics for both evidence and argument. In this presentational style, most often the numerical data are gathered and analyzed statistically, but sometimes the argument takes the form of abstract mathematical reasoning, as when game theory is employed. Articles assuming this mode of discourse may be more fully characterized as opening with a problem expressed through a review of literature that orders the existing knowledge in a coherent system of findings and issues. The article then proposes a hypothesis or solution to the problem, presents (and perhaps justifies) a methodology, then tests the hypothesis through mathematical data and argument. At the end only narrow conclusions are formed, limited to what can be documented by the mathematical argument. In 1979 over 70 percent (30 out of 42) of the articles in the *American Political Science Review* could be so characterized. The remaining articles, other than a presidential address statistically examining one aspect of quantitative political studies, are devoted to historical narratives about political movements and philosophical discussions of new and classical political theory. Moreover, the articles in the natural scientific mode averaged 1.63 authors per article compared to an average of 1.08 authors per article for the historical and philosophical texts. The multiple authorship implies a research team practice resembling that of the natural sciences, resulting from the complexities of data collection and analysis (e.g., Physics Survey Committee, p. 1368).

Such a textual organization is a direct correlate of the model of scientific activity presented in scope and methods books that explain to students of political science how to go about their intended profession (e.g., Greenstein and Polsby; Hayes and Hedlund; and Isaak). These books emphasize hypothesis testing and data collection as the core of science. Isaak, for example, discusses induction in the following terms:

> We test a hypothesis by seeing if it fits the world of observation. Suppose we want to test the hypothesis, "Businessmen tend to be conservative." A sample of businessmen would be questioned . . . to determine their ideological orientations. On the basis of this sample—and the confidence we place in our conclusions depends upon its size and randomness—we accept or reject the hypothesis. (91–92)

The task of the political scientist is to compare claims to empirical reality; the function of political science writing is to communicate the findings of these comparisons.

A closer examination of political science articles, however, reveals difficulties and complexities in this straightforward aspiration to a scientific ideal. Arguments do not fit together as crisply as the ideal would have it, and the political scientist as author inevitably finds himself in explanatory, justificatory, reconciliatory, and persuasive tasks that are not part of the idealized version of the scientific report.

Analysis of Political Science Texts

The following analysis is based on examination of all articles in the *American Political Science Review (APSR)* of 1979 (volume 73). Three articles, selected for their range of topics and styles, are analyzed in detail: Edward T. Jennings, Jr., "Competition, Constitutencies, and Welfare Policies in American States," 414–29; Diane L. Fowlkes, Jerry Perkins, and Sue Tolleson Rinehart, "Gender Roles and Party Roles," 772–80; and Benjamin I. Page and Calvin C. Jones, "Reciprocal Effects of Policy Preferences, Party Loyalties and the Vote," 1071–89.

The most obvious characteristic of the papers in *APSR* is their length. The mathematically developed articles in *APSR* in 1979 run from seven thousand to fifteen thousand words in length, with a mean of about twelve thousand words. The articles each occupy from nine to twenty pages of closely packed, double-column pages. In comparison, Watson and Crick's famous paper (examined in the second chapter of this book) is under one thousand words; most of Compton's papers on x-radiation (discussed in chapter 7) are between one and two thousand words.

The only groups of papers averaging a comparable length that I found in the course of my researches were late eighteenth-century articles in the *Philosophical Transactions* (see chapter 3) and recent articles in physics (see chapter 6). The late eighteenth-century articles gained length through the long series of experiments (as many as ninety-five) reported in a single article. The recent articles in *Physical Review* have reached an average length of about ten thousand words through the embedding of arguments within complex theoretical contexts. However, neither of these reasons accounts for the length of the *APSR* articles. As the following analysis suggests, the reasons are rather to be found in the kinds of rhetorical work that must be accomplished within the political science article. The amount of that rhetorical work appears comparable to that required in the twelve nonmathematical essays appearing in volume 73 of *APSR*. Although developing arguments through political theory or historical accounts, and although adopting overtly different styles, the nonmathematical essays run about the same length as the mathematical ones.

Establishing the Literature

One of the kinds of work taking substantial space in the political science articles is discussing the prior literature. The typical bibliography of an *APSR* article in 1979 has from twenty to forty items, whether the article is mathematical, historical, or theoretical. The citation method obscures the actual number of textual references; the three articles examined closely each had from thirty to fifty mentions, discussions, or characterizations of other sources. In comparison, Watson and Crick had six footnotes, Compton typically referred to less than ten sources, and through 1960 articles in *Physical Review* averaged under a dozen references per article. Only in the most recent theoretically embedded articles in *Physical Review* has the average number of references grown to around twenty-five.

Unlike the references in recent physics articles, however, the references in *APSR* do not reflect embedding in a highly codified literature. Rather than infusing all parts of the argument, the references are concentrated in extensive opening reviews of the literature (in one case comprising half the article) and in the last few pages of conclusions. These reviews of literature, rather than discussing selected recent articles with direct bearing on the subject at hand, instead assemble and discuss all the literature in the problem area. Unlike articles in codified sciences where older texts have developed stabilized meanings and have been incorporated into the tacit assumptions of shared knowledge (Cozzens, "Taking"; Messeri) so that only recent articles tend to be explicitly mentioned and discussed (Price, *Little Science*), the political science articles reassemble, reinterpret, and discuss anew wide ranges of the literature, dating back into the discipline's history.

The article by Jennings begins with V. O. Key's seminal comment on welfare policies (1949) and then discusses every major test of Key's hypothesis (1959, 1963, 1969, 1970, 1976). The discussion then reinterprets Key's original comments. Jennings obviously cannot rely on the audience identifying and understanding the background literature in the same way he does; in his extensive discussion of the literature he establishes his vision of the prior work. Similarly, Page and Jones review thirty years of voter studies in nine pages and over seven thousand words; Fowlkes, Perkins, and Rinehart mention all work they consider important on women in party organizations and on differentiation of party membership—much of this work between ten and twenty-five years old. Whereas Price calculates that 72 percent of the references in recent volumes of *Physical Review* are to papers published in the preceding five years, a similar calculation for these three political science articles reveals that only 30 percent of the cited sources are from the past five years.

This extensive reinterpretation and reconstruction of the literature requires a broad-stroke treatment of a large number of sources. Works are frequently categorized as part of a group, with only representative articles discussed in detail. In their seemingly detailed discussion of prior voting studies, as an extreme example, Page and Jones actually discuss their own versions of typical arguments and then list sources which they claim take these approaches. Brief general characterizations, group characterizations, and simple lists of sources are common in all three articles.

Such patterns of generalization rely on the audience's faith in the author's judgment for their persuasiveness. Little compelling evidence can be given to justify interpretations and evaluations or eliminate alternative judgments. Selected detailed discussions of some sources do provide details of some interpretations, but even here justification for the readings is rarely provided. For example, Jennings summarizes two studies which he claims "can be interpreted to support [the preceding] analysis" (416). Each summary is about a hundred and fifty words long; however, Jennings never explicitly identifies the issues open to interpretation or the justification for his interpretation.

Four: *The Reinterpretation of Forms in the Social Sciences*

Establishing One's Contribution

The lack of codification of the literature offers the political scientist large opportunities for putting his or her current work in the most advantageous light. With prior work regularly open to reinterpretation and criticism, each new contribution can be represented as a radical new departure or a fundamental solution to ancient gordian knots. All three articles, in fact, claim all prior work misses the boat; the reviews of literature, consequently, are critiques of the fields in question.

Fowlkes, Perkins, and Rinehart suggest that gender roles in politics have been incorrectly conceptualized. Jennings argues that "the logic underlying standard formulations of the interparty competition (IPC) hypothesis" (415) is faulted and needs reformulation. Most totally rejecting the literature, Page and Jones suggest "that virtually all past voting studies have erred by ignoring the possibility of reciprocal causal effects among the central variables of the electoral process" (1071).

The problem of each paper is simply to rectify the earlier mistakes. Without strong codification of the literature, more precise forms of contribution (such as the solution of recognized problems, the reconciliation of anomalies, proposing a new account of previously identified phenomena, or extending previous work to new domains) are difficult to identify. Moreover, the consequences of the current contribution for related work are also difficult to pinpoint. Within the loosely connected, personally interpreted and evaluated political science literature, any particular new finding, though interesting or striking, may not suggest immediate follow-up work.

Emphasizing the methodological innovations of a study is a way of increasing its consequentiality and importance. A new way of seeing creates a clear imperative for future studies, even though facts and hypotheses may not reverberate strongly with the work of others. Two of the three articles analyzed emphasize methodological innovations. Page and Jones offer the most pronounced case. They open with one-page review of the literature, followed by an eight-page methodological critique of the literature and a three-page description of the authors' methodological innovations. Less than five pages are devoted to the actual presentation of data and discussion of findings. Of the ten paragraphs of conclusion, nine are devoted to methodological issues and only one to empirical discoveries.

This is a distinctly different function for method discussions than we have seen elsewhere. In the early *Philosophical Transactions* the growth of methods sections served to identify the conditions of the experiment, establish verisimilitude, and argue for the results. In experimental psychology, we saw the importance of methods sections in protecting the acceptability of results. Compton described methodological innovations (such as the cloud chamber) as a means of obtaining new data. In most cases methodological innovations were not seen as invalidating previous results unless they revealed a serious flaw in prior work, as when Duane challenged Compton over the geometry of the box surrounding the target. More usually the results of prior methods are preserved as valid, although perhaps limited or crude.

Technical Studies and Real World Meanings

In political science, uncertainties over the consequences of findings and methodological propriety lead to an uncertainty over the reality and meaning of results. The specialized technical study seems not able to stand purely on its own terms, as technical discussions alternate with ordinary language accounts of historical cases, hypothetical situations, or traditional political theory. The studies seem to be hanging under the question, "What does this all have to do with the real world?" Even though the data of molecular biology or spectroscopy are much further removed from everyday experience than voting statistics or per capita welfare expenditures, authors in the natural sciences do not seem to need to defend the reality of their data beyond presenting acceptable technical methods for the data production. Jennings, however, begins with a commonsense paraphrase and quotation from Key and in the later statistical passages keeps converting the statistics into historical descriptions. Page and Jones rely on commonsense observations about recent presidential elections to reinforce and interpret their data manipulations, and they let a series of plausible hypothetical statements carry an argument. Finally, the gender role article steps back from a specialized statistical approach to offer general speculations in ordinary language.

The insecurity about the force of a purely technical argument is related to difficulties in identifying just what is being indicated by statistical indicators and what real world behaviors are identified in the nomenclature. Although a vote is an isolatable measurable action, vote decision making is, by its nature, invisible to the outside observer; at the same time we have a wealth of anecdotal, testimonial, historical, and introspective data about the phenomenon. Any model we put forward is a speculation about an internal process that begs for comparison with our experience, knowledge, and intuition on the subject. No matter how

Four: The Reinterpretation of Forms in the Social Sciences

detailed and concrete the data manipulations one can perform, one may suspect that the model does not really reflect the way things are. Even more of an indicator problem arises when you try to connect survey responses to actual behavior. The practical meaning of the terminology in the gender roles article is particularly befuddling. Suggestive psychological terms, although vague and not based on widely accepted theory, are made the basis of rather concrete distinctions. Thus, although the conclusions are intelligible in general terms, some of the specifics about how the conclusions refer to actual political behavior are elusive. Page and Jones directly address a similar issue when they try to identify the factors and relationships in their model of voter decision making:

> We cannot estimate any of the coefficients in Figure 6, as it stands, because the model is hopelessly underidentified. That is, there are only three empirically observable relationships among the central endogenous variables available to estimate the six causal processes of theoretical interest. (1079)

The observable behaviors are not rich enough to tell them about the internal processes they are interested in. Page and Jones then try to define the internal machinery, but they run into further obstacles:

> It is in the search for suitable exogenous variables that difficulties mount, for most of the *pertinent social theory is either not very powerful or not universally accepted.* The grounds for specifying that a given variable theoretically cannot affect or be affected by another are seldom overwhelming. The situation is *worse than usual when one deals with psychological measurements or attitudinal variables,* since practically any attitude might conceivably affect any other. There are times when we seem to be studying *relationships between mush and slush.* (1080, emphases added)

The authors escape from their dilemma only by an eclectic synthesis of plausible factors suggested by the literature, history, and common sense. But no grounding theory or unifying approach make the factors and relationships anything more than assertions. To their credit, Page and Jones recognize their conjecture.

The Authorial Vision

This last case exemplifies the exposure of the author's intellectual processes, typifying an authorial role for political scientists that both resembles and differs from the authorial role of natural scientists. In political science papers, as in natural science papers, the first person frequently is used to express the author's active role in constructing ideas and collecting data as well as to claim credit for the research process and results. For example, Page and Jones use such phrases as, "we intend to specify and estimate," "we first consider," "we conceptualize," "we prefer to analyze," "we measured reactions to candidates' personalities by counting the net number," "in short we are suggesting," "we can with some confidence specify," "to us the most striking aspect," and "perhaps the theoretically most important of all our estimates." The authors of the other two articles also represent themselves as the doers, interpreters, and owners of the research.

Yet in the natural science articles, the results tend to rise above all the separate doings of the authors. As Latour and Woolgar note, the claim seeks to rise above the condition of its begetting. The claims of political science may have the same ambitions of disembodied knowledge, but because of all the problematic conditions discussed earlier, the claims cannot easily rise above the author's perception of the literature, definition of problem, choice of methodology, naming and division of the phenomena investigated, and development of the argument. The author as conceiver, doer, and owner of the claim cannot so easily shift responsibility to nature for the truth of the claim. The authorial stance can be no more than "I have an interesting and revealing way of looking at political behavior and institutions. Look at them my way." Some readers come to share the vision and others do not.

In this way the discourse of political scientists still bears some resemblance to the discourse of political philosophers, who also ask the readers to see it their way, although the philosopher's vision is less constrained by empirical methodology. Rhetorically, political science is somewhere in the middle—whether that middle is part of a historical development or of a permanent dilemma I leave to epistemologists and future historians of knowledge. In the meantime political science needs the resources of both forms of discourse.

The gender roles article suffers from sidestepping its need for traditional discourse and using the stereotype of the scientific paper as a persuasive resource; we can see the rhetorical ambitions from the section divisions—untitled introduction, "Methodology," "Data Analysis," "Findings," "Discussion."

The welfare policies article shows greater concern for the problem of translating the terms of ordinary political discourse into mathematically more solid form, as evidenced by the broader conceptual discussion preceding formulation of hypotheses and by interplay of historical descriptions and statistical indicators. Again the section headings reveal

Four: The Reinterpretation of Forms in the Social Sciences

the stance; although the underlying structure of the paper follows the typical pattern of introductory review of the literature, hypothesis, methodology, data analysis, discussion, and conclusions, the division titles are more discursive: untitled introduction, "Party Competition and Welfare Policy," "State Welfare Policy and the Lower-Class Electorate," "Changes in Politics and Policy in Eight States," "System Differences and Policy Differences," "Electoral Support and Change in Policy," "Further Considerations," and "Conclusions." Thus Jennings preserves the appearance of a commonsense political discussion even as he moves the argument into mathematical terms.

Finally, the study of voter decision making treats the scientific mathematical discourse it relies on as problematic. The underlying structure remains the typical one, but the review of literature is expanded into an extended theoretical methodological discussion. The division titles are then drawn from the methodological problem: untitled introduction, "One-Way Causation: Recursive Models of Voting," "Two-Way Causation: Non-recursive Models of the Vote," and "Conclusions."

Each of the three political science articles discussed employs a strategy to maintain a stable rhetorical base on which to frame statements about real world political behavior and institutions. The articles share some points of strategy, but the overall stances toward the discourse differ. This rhetorical variety suggests that political science has yet to forge a consistent rhetoric. Whether such a consistent rhetoric that addresses all the relevant dynamics of political studies is possible or advisable will only be decided by the collective wisdom of the discipline over time. At the moment, the one certainty is that mandating a rhetoric, borrowed (and reduced) from the practices of a different community does not make the real rhetorical complexity of a community vanish. The ambitions expressed in the transplanted rhetoric only add to the complexity of the rhetorical task. Writing a science of politics may be a worthwhile task, but it is no easy task.

PART FIVE

SCIENTIFIC WRITING AS

A SOCIAL PRACTICE

11 HOW LANGUAGE REALIZES THE WORK OF SCIENCE

SCIENCE AS A NATURALLY SITUATED, SOCIAL SEMIOTIC SYSTEM

> *There, in front of us, where a broken row of houses stood between us and the harbor, and where the eye encountered all sorts of strategems, such as pale-blue and pink underwear cakewalking on a clothesline, or a lady's bicycle and a striped cat oddly sharing a rudimentary balcony of cast iron, it was most satisfying to make out among the jumbled angles of roofs and walls, a splendid ship's funnel, showing from behind the clothesline as something in a scrambled picture—Find What The Sailor Has Hidden—that the finder cannot unsee once it has been seen.*
> Vladimir Nabokov, *Speak Memory*

The chapters of this book have projected a few short moving pictures of language being used in science. Like all texts, these chapters have been constructed with as much intention and art directed toward the anticipated readers as the struggling writer can muster. The intention has been to share parts of a pattern, an understanding, which I have increasingly seen through contact with materials examined in the course of research. This pattern, although incorporating many patterns pointed out by previous authors, seems somewhat different in form and total mass than that perceived by others considering related problems. Why I think the pattern I have seen is important will, I hope, emerge in this and the next chapter, but first the entire pattern must be exhibited, by juxtaposing it with some other patterns, familiar and less familiar.

Put most baldly, the pattern I see addresses the problem of how language accomplishes the work of science. Such a discussion could be simplified if we could independently define the work of science; however, for reasons I hope to make clear, we cannot separate our view of the work of science from our view of the praxis by which the work is realized. Thus, we can best get an understanding of the various views of science and language by seeing them as unitary relations.

The Difficulty

From an everyday point of view, how language accomplishes the work of science is hardly a problem at all—or only a problem in the most practical sense of the word. From this perspective, language represents the objects of nature and their relations. As we discover new things we invent new words and we put those words in relation to represent the relations of the real world. Science tells us about nature; words and numbers are the symbols it uses to tell us. By representing nature symbolically, we can understand, predict, and manipulate it. The symbols give us a picture of the way things are. The only problem is the most practical one of making the symbols precise, unambiguous, univocal, to create a clear one-to-one correspondence between object and symbol. The prescriptions of technical writing manuals largely reflect this everyday perspective (see, for examples, Day; Fear; Houp and Pearsall; and Mills and Walter).

From a commonsense point of view we have many reasons to credit such an account. The formulations of science—rules, laws, descriptions, knowledge—have provided us with detailed accounts of many natural events, accounts that seem tightly congruent with repeated experience and precisely predictive for future experience. Moreover, these formulations have given us unimagined dominion over the objects and creatures that surround us. These formulations allow us to conjure great forces, quicken those at death's door, and create new forms of life. Our trust in the congruence of these formulations with the ambient world goes beyond appreciation and spectacular display. We regularly trust our lives on airplanes and feel ourselves distinctly disadvantaged when our television or computer breaks down.

When we look at scientists themselves, we see so many of them working so intently to create new formulations and to create evidence for the correspondence between their claims and the phenomena they are exploring, that it is difficult not to share their conviction that they are describing something. Indeed, hard-headed corporations and realpolitik governments have invested heavily in science's ability to create bottom-line economic power.

When we look to the formulations created by science as reflected in symposia and published articles, we certainly see a very specialized development of language, distinct from our everyday conversation and newspaper reading. Unfamiliar words signify objects and phenomena from the microscopic and macroscopic limits of the universe, objects distinguished from each other and classified with a precision and taxonomic care having little to do with our everyday fuzzy naming of the objects of domestic life. Moreover, this specialized language of science seems constantly filled with evidence, numbers, observations, pictures, to ensure that the formulations correspond to real things. Fat scientific dictionaries, histories of the rise of scientific vocabulary, detailed handbooks of scientific writing, and the teaching of technical writing and scientific German as special subjects all reinforce our notion that scientific language is something special and privileged. Even such varied and opposed reductionists as Garfinkel (*Studies in Ethnomethodology,* chapter 8) and Skinner (*Verbal Behavior,* chapter 18) afford scientific language a special status separate from the turbulent, murky, and illusion-ridden language of the rest of the human world.

Yet from the perspective of our murky, deluded human world, we have always had good reasons to doubt such simple accounts. The Sophists early saw the fluidity and uncertainty of symbolic representations and thus the questionableness of whatever formulations we see as knowledge. Plato shared this perception despite his being cast as the Sophists' first and most formidable enemy in the saga of philosophic history. The cave allegory in the *Republic* is a critique of the shadowy representations by which we know the world; Plato only adds the difficult possibility of escape from the cave (514a–517c). This is the same problem Bacon grappled with in considering the idols that obscure our language *(The Advancement of Learning).* Although some Baconians—notably Sprat and Wilkins—may have believed in the possibility of a pure philosophic language totally expurgated of the idols, Bacon himself seemed to see the cleansing process as always a partial and incomplete process, so that we would always be burdened by the constraints of language. Nor could the naive linguistic realists identified as Baconians have held unquestioned sway after Swift's damning parody of the Royal Society in the third book of *Gulliver's Travels*. In the eighteenth and nineteenth century eminent scientists and philosophers of science repeatedly warned of the uncertainty of language and symbolic representations (Bellone).

Reasons for distrusting the direct correspondence between scientific formulations and nature have been in recent years rearticulated with great force, and with persuasive empirical evidence. The faint irony of empirical evidence being used to undermine naive empiricism has not escaped the attention of a number of authors making the argument for the opacity of scientific discourse, and they have dealt with this awareness variously, with some considering it a great paradox and difficulty (see, for example, Woolgar, "Irony"; Mulkay, "The Scientist Talks Back"; and Oehler and Mullins, "Mechanisms of Reflexivity"). Yet, from the perspective to be sketched below, recognition of the opacity of language does not necessitate disowning empirical constraints on what we say.

The reasons to distrust scientific language are of several kinds: ·

1. All languages are semiotic systems, incorporating basic assumptions about the nature of reality (for example, Bloor). These assumptions color not only representations made within the language, but sensory perception about the ambient world (see, for example, Hanson). From this perspective it would seem that the work of science is to maintain and elaborate the existing semiotic system

2. Scientific formulations embody ideological components from outside the realm of science. From this point of view the work of science is to advance or provide foundation, legitimacy for larger social programs which themselves may simply be the result of class interests (see, for examples, the various essays in Barnes and Shapin, *Natural Order*).

3. Scientific language serves to establish and maintain the authority of science, largely through exclusion and intimidation. By establishing the special and elevated character of science, scientific communications accrete power to the scientific community (see Knorr and Knorr, "From Scenes to Scripts"; Gieryn, "Boundary Work"). Here the work of science is to advance itself.

4. Within the scientific community, scientific language serves the competitive interests of separate individuals and research groups. The language is partisan, argumentative, and manipulated for individual gain rather than an objective, dispassionate representation of things as they are (see Latour and Woolgar; Yearley; Pickering). Under this rubric the work of science is to advance the careers of individuals.

5. Scientific language is often fuzzy, incomplete, undefinitive. In particular the reference to actual events is obscured if not made fully obscure by the inadequacy of methodological description, the importance of inarticulate craft knowledge to produce results, the lack of precise replication of results, and the selectivity and emphases in the representation of results (see Knorr, "Tinkering"; Collins, *Changing Order*). This fuzziness leaves room for many kinds of social activity, with the apparent work of scientific discovery being only a screen.

6. In sum, scientific formulations are a human construction and thus are heir to all the limitations of humanity. Scientific formulations, giving

us no direct access to things in themselves, seem to do all the social work of being human with no overt means of doing the empirical work which has been considered the work of science. The appearance of reality projected in scientific texts is itself a social construction.

I have cast the modern formulations of the problem of language in the most radical form, and there are many who present less extreme positions. Some, claiming interest only in the social processes, simply postpone considering the empiricist issues. Others see the social processes somehow embedding empiricist procedures. Kuhn, for example, despite the rather radical uses he has been put to, insists he is a rationalist and empiricist. Yet, he has been unable to make that case forcefully enough to harness the widespread radical interpretation of his work. Currently, the radical positions put the issues most powerfully.

The Conceptual Source of the Difficulty

Our current inability to forge a convincing link between the socially constructivist critique of scientific formulations and the empiricist project has roots in how we have become accustomed to think about language in this century. When socially minded observers of scientific activity come to think about the role of language, our current concepts of language offer no strong clues about how language talks about anything other than itself. The main lines of twentieth-century linguistic inquiry have turned away from issues of how language interacts with the world of experience, although in recent years some linguists have shown increasing interest in how language constitutes the social world. Thus on the question of the nature of linguistic representation of the experienced world, linguists have only to offer some version of correspondence theory (that words do in some fairly direct way correspond to the objects of nature) or of social relativism (that every society creates its own reality through its symbol system.)

More specifically, what has been lacking is a unitary concept of signifying events simultaneously contexted within and realizing linguistic code, social relations, psychological cognition, and perception of the ambient world. Only in the recent attempt to elaborate the work of the Russian psychologist Lev Vygotsky and his followers has a strong enough model of language activity developed to encompass all these elements, and to enable us to see how in making statements we bring together many elements—cultural, social, psychological, and material—to accomplish our activities and create cognition, a cognition that can be

Five: Scientific Writing as a Social Practice

empirically conditioned. That empirical conditioning of cognition is highlighted by Ludwik Fleck's vision of scientific activity. The prescient work of both Vygotsky and Fleck was buried by the politics of the 1930s, but their recently rediscovered ideas point the way toward the understanding of scientific knowledge as a socially and individually constructed, semiotic, cognitive, empirical activity—a practical part of our being human in the world.

In his *Course in General Linguistics* Saussure, rightly considered the founder of modern linguistics, admits the complex reality of language, but finds this complexity far too much to contemplate with any clarity within any discipline (24–25). For the sake of analysis and the sake of establishing linguistics as an autonomous discipline, he separates *langue*, the linguistic code, from *parole*, the use of language in particular circumstance for particular purposes. He considers only the former, linguistic structure, as the proper study of linguistics. In so doing, he separates code from meaning, even though he recognizes that the sign is not an independent linguistic entity, but is a dialectical unity of signifier and signified (99–100). That is, sign systems not only embody meanings, they are embodied out of meanings. Words and meanings dialectically define each other. The immediate implication is that one cannot understand language without looking at the contexts in which it is used to convey meanings. Yet by distinguishing *langue* from *parole*, and limiting linguistic science to *langue*, Saussure has effectively ruled the fundamental questions of language out of bounds.

Three other Saussurean gestures heighten this context-free code orientation. First, to isolate the study of code from the study of the historical evolution of particular features (as characterized nineteenth-century philology), Saussure distinguished synchronic from diachronic study. Systematic linguistics would consider language only synchronically (40–43; 114–40). By ruling history out of bounds to systematic study, Saussure not only eliminates large-scale evolutionary studies, but also the examination of the brief historical moments in which code interacts with context to realize meaning and during which code evolves to meet communication needs. This antihistorical gesture effectively keeps the code orientation clean, at some distance from challenging data.

Second, in discussing the form of the sign, Saussure calls the sign arbitrary (100–102). The argument and examples that follow the designation of arbitrariness suggest only that the phonetic realization of the sign—the sounds—are abitrary. Roosters go *cock-a-doodle-doo* in English and *kiekeriki* in German. Nonetheless, the slogan that the sign is arbitrary has been taken as justification for the divorce between code and meaning (or use in context). From the text it is unclear how much Saussure himself was willing to use this more general claim to buttress his strategy of excluding *parole*, but certainly the claim of abitrariness has eased the conscience and consciousness of many linguists to follow.

Finally, through an imaginative gesture, Saussure brings into creation an as yet unestablished but broader field of semiology, the study of sign systems (33-35). Semiotics was thus grounded in the model of the study of linguistic code separated from context of use and meaning, even though Saussure proposes that semiotics would study "the role of signs as part of social life" (33). This founding heritage has directed semiotics to consider sign systems as having autonomous structure and power.

Saussure's judgments about how best to make progress in the study of language have turned out to be quite shrewd. In looking closely at synchronic codes descriptively and structurally, linguistics has made great conceptual and concrete empirical advances, particularly at phonetic, grammatical, and syntactic levels. And this orientation was reinforced by such different kinds of linguists as Hjelmslev and Chomsky, who saw in the synchronic system not just an analytical fiction (an artifical cut to allow some clarity), but hope of a more substantial explanation of realities beyond the code. In explaining the rules that govern the code we might find the rules that govern meaning (in Hjelmslev's *Prolegomena to a Theory of Language*) or the rules that govern the mind (in Chomsky's *Language and Mind*). That is, code separated out and elaborated as an autonomous object has come to be seen as dominant. This tendency has also generally been followed in semiotics, where sign systems are seen to be determinant of consciousness, perception, and social behavior, rather than interactant with them.

This is not to say that there haven't been contrary observations, hybrid ideas, and minority traditions, but these have until recently tended to remain either vague or underdeveloped. Malinowski, Whorf, Sapir, and Firth got little beyond programmatic statements and/or preliminary investigations into the social embeddedness of language. Their undeveloped work was too easily reinterpreted in code-oriented ways, as forms of code determination of social/psychological realities. After all, the synchronic code seemed to have an elaborated, solid structure—something a linguist could analyze—while social and psychological phenomena seemed inchoate, and therefore open to be shaped by the structured linguistic or semiotic codes.

Thus from language and sign studies we tend to get either of two attitudes toward reference and meaning. First, within the majority code-oriented tradition, because the study of language structure is cut off from problems of meaning and use and thus the relationshp not looked into, it is simply assumed that there is some sort of not very interesting correspondence between words and meanings. Or alternatively, from a code-oriented reading of the minority tradition, since meaning and use seem to have no grounding equivalent to that found in synchronic code, they are free to be pushed around by the code—leading to a simple relativist position.

Linguistic studies of scientific language (or scientific register or scientific sublanguages) have come rather directly from the code orientation. They have been looking largely for the subset of syntactic and grammatical features used in scientific communications, considered fairly independently of use, context, or meaning (for example, Gopnik; Lee; Huddleston; Kittredge and Lehrberger). Relationships to meaning, use, and context are just not problems, and the implicit acceptance of some sort of correspondence theory of meaning need not even be raised.

For obvious reasons, these studies have been of little interest to the social relativist critics of scientific discourse, who have been concerned precisely with the social, ideological use of scientific language, but apparently reflected in syntactic, grammatical code. They have, however, found some greater affinity with literary philosophic work developing out of semiotics and transmogrifying into deconstruction—revealing the text only as a linguistic structure, a contrivance, having no inherent meaning, but creating sociopsychological realities out of its semiotic code. Both Knorr and Latour have shown particular interest in semiotics and deconstruction.

On the Way to a Solution

Recent developments in linguistics and related social sciences, however, have loosened the strict code orientation, thereby undermining linguistics as an autonomous discipline, having a separable matter for study. Sociolinguistics at first addressed the code descriptive task of identifying variation in the code and/or alternative codes among different groups distributed geographically and/or by class, but the variation found was so extensive as to call into question the notice of a stable/coherent code. Codes just ran into each other with no distinct boundaries (Hudson, *Sociolinguistics*, provides a critical review). Even more distressing, individuals seemed to speak no one code but have a repertoire of codes, with their choice of codes to use at any moment itself being a meaning-creating act (see, for example, Gumperz, *Discourse Strategies*; and LePage and Tabouret-Keller, *Acts of Identity*).

Similarly, the recently developed linguistic specialty of pragmatics has been fraying the edges of a firm code. Pragmatics is the study of how people use language in real life to do things, a topic seemingly beyond the edge of Saussurean linguistics. The topic first had to be domesticated enough to be brought into linguistics. This was done by Austin, who located Wittgenstein's concept of language in use (not far from Malinowski's observations on the social use of language among the Trobrianders) within certain sharply definable speech acts, which Searle further reduced to a series of rule-governed procedures. Thus framed, the concept of doing things with words seemed a code-consistent issue, opening up the new domain of communicative competence to parallel other code-based competences (see Searle, Kiefer, and Bierwisch; Leech). But this open-ended issue would not remain domesticated for long, as the observations of what people did with words started extending beyond crisp examples such as christening a ship and making a bet. Moreover, the action taken was not always crisply related to the linguistic forms used to realize the action. Social activity in language was seen to be a complex and creative force, not easily reduced to rule-governed behaviors.

Searle himself planted a major surprise when he argued that making reference itself was a speech act *(Speech Acts)*. This problematization of reference impelled the study of deixis—that is, how one attaches one's talk to the surrounding world. At first deixis seemed a fairly containable subject, dealing with simple words like "this" and "that," but deixis too has been discovered to infuse all aspects of the language in complex ways (see Lyons, chapter 15). Thus the code again seemed unintelligible and uninterpretable and even unsystematic when separated from its contextualized use.

Increased attention to detailed developmental data, in part motivated by Chomsky's strong claims about the psychological implications of code structure, has as well revealed that language develops as part of the child's increasingly complex interaction with the world and people. Cognition, experience, and social interaction are all significant variables in language development, which can no longer be seen as an autonomous linguistic phenomenon.

And, finally, the great success of code-oriented linguistics in phonology and syntax has encouraged consideration of larger orders of organization, in the specialty at first called text grammars and then discourse studies. The change of nomenclature itself indicates how little the phenomena could be contained within a formal code-based model. Questions of textual interaction with cognition (schema, story grammars), social interaction (ethnomethodological approaches), and social

Five: Scientific Writing as a Social Practice

history (genre approaches) currently seem more promising in understanding textual organization. Even formal models seemed to require awareness of how texts were situated within task, social relationships, and communication channel (field, tenor, mode) to begin to account for variation in discourse patterns attributed to various subcodes or registers.

This exciting rediscovery of language's intimate dialectic with the lives of people in the world has hardly settled into any clear picture, appearing to reconfirm Saussure's warning that linguistics must isolate itself from these variables to gain any rigor. Thus while we are now starting to get much more detailed and vibrant pictures of separate linguistic phenomena, linguistics has not developed any sharply articulated model of language activity that could guide social studies of science.

The boldest and most influential attempt by a linguist to form an overall view of language activity has been by Michael Halliday, who argues that linguistic features are only surface realizations of larger social activities *(Language as Social Semiotic)*. In his study of child development of language, for example, he sees the developing language system of the child as part of the child's growing system of social interaction *(Learning to Mean)*. Only once that social system is formed is the child ready to adopt the socially given model of adult language. Accordingly, he interprets features of the code as realizations of communicative impulses and social interactions *(Functional Grammer)*. And he argues that any communicative impulse may be realized in a variety of apparently different surface forms, which we cannot properly understand unless we see the connection to the underlying impulse. For example, in some contexts the question "How are you doing?" may be more closely related to the command "Have a good day!" than to the more superficially similar question, "How are you traveling?" ("Language as Code and Language as Behavior").

Despite Halliday's boldness in reestablishing meaning-making without context as prior to code, his formulations (which he considers within the Malinowski-Firth tradition) fall short of solving the puzzle presented by scientific use of language as currently perceived in social studies of science. First, while recognizing the evanescence of linguistic code, Halliday seems to have a much greater confidence in the firmness of social stucture and culture as a priori frameworks from which to derive language behavior. When he talks of the influence of social structure and culture on language he has presented a synchronic vision of a well-ordered system, as though society—rather than logic and brain structure, as Chomsky might claim—offered a deep structure one could rely on (for example, *Lanquage as Social Semiotic*, chap. 10). In this way he

How Language Realizes the Work of Science

not only elevates and reifies society as a primary principle more than current sociological thinking might support, he seems to be running contrary to sociological interest in how society is constituted through language.

No doubt, regularities and structured elements appear in both language and society, but I know no reason to believe that either is prior to or privileged over the other. Until we have positive reasons for believing otherwise, we must assume society exists no more firmly, nor no less firmly than language (and other symbolic and physical means of coordinating activity). They are simultaneously realized in the social language act. Linguistic and sociological regularities—realized and institutionally structured in successive acts—might be best accounted for as parts of mutual realization.

The second area insufficiently addressed to this date from the Hallidayan perspective is the influence of the material surroundings on the sociolinguistic interactions and activities—that is, in what way, if any, language can talk about the world or influence doings in the world. This issue has just not been raised within Hallidayan linguistics, as far as I know, although there is no reason why it should not. Until that is addressed we are left with a vision of language activity floating somewhere above the world, as in a middle-class living room, with attention only on social coexistence. The mutual construction of reality seems only a matter of free choice and social imagination, with all the work of the world handled by machines behind the woodwork.

My intent here is not to privilege practical boiler-room language over the elaborate imaginative constructions of the drawing room, but rather to avoid a separation of the two. Certainly consideration of how language is used in science brings questions of the connection between elaborate human intellectual constructions and material activity to the fore, if for no other reason than science has allowed us such unimagined mastery over nature. Yet the issue is not limited to discourse areas which take the natural world as their overt topic, as science does. Much is to be gained by seeing all forms of language as practical activity in the material world, no matter how complex and apparently removed from the production of goods and services. Even play—both child's and adult's—is an important part of our material existence, as psychologists, sociologists, ethologists, historians, and critics of the arts have often reminded us. Any attempt to understand language that does not pay sufficient attention to how language works as a social tool in the material world invites the extremes of materialist and antimaterialist reductionism that see potatoes as more real than books or books more real than potatoes. Whether one sees human constructions as arbitrary and immaterial because they are just epiphenomenal by-products of less culturally conditioned material objects or arbitrary because society seems to ride above the material, one loses sight of the way human constructions provide our means of living in the world.

A Vygotskian Model of Practical Social Semiosis

A more crisply defined, and I believe ultimately more powerful, model of the role of language in human activity, society, and consciousness can be developed out of the work of the Russian psychologist Lev Vygotsky. This line of work, at first carried on in the Soviet Union, has in recent years also been carried on in the West. The following account of the practical use of language in science borrows deeply and freely from the work of Vygotsky and his followers. However, in applying these ideas (which have been largely elaborated through study of the development of higher cognitive functioning in children) to the problems of science's advanced system of literacy, I have transformed some of them, perhaps beyond recognition. But the influence of these ideas upon me has been so deep, I am no longer capable of offering a full archaeology of the sources of the model I am about to propose. In the following discussion I will identify and describe some discrete Vygotskian concepts, but in general I will not attempt to disentangle my own elaborations and transformations from ideas previously proposed in the Vygotskian literature, nor will I attempt to give a coherent account of Vygotsky's theories. For a less idiosyncratic exposition of the ideas, you may refer to Vygotsky's two books translated into English, *Thought and Language* and *Mind in Society*; Kozulin's history of Soviet psychology, *Psychology in Utopia*; Wertsch's commentary, *Vygotsky and the Social Formation of Mind*; or Wertsch's two edited volumes offering work in the Vygotskian tradition, *The Concept of Activity in Soviet Psychology* and *Culture, Communication, and Cognition: Vygotskian Perspectives*.

The following model of scientific use of language will suggest how the work of science can be accomplished through the unfolding social and empirical activity of individuals coordinated (cognitively and behaviorally) within groups. To start, language is a tool that helps us carry on cooperative activities (a frequent theme in Vygotsky's writing; see for example, *Mind in Society*, 19–30). But in order for cooperation to be successful, we must already share much, not just the meaning of words and the syntactical operations but how those generalized words apply in this situation and how they are to be realized in action. (Wertsch offers a preliminary discussion of presupposition and intersubjectivity in *Vygotsky and the Social Formation of Mind*, chap. 5.) Written directions on observations to be shared through a microscope require congruence between the direction writer's and the direction follower's apparatus, defined by common terms and perhaps aided by standardization in design. But also it requires congruent craft skill in manipulating machinery, dies, slide preparation—a craft knowledge that can only be to some extent spelled out in print. A joint language and organization of the visual field is necessary for one observer to be able to see what the other sees, to identify designated patterns and salient features. Much shared background knowledge and shared experience are necessary to create the shared perceptual schema. And finally the shared observation is aided by standard observational routines that organize the activity.

In the literary economy of scientific articles, much of this shared background is relied on—not just the shared technical words, but shared conceptual, practical, and social worlds. In books for neophytes more of these shared elements are made explicit, but still much that is tedious, difficult or perhaps impossible to reduce to shared print symbols is left unsaid. Similarly, in the realm of research, which by its nature lies just beyond the edge of the familiar and communally certain, the symbolic reduction of the world and action conveys less firm and stable meanings, for just those elements necessary for shared understanding have yet to be established.

Another kind of shared knowledge required is of the social interaction being engaged in through the language. Often, for example, students socialized into the authoritarian relationships of textbooks (which dictate the student's experience, perception, and general claims) have difficulties entering into the more active engagement offered by educational materials emphasizing student observation and the development of individual perceptual schema. Perhaps even more to the point for scientific research, research communication requires practical social understanding of cooperative endeavor, aggressive assertion, and agonistic competition. As in any competitive activity, one must grasp the limits of violence and cheating and understand the forces that would bring the game to the edge of disintegration or transformation to a different kind of activity. Only under certain conditions and certain mutual understandings can the mutual activity flourish, just as ice hockey can flourish as ice hockey under certain conditions and understandings; when other conditions and understandings reign, the game transforms into a public display of team street fighting.

Given personal investments of all kinds that scientists have in their published claims, the maintenance of a cooperative, honest, problem-solving endeavor may often be threatened. Appeals to the rules of the game are almost necessarily self-serving resources (Would you complain to an umpire unless you had some interest at stake?), but mastering and developing allegiance to the interactional rules are an important part of socialization into scientific activity. Different individuals have different understandings of the rules of the game and make different adjustments to them. Different subcommunities vary or elaborate the interactive practices differently, with perhaps greater passion, cynicism, or avoidance of severe struggle. But whatever the interactive pattern is, the scientist must come to understand it. What fascination working scientists have for the sociology of science may come from the need to come to terms with this aspect of the communication system.

The Material Bases of Shared Understandings

Since communication depends on shared knowledges of so many kinds, we need to identify the source of shared understanding to establish the grounds of the communication and to identify the social range and cognitive degree of the sharing in any interaction. That is, who shares and with what degree of congruence? Which individuals are brought into a social understanding and how fully is intersubjectivity established? We need to unpack the mechanisms by which shared understanding is achieved locally, and then by which local sharings spread and maintain stability over larger collectivities.[1]

The achievement of shared understanding can be examined in two different kinds of situations, both of relevance for scientific communication. First is of the neophyte becoming familiar with knowledge already shared within a community. Through interactions, such as with the mother-child dyad now so energetically being studied in developmental psycholinguistics (for example, Bruner), the neophyte's utterances are interpreted and recast so as to fit within the interactive patterns and linguistic formulations accepted within the adult community. A kind of negotiation goes on between the beginner (with some kind of expressive or interactional motive) and an accomplished speaker, until the beginner produces an utterance recognized as bearing meaning within the socially shared system. Often within such socialization situations

1. For another account of how shared understanding is achieved locally in the laboratory, see Lynch.

the neophyte's comments are interpreted through a broader and more charitable interpretation of the comunicative system than would be granted to a speaker recognized as fully socialized.

Significantly, these activities usually embody some aspect of the material world that provides a reference point, constrains the language negotiation, and often defines successful completion of the activity. With a child, the material considerations may involve food and dry clothing to be obtained or a jigsaw puzzle to be assembled or a series of sounds to be played with; with a student of science, the material considerations may be of a textbook experiment to be carried out or a functioning machine to be explained or a printed equation to be explicated. In both sorts of cases, objects, which stand independent of the conversation constructed around them, take an essential part in the activity of the conversation and allow the neophyte to associate the symbolic interaction with concrete operations on concrete objects. (For the importance of active engagement with the material world for Vygotskian theory, see *The Concept of Activity in Soviet Psychology,* 37–71.)

The symbolic interaction shapes perception and meaning to be taken from these concrete objects by calling attention to particular features and placing them in symbolic relations to other features foregrounded as salient, as when an instructor identifies a piece of paper as litmus and tells the student to pay attention to color change when placed within various solutions. Such use of language establishes categories of significance; dialectically, the presence and character of objects make such categories of significance possible, constrain appropriate comments to be made, and provide meaning to the interaction. If there were no paper, or there were no chemical solutions to dip the paper into, or the colors did not appear to change in the predicted way, the interaction would go differently, have different meanings for the participants, and would provide a different kind of learning experience for the neophyte.

Language use in the communal enterprise of chemistry is taught and learned in textbook diagrams and charts to be memorized, in classroom discussion of the previous night's reading, in pencil problems to be solved, in the teacher's commentary on demonstration experiments, in getting particular bottles down from the shelf, in student groups with lab book on the table attempting to set up an experiment, in the teacher's comment on the experiment's write-up. Students learn not just names of chemicals, but when to use such names, how to label the results of experiments, how to determine whether their results fit the standard description, how to answer questions.

Even the well-known forms of laboratory fiction-making practiced by students—such as the fudge factor—require that the students understand the discrepancy between the symbolic representations constituted by the students' activity (that is, the recorded results of the student experiment) and the representations the students are expected to reconstitute based on the prior experience of the expert community, codified in the textbook experiment (that is, the "right" results). The clear intent of student fudging is to hide their apparent manipulative incompetence in reconstituting the symbolic object according to the shared procedures and perceptual schema of the disciplinary community. By fabricating expected results through calculations based on textbook theory, students hope to hide their inability to do the experiment "correctly."

As students move up the hierarchy of expertise in their scientific communities not only do their technical vocabularies expand, but so do their ranges of contact with the subject materials, their abilities to manipulate these materials in congruence with the formulations of their disciplines, their abilities to formulate symbolic expressions in less teacher-constrained situations (that is, taking their linguistic constraints from the materials rather than from sentences fed them in class), and the ranges of interactive processes they are expected to handle with peers and mentors.

In the course of these interactions students gradually expand functional competence in language activity through what Vygotsky calls the zone of proximal development (*Mind in Society*, 84–91). At any stage of development, an individual can accomplish certain things on his or her own, whether uttering babble syllables or boiling a liquid. But that same individual can accomplish a broader range of activities with the cooperation of a more skilled individual, such as associating certain of those babble sounds with meanings, or boiling the liquid within a distillation apparatus. The expert intervention provides a scaffolding into which the neophytes' behaviors can grow. By actual physical manipulations, giving instructions, asking questions, or responding appropriately, the skilled partner provides a framework of meaning into which neophytes' impulses, behaviors, and language can shape themselves.

As the neophyte gains control of the structured meaning/behavior system transmitted through the scaffolding, she starts to incorporate parts of the scaffolding in her own behavior. She starts to repeat the phrases the adult utters, starts to grab toward the picture the adult points to in association with an appropriate word, starts to repeat to herself the instructions provided by the instructor or the lab manual (e.g., "First you connect the rubber hose to the glass tube. Make sure that . . ."). An important moment in the child's development for Vygotsky is when the child starts to develop an internal language so that these self-instructions, regulating the child's behavior, go underground becoming invisible to observers and even eventually to the child.[2] In this way, gradually the neophyte becomes socialized into the semiotic-behaviorial-perceptual system of a community with language taking a major and multivalent role in the organization of that system, but with that system also shaped around concrete worldly activities. In terms of contemporary cognitive psychology, she will have developed the scripts, schema, and plans appropriate to participation in the community.

Thus the apprentice chemist learns to think and behave like a chemist, such that when she walks into a laboratory, she will perceive the surrounding material through the acquired framework of chemical formulations and will behave with respect to the material so as to reliably reconstitute phenomena accepted by chemists as reliably reconstitutable. She knows how to make recognized chemical phenomena appear to those who have the appropriate chemical perceptual framework. And finally she knows how to interact with chemists—to discuss the happenings in chemical laboratories in terms of significant chemical issues and so as to make an appropriate contribution to a communal endeavor.

But all this requires the cooperation of the material she is working with. If someone switches the bottles or the chemical nature of the universe changes without her awareness, she cannot make the anticipated phenomena reappear reliably, nor can she carry out the day's work with colleagues. Her language will break down into the common language of bafflement, where referrents no longer seem to refer, anticipations do not hold, and symbolic relationships do not wrap tightly around ambient conditions. The language withdraws from intimate interaction with the control of the processes—one literally does not know where one is. Under such conditions language moves to questions such as "Why isn't X happening?" and "What is going on here?"

If only deception is involved, standard chemical tests can reassert order by putting the right labels on the right bottles. But if the material of the universe changes, the chemist will have to begin chemistry from scratch, with all previous knowledge serving at best as an uncertain analogy. That is, the semiotic-cognitive-behaviorial system ties language use procedurally to specific manipulations of materials, and if those ties do not hold, our language use in concrete situations breaks down.

2. Vygotsky's concept of internal language (as elaborated in *Thought and Language*) is a conceptual precursor to Polyanyi's tacit knowledge.

Five: Scientific Writing as a Social Practice

Constituting New Reliably Reconstitutable Phenomena

The second kind of situation in which shared understanding needs to be established is when change, growth, or instability occurs within the system of understandings already shared by fully socialized members of the community. This kind of situation is particularly central to the activity of the scientific community. Unlike some other social systems that seek stability and ritual regularity in their communications (such as churches, island tribes, or lower echelons of bureacracies) and change only when forced by exigencies (such as climactic change, new populations to proselytize, or political revolution) scientific communities are by their nature committed to new formulations, new knowledge. If they have no new knowledge to create, they cannot be legitimately maintained. In that respect they are like legislatures; with no laws to be made, they would be adjourned or turned into shams.

Change in scientific formulations can come from many sources. Some sources can be from outside the scientific community such as political ideological movements (state Marxism has served as both a stimulus and a constraint within Soviet sciences), changes in other forms of communication (such as the rise of a periodical press), new means of communication (whether printing press or modem), or idiosyncratic individuals with complex personal histories that import foreign styles (as when physicists went into biochemistry, or Newton perceived physics as mathematics). Or the sources of change may come more directly from within the activity of a science—as when phenomena refuse to fit formulations or when a new idea developed for a narrow problem is seen to have much broader power, or when an individual, whose work is rejected, discovers new and compelling means to assert his position.

Whatever the source of the new impulses and new forms the acceptance of these new formulations and styles of formulations into the common stock (or disciplinary matrix, as Kuhn calls it) depends on the community. The community itself must see these formulations as more useful, productive, promising for its current set of problems as currently perceived and formulated. The new formulations must be perceived as realizing desirable lines for the group activity (that is, as part of a progressive research program, in Lakatos' terms). In this competition for intellectual survival (as Toulmin has elaborated in *Human Understanding*), formulations must be cast persuasively, and preferably compellingly.

But general and immediate capitulation is rare, for the new formulation represents at first only the realization of the experience and cognition of one individual or small working group within the larger research community. Other members of the community would likely have interests in seeing, thinking, and talking about the phenomenon (or related phenomena) in other ways. Resistance to new formulations exists for reasons beyond narrow-mindedness and bull-headedness. Persuasion, rather than being a single, sudden event, can be a lengthy process of negotiation, transformation, and growth of the central formulations and related arguments. Other researchers with their own perceptions, experiences, and research goals are enlisted not by checking off an approval rating in a Gallup survey, but by somehow taking the new formulation into account in their own work, if even only as a target of criticism. Formulations survive only by entering the living body of scientific activity, influencing behavior, cognition, social relations, future experience, and new formulations.

New formulations entering the common stock of formulations influence future activity and thus enter into a dialectical process with experience through the medium of working scientists. A successful incorporation negotiation ends up with a symbolic representation of an object or phenomenon that can be reliably reconstituted by members of the community under appropriate conditions in appropriate relation to activities and other reliably reconstitutable phenomena as perceived through the shared perceptual screen of the field. The two most immediate points of contact between active experience and formulation—the experiment or observations reported in the article and replication attempts—have been most criticized as having a loose correlation between events and formulation, but in the long run they may not be the most decisive in incorporation or rejection. They are only the most obvious first steps, and there is no reason to assume a stable reliably reconstitutable object will emerge from such first attempts at formulation.

The original report of an experiment or observation will not necessarily establish for all lookers the existence and character of a phenomenon, though the authors might wish so. Rather it will only indicate that these authors have been able to constitute an object for themselves with enough conviction that they will hold it up for public inspection. Since they are holding up for inspection a previously unconstituted phenomenon and since their formulation is a new one, one would expect neither that such a formulation would be stabilized in its final form nor that the object would be easily reconstitutable. The authors, to give the impression that their formulation captures a robust and reliably reconstitutable phenomenon, may be selective in their report, telling only of those occasions when they were able to constitute the object and telling

Five: Scientific Writing as a Social Practice

only those key behaviors in their belief necessary to constitute the object. Moreover, they will be talking through a cognitive/symbolic framework that already incorporates the possibility (if not the reality) of the existence of such an object of the precise kind represented. Their own description may not be useful in helping others (or even themselves at a later time) in reconstituting that object.

The difficulties with replication, as pointed out by a number of observers, include that there is often little incentive to attempt an exact replication. Where replication is attempted, local differences.in behavior, experiences, and craft knowledge influence the outcomes; that is, the active attempts to reconstitute phenomena represented symbolically may lead to different (or differently perceived) results. Further, the replication attempts might not be carried out by people with the same commitment to the claim/representation and the implicit perceptual/behaviorial world as that of the originators of that claim. In fact, finally, it would seem those most motivated to attempt replication may well be those who most distrust the reported results and would have least shared in this activity/language/perception matrix. That is, while certain stabilized framing elements of the disciplinary matrix may be shared, unstabilized elements will lead to variations in the created and perceived event. Replicators will understand the words differently, do the experiment differently, and see the results differently.

Great intersubjective fuzziness may therefore surround a newly proposed phenomenon. Much negotiation may be needed before a communally accepted formulation emerges that defines a reliably reconstitutable object. This negotiation may involve many different kinds of empirical experiences, and not just attempts at immediate replications. In cases of direct opposition, other kinds of experiments and observations may be offered, putting the phenomena in different contexts of activity and representative framework. Not only more sensitive tests or new equipment or experimental variations may be involved, but new ranges of data may be deemed relevant to determine the character of the phenomenon, as well as new kinds of formulations. In the course of this debate, the object may turn out to vanish from sight, turning out not to be reliably reconstitutable in the emerging terms of the discussion. The noose of language and activity may pull closed and discover it holds nothing, or the phenomenon may slip out of the noose. Or the stabilized phenomenon may turn out to be a somewhat different thing than first formulated. Or the negotiation may never be resolved, with the community splitting into subcommunities based on acceptance of the object.

Stabilizing of a reconstitutable phenomenon may occur in ways other than direct conflict and negotiation. Competing scientists may carry out

How Language Realizes the Work of Science

brief, successful, and unreported replication. They may accept the new claim as consonant with their own or the discipline's previous experience and current conceptual frameworks. They may find the formulation of the new phenomenon powerful in solving problems in their own work. In such events they may quietly accept the phenomenon as reliably reconstitutable, and will employ it in their own future work. If this new work, however, proves troublesome they may have cause to look back upon their reliance on this phenomenon as reconstitutable. They may be forced to reconceive their experiments and observations under alternative assumptions in order to have them work out. On the other hand, if the formulation proves a reliable assumption, it may be used in a wider range of changing theoretical contexts and empirical situations, thus transforming the object by making it part of different activities. Similarly, it may become more and more a fundamental assumption built in almost invisibly to activities and formulations at great remove from its original use.

A phenomenon may become so regularly used and so reliably reproducible that it is built into a machine. Every time the machine works as anticipated it reconfirms the reliable reconstitutability of the phenomenon. Every time I drive my car I am reconfirming the reliable reconstitutability of physical and chemical formulations about such things as explosive combustion, friction, and electrical current flow. Every time an oscilloscope is incorporated into an experiment, the success of the experiment relies on the reconstitutability of many electromagnetic phenomena.

Alternatively, the formulated phenomenon may prove of no interest to anyone else so that it is not reconstituted at all. It is not reliably reconstitutable, not because nature might not cooperate, but because scientists do not. Scientists must see the phenomenon as a significant one for it to enter the living body of scientific activity.

Active and Passive Constraints

Thus within the negotiation of meanings that turn individual proposals into intersubjective realities, we find ambient nature passively constraining possible meanings through the active experience that is inseparable from the language use. Claims that may appear crisp and certain to their proposers will only be fuzzy intersubjective speculations until they settle into a regularized use within repeated activities, and these activities will only be repeatable if they are conso-

Five: Scientific Writing as a Social Practice

nant with some regularity in the appearances or operations of the natural world.

Similar constraining processes occur in all discourse communities. Cult leaders' claims that the world will end tomorrow must contend two days later with cult members' perceptions of the continued existence of the world. Literary critical claims that a particular theme is central to a novel must contend with the words inscribed by the author as read by a reader. Various discourse communities appeal to various kinds of experience as touchstones for their negotiations of communal meanings. In some religious communities, for example, particular emotional states, identified and interpreted appropriately, serve to confirm and define the reality of a cluster of essential meanings. Such states are in fact encouraged through architecture, music, ritual activity, and rules for regularized prayer and group interactions.

Science, however, has taken empirical experience as its major touchstone, so that in the process of negotiation of meaning, empirical experience not only constrains the range of possible meanings but is actively sought in the attempt to establish stable meanings from the negotiation. Thus, whatever may be the source of statements, the fate of statements depends on the experience generated by them. In this way science has made nature its ally. The claims that endure do so precisely because (within the particular set of problems and activities considered important) they have been able to ally themselves closer to nature than their competition, so that in the long run, one set of terms rather than another proves more fundamentally useful in carrying on activities.

In the last three paragraphs I have been elaborating a Vygotskian perspective on cultural/semiotic evolution through concepts borrowed from Ludwik Fleck. In *The Genesis and Evolution of a Scientific Fact*, Fleck proposed that formulations of knowledge within a community (or thought collective) were influenced by two types of constraints. The first, active constraints, consisted of the elements of the thought style of the thought collective. In his analysis these elements of thought style actually turned out to be habits, patterns, and available means of representation—through language, drawing, or other symbolic media. This seating of thought within a collective drawn together through semiotic means places his ideas in the same general area as Vygotsky's.

Moreover, in proposing a second kind of constraint on formulations, what he calls passive constraints, he comes even closer to Vygotsky. Natural phenomena passively constrain the kinds of formulations you can make in the sense that once you begin formulating statements in whatever style of your thought collective, certain behaviors or features of nature will limit what you can properly say. Once you have established, for example, a procedure for identifying the hardness of rocks and have developed a taxonomy of rock types, which rocks are labelled as harder than which others is no longer a matter of cultural discretion. Thus, formulating practices are constrained by the activities that bring the language user into active contact with nature.

According to Fleck, a scientific fact for a thought collective is the representation of that passive constraint within the stylized representational manner of the thought style. Moreover, Fleck suggests that the scientific community is marked by the active pursuit of passive constraints. That is, the thought style of science actively seeks to increase the relationship between representations and empirical experience.

This Vygotsky-Fleck model of formulating practices seems to me most fruitful for the issues I have investigated in this study and the data I have examined. By seating language use in a social/empirical/cognitive activity, this model allows us to see the multivalency of symbolic formulations and to give a plausible account of the kinds of work we know through our daily experience that science does. But it does not give undue status to the statements of science, which by their own nature can be nothing more than constructions of the humans who use them. Scientific formulations embody all the complex impulses and limitations of any human product. Such a model allows us to accept the deep insights of the recent social analysis of language use within science without being driven to the absurdity of considering scientific activity cut off from its concern with the natural world.

The Historical Analysis of Language Use

By situating scientific language use and cognition within specific social/empirical moments, this model suggests that scientific language needs to be studied as a historical phenomenon. (Vygotsky argued similarly for a historical/genetic analysis of language; see Scribner, "Vygotsky's Uses of History" in *Culture, Communication, and Cognition*, and Wertsch, *Vygotsky and the Social Formation of Mind*, chap. 2). To understand what scientific language is and does, we need to look at what kind of tool it is. We need to see when, how, and to what purpose it is employed in the concrete settings of human history. History is not just kings quarrelling, but apparatus being built, balls being released down ramps, astronomers looking at the moon and arguing over the different things they claim to have seen, political scientists interviewing southern voters, articles being written, articles being read. Thus in this book, I have offered accounts of what forces constrained and impelled

Five: Scientific Writing as a Social Practice

Newton, Compton, Oldenburg, Wundt, and a host of other scientists to use language (both reading and writing) in particular ways at particular moments. I have looked at their linguistic inventions as creative responses to their situations, investigations, and goals as they evolve in historical settings. But this kind of narrative of rhetorical moments only displays the first level of history, the single living moment.

But the model proposed and the data examined suggest that history makes history, so that I have looked at a second, third, and fourth levels of history. The second level is the history of an individual that defines the symbolic resources, experience, and perceptions of that individual coming to any particular moment. This corresponds to Vygotsky's autogenetic analysis. Accordingly, chapter 6 of this book describes how Compton's underlying conception and formulations in one article had been shaped by his history in trying to come to terms with a problem—both in his laboratory and in communication with the ideas and opinions of his colleagues. Chapter 4 similarly reveals how Isaac Newton had to work through many formulations in many situations in order to find the final public form in which to express the scientific meaning of some empirical experiences of forty years before. In the process of finding a satisfactory mode of public discourse, both Compton and Newton were creating intersubjective, reliably reproducible phenomena for their disciplines out of what first were only private experiences.

This creation of community-wide, intersubjective realities brings us to our third level of history—the genetic account of the community as a whole—those events that have lead to the momentary state of the debate or communal activity. This corresponds to Vygotsky's cultural history. We have seen this in the issues and claims and counterclaims, the negotiations going on in almost every chapter in this book. We see the development of arguments, the mutual construction of theoretical perspectives, the populating of the experiential/conceptual world with reconstitutable phenomena of varying reliability, states of negotiation, and intersubjective congruence. We have the emergence of procedures and formulations.

But out of this fluid world of ever-new and ever-different social action, interaction, and symbolic realizations certain regularities develop in the social forms—what Fleck would call the distinctive features of thought style, but which Vygotsky might see more broadly as the characteristic cultural forms. These regularities encompass when and how one would approach a test tube or a colleague, how one would go about reading a text, as well as how one would draw a diagram or frame an argument. An account of the emergence, evolution, and extinction of these regularities comprises the fourth level of history: the history of cultural forms. The evolution of cultural forms shape, constrain, and create opportunities for the historical events seen through the previous three perspectives.

Previous structural examinations of scientific language have been at this level of analysis, but without recognizing the historical/cultural character of the forms studied. By identifying certain regularities appearing currently within certain limited locales and activities as characteristic of scientific language, linguists have given the impression that these regularities are timeless expressions of the essential character of science, and that these regularities give a grasp on the whole of scientific use of language. But when we view these regularities through the model proposed here, we become aware that we must account for the functional emergence of such regularities to understand what they are and what they do. We must see them as fluid to varying degrees and in relation to even more fluid elements, and must see them in relation to the complex activities that employ these regularities. Thus broad, ahistorical, static identification of features such as the standard five-part structure of the experimental report or the use of the passive voice and avoidance of the first person, are found inaccurate with the slightest amount of historical digging; moreover, such investigations tell us very little about how and why to use these features.

Rather, we need to understand why regularities emerge, evolve, and vanish; what the writers accomplish through the use of these features within the activity of the discipline; why these particular symbolic choices have seemed advisable to so many members of the community that they become regular practices; whether these habitual practices have become institutionalized; and what the effect is of regularities and institutions on science's ongoing work.

The Cultural Form of the Experimental Article and Its Impact

The studies represented in this book have looked at all four levels of history realized in the linguistic moment. But the central focus has been on the fourth—the history of cultural forms. In this case, the cultural form is the genre of experimental research article. In its emergence and continuing fluidity we see the impact of the other three levels of historical analysis, and in its normative stabilization and institutionalization we see the consequences of the genre for the other three levels. That is, cultural forms emerge and evolve through individual and

Five: Scientific Writing as a Social Practice

communal activity; in turn, cultural forms give shape and focus to continuing activity.

The framing themes of the historical narrative of this book are (a) that the features of the modern scientific article emerged as responses to (and realizations of) social and intellectual history within the emergent scientific community; and (b) that these larger communal regularities emerged out of the activity of individuals, attempting to accomplish their goals within their perceived situations. The growth of the scientific periodical press and the rise of scientific societies (both in seventeenth-century England and late nineteenth-century America), and the emergence of new disciplines and reformulations of fundamental problems (as with experimental psychology and political science) have created major shaping pressures on the genre. But it is the individuals (both towering figures and lesser souls) who perceive and respond to these pressures to remake the genre at each act of reading and writing.

Regularities occurred because individuals perceive situations as similar and make similar choices. Institutionalization and codification occurred because repeated choices appear to the collective wisdom (or wisdom of a few powerful actors) to be generally and explicitly advisable. The agonistic forum of the scientific journal made special demands on communication that made exploring the rhetorical possibilities of empirical representation a particularly attractive rhetorical resource. As the genre and the consequent literature took shape, they themselves became increasingly important social facts to be addressed in new texts. References, citation practices, and embedding of contributions in theory gave textual form to the increasing explicit intertextual activity of each individual author. The success of the genre in carrying out the business of the scientific community has also turned the genre into another kind of social fact, as an authoritative model to be emulated by other disciplines, interpreted through their own perceptions and problems.

Institutionalized patterns of representation not only shape the form of the utterance, but all the activity leading up to, surrounding, and following after the utterance. We have seen some of the argumentative assumptions built into generic features of the research article. In the case of Compton we have seen how his activity, his normative behavior, and his basic perception of the cognitive task he was engaged in were shaped by the form of the answer that he was seeking. We have examined how the patterns of argumentation impel the strategies of argumentation and the surrounding activity. Good science—both experimental psychology and physics—seems in part defined by the form of one's claims, and that desired form provides a goal for the activity. The history of the APA style sheet reveals it not only as an attempt to regulate form, but as a way to socialize neophytes into acceptable scientific practices and appropriate communicative relationships among professional researchers. We have also seen how the necessity to produce new statements even influences the reading of prior statements in fundamental ways.

But of all the stories recounted in this book, the most poignant one, to my mind, for revealing the utterance as crystallization of experience, realization of social action, and shaper of personal and social cognition is the story of Newton's search for the most persuasive, compelling form to create shared appreciation of his perceived experience—what he saw with his prism. In eventually finding that his material was amenable to a tightly sequential form, constraining and constructing the reader's reasoning, experience, and perceptual framework, Newton not only quietened his critics and won the argument; not only did he establish his "facts" as reliably reconstitutable phenomena for all to see; not only did he create a perceptual/behaviorial/empirical complex so strong that he closed off serious investigation of alternatives for a century; not only did he invent a way of arguing that led to the even more mighty *Principia* that seemed an immovable mountain for two centuries; but, most powerfully, he provided a model for the form of scientific argument that influenced all of scientific practice.

The evolution of scientific use of language hardly ended with Newton, nor had it begun with him. But given the contemporary means, problems, social relationships, and activity of science, he organized them to create a shared, relatively stable semiotic universe which has only in this century been displaced by a communal creation. He dominated the history of science not just because he discovered a few major laws, but because in finding the way to articulate those laws he found a powerful, long-lasting (though ultimately and necessarily temporary) solution to the problem of how one should talk about the subject.

That debate over how to talk about one's subject continues in all disciplines today, and cannot be separated from the fundamental practices of those disciplines. If there is any essential message of this book it is in precisely this: in those communal endeavors whose goal is symbolic knowledge, the more we understand the way symbols are used in the activity, the better we can carry out that activity. In Vygotskian terms, ability to talk about our language behavior offers us a higher form of self monitoring and regulation of behavior.

12 WRITING WELL, SCIENTIFICALLY AND RHETORICALLY

PRACTICAL CONSEQUENCES FOR WRITERS OF SCIENCE AND THEIR TEACHERS

The forms of writing are historical phenomena—created, recognized, mobilized, and given force within the mind of each writer and reader at specific social-historical moments, but transmitted in the accumulation of texts. Accumulated, socially contexted, textual experience increases the formal repertoire and procedural command of each writer and reader. This book has explored the changing repertoire within the domain of scientific writing and the social, empirical, and epistemological consequences of that repertoire in use within changing contexts.

The repertoire has grown and changed as individuals have confronted specific rhetorical problems within specific rhetorical situations. In adopting the role of scientist, individuals commit themselves to creating novel claims persuasive to other scientists knowledgeable and experienced in their specialty. They must draw on their reading, their empirical experience, and their interactions with their peers so as to use the existing symbol system to point to phenomena previously uncontained by symbols but reliably reproducible, recognizable, and persuasive to peers. In cases presented here and elsewhere in the literature on the rhetoric of science, we have seen individuals use, transform, and invent tools and tricks of the symbolic trade.

Genre as a Sociopsychological Category

Some of these tools and tricks have proven so useful and forceful as to become regularized and even institutionalized mandatory features (both formal and procedural) of particular types of scientific communication. What we recognize as the genre of the experimental article embodies many such regularized formal and procedural elements. Genre, then, is not simply a linguistic category defined by a structured arrangement of textual features. Genre is a sociopsychological category which we use to recognize and construct typified actions within typified situations. It is a way of creating order in the ever-fluid symbolic world.

The textual features we may associate with any particular genre have no necessarily fixed definition. Even attempts to hold features firm by social processes of institutionalization lead only to a temporary stability; despite the great influence of the APA *Publication Manual*, a quick scan of psychology journals in 1987 will reveal a wide range of rhetorical innovation, hardly contained within the bounds of the idealized model (chapter 9). Nor are the textual features that we associate with a genre all of the same order. Some are large organizational features, such as the presentation of method after the introduction and before results in many versions of the experimental article. Others are associated with citation practices (both in terms of citation format and quantity and in terms of the role of citation within the argument). Others are matters of quantity and location of detail. Still others have to do with the level, function, and placement of generalization. The use or absence of transitions also characterizes the genre at different moments in different disciplines—and so on, through all the myriad kinds of features discussed in the previous chapters.

Most important, the features we may associate with genre are hardly contained in their formal appearances on the page. The formal features are only ways more fundamental relations and interactions are realized in the act of communication. In recognizing and using genre, we are mobilizing multidimensional clusters of our understanding of the situation, our goals, and our activity. Some of these relational themes we have seen expressed at various times within the genre of experimental article have to do with the agonistic structure of discussion within journal forums, the desire to compel assent, the emergence of a domain of general claims separate from a domain of specific claims, the attempt to construct empirical experience through experimental intervention into nature and to represent that experience, the enactment of the emerging role of scientist within a changing structure of the community, the mutual construction of a shared knowledge within the community, and the changing relations with communities involved in more applied endeavors. These relations are played out on social, psychological, empirical, epistemological, as well as textual gameboards. Understanding the genre one is working in is understanding decorum in the most fundamental sense—what stance and attitude is appropriate given the world one is engaged in at that moment.

Because genre is such a multidimensional, fluid category that only gains meaning through its use as an interpretive, constructive tool, the reduction of any genre to a few formal items that must be followed for the sake of propriety (decorum in its most restricted sense) misses the life that is embodied in the generically shaped moment. As writers, we find a list of formal requirements of any particular genre gives us only weak command over what we are doing and gives us no choice in mastering or transforming the moment. As teachers, if we provide our students with only the formal trappings of the genres they need to work in, we offer them nothing more than unreflecting slavery to current practice and no means to ride the change that inevitably will come in the forty to fifty years they will practice their professions. We do better to grant ourselves and our students means to understand the forms of life embodied in current symbolic practice, to evaluate the consequences of the received rhetoric, and to attempt to transform our rhetorical world when such transformation appears advisable.

Rhetorical Self-Consciousness and the Invention of Science

Sometimes individuals who have significantly transformed scientific writing have had some degree of rhetorical self-consciousness, as we have seen in the cases of Newton and Oldenburg. Elsewhere individuals seem to veil their rhetorical awareness behind other sets of beliefs, as in mid-twentieth-century experimental psychology. There rhetoric is denied even as it is practiced, because the practitioners feel they have no other alternative; as I have heard a number of experimental psychologists say in response to my chapter on the writing of their field, "the practices you describe are not rhetoric; they are simply good science." And some individuals with little self-consciousness about their formulating practices just keep doing what seems demanded by the situation, what is rewarded by persuasive success, as seems to be the case of many of the contributors to the early *Philosophical*

Transactions. When elaborate practices are deeply embedded in the training and socialization of scientists, as among twentieth-century physicists, innovations in symbolic process are likely not to be perceived as either rhetorical or innovations, but rather just as continuing business as usual.

No matter the degree of self-consciousness accompanying the innovations and emergence of regularized procedures, these transformations of rhetorical practice matter. They matter significantly, for they create the symbolic ground on which scientific formulation and argument occur and they shape the communal action and the structured interaction of the scientific community. The regularized symbolic practices define the symbolic universe within which the community operates; and the degrees and kinds of restrictiveness within these practices define the directions and dynamics of growth for the knowledge to be produced by the field. As we have seen, the symbolic practices even deeply influence the empirical experience of individuals and the identification of communally reconstitutable phenomena.

The tools and tricks of the symbolic trade are what make possible an empirical science that uses symbols to formulate knowledge about natural phenomena. The various cases studied here all reveal a history of symbolic practices defining phenomena of substantive and evidentiary interest, then drawing closer to the phenomena within the stylized communication of the research community, driven by the difficulties of persuading motivatedly agonistic peers. Persuasion is at the heart of science, not at the unrespectable fringe. An intelligent rhetoric practiced within a serious, experienced, knowledgeable, committed research community is a serious method of truth seeking. The most serious scientific communication is not that which disowns persuasion, but which persuades in the deepest, most compelling manner, thereby sweeping aside more superficial arguments. Science has developed tools and tricks that make nature the strongest ally of persuasive argument, even while casting aside some of the more familiar and ancient tools and tricks of rhetoric as being only superficially and temporarily persuasive.

Scientific Writing and the Rhetorical Tradition

Skill in scientific writing, as with most human arts, is knowing what you are doing and making intelligent choices. This is hardly a startling pronouncement and firmly within the rhetorical tradi-

Five: Scientific Writing as a Social Practice

tion. Classical rhetoric is an art of oral performance built on the analysis of the kinds of rhetorical situations, goals, and tools that resided within the legal and political world of ancient Greece and Rome.[1] The basic goals of the study of political and scientific language, as of all language uses, share a fundamental concern: to understand and control the symbolic actions in order to achieve desired communal ends.[2] But the rhetorical situations, goals, and tools of contemporary journal science are quite different from those of the Athenian agora, and as the symbolic means for science developed they were consistently distinguished from arts of political oratory identified as Rhetoric. Quite appropriately the two forms of symbolic art developed different conceptual vocabularies and analyses. However, in the search for certainty of statement and compellingness of argument, the constructed, socially active character of the scientific symbolic system seemed to be forgotten. Scientific language began to seem an escape from language, and thus not a matter for conscious control. Propriety and clarity, not letting errors of language get in the way, were all the scientific writer needed to worry about. Where this book diverges from tradition is only in explicitly recognizing that scientific language is of our own making and used only in human, social contexts; therefore it is a matter for our conscious control.[3] And the levels of our conscious control can extend as deeply as we can come to understand the communication process.

The historical overt disavowal of the socially active, rhetorical character of scientific use of language did not, however, mean that individual writers confronting blank pages to be filled and filled pages to be read did not implicitly have an understanding of what written texts could do. They expressed various kinds of conscious and unconscious practical control over their language and the complex practices in which the language was embedded. The detailed analyses of the preceding chapters

---

1. George A. Kennedy has written the standard surveys of classical rhetoric: *The Art of Persuasion in Greece, The Art of Rhetoric in the Roman World: 300 B.C.–A.D. 300,* and *Classical Rhetoric and Its Christian and Secular Tradition from Ancient to Modern Times.* Useful introductions to the field are also provided by James J. Murphy, ed., *A Synoptic History of Classical Rhetoric* and Winifred B. Horner, ed., *The Present State of Scholarship in Historical and Contemporary Rhetoric.*

2. In composition and the teaching of writing, research and theory have recently turned toward an examination of the social bases of writing, thereby coming closer to the concerns of classical rhetoric in understanding statement making as a socially embedded form of social action. See, for example, Bizzell; Cooper; Ede; Faigley; Herrington; LeFevre; North; Nystrand; Odell and Goswami; Perelman; Rubin and Rafoth.

3. In the last decade within the discipline of rhetoric some limited attempts have been made to address the rhetorical character of scientific writing. See, for example, Fahnestock; Halloran; Overington; Weimer.

Writing Well, Scientifically and Rhetorically

serve exactly to make explicit the complex rhetorical concerns embedded within the emerging practices. These kinds of rhetorical analyses help us to understand the meaning of our choices and raise the possibilities of alternatives.

A rhetorical approach to writing well in science would not set forth a set of formal prescriptions to be followed for propriety's sake, nor would it suggest a set of universally advisable procedures. A rhetorical approach would attend to the range and meaning of current practices and then suggest how to deploy them appropriately and effectively within specific contexts. The current practices, properly understood, within themselves contain their own recommendations for appropriateness and advisability, for they embody a history of inventions and choices by prior writers addressing and shaping similar situations. The following practical morals of the analyses of this book neither identify a set of rules nor define a limited linguistic technology of responses for all of science. Science is no one single thing, and rules and language technology are continually changing in form and meaning. The advice I offer, rather, is to hold up for reconsideration the concerns embodied within the historical development and current practice of scientific writing. Reexamination of fundamental concerns gives us a position from which to reconsider our current choices.

- CONSIDER YOUR FUNDAMENTAL ASSUMPTIONS, GOALS, AND PROJECTS

The underlying epistemology, history, and theory of a field cannot be separated from its rhetoric. The rhetorical action is mounted within a conceived world and in pursuit of ultimate as well as immediate goals. The more you understand the fundamental assumptions and aims of the community, the better able you will be able to evaluate whether the rhetorical habits you and your colleagues bring to the task are appropriate and effective. Much of the rhetorical change we have observed in various periods has been driven by the gradual realization of the rhetorical consequences of epistemological commitments and communal goals. The realization of the empiricist project (as embedded in an agonistic social structure) lies behind much of the movement of the experimental article in the seventeenth and eighteenth centuries. Similarly behaviorism, as we have seen, has had a deep effect on the institutionalized rhetoric of experimental psychology. Greater rhetorical self-consciousness may not have changed the overall shapes of the rhetorical practices that eventually emerged, but it may have led to those results more rapidly and with greater precision. Indeed, some of the current problems of

Five: Scientific Writing as a Social Practice

writing in political science seem to come from inadequate consideration of the epistemic consequences of the rhetoric adopted.

And conversely, epistemological change and reformulations of goals have come in the wake of rhetorical change. The ideal of Newtonian science structured as a comprehensive deductive system of great generality can be seen as fostered by Newton's discoveries of the most advisable procedures for winning his arguments. Newton's abilities to recognize and heighten the epistemic consequences of his rhetorical struggles presented him with powerful tools to transform science.

More locally, it is useful to understand how your individual assumptions and goals fit in with the epistemology and goals of the community you are participating in and contributing to. If your work is simply harmonious with disciplinary assumptions and projects, and if the discipline has forged a rhetoric adequate to its beliefs and tasks, you can adopt the local rhetoric with a fuller understanding and commitment. If, however, you find yourself in some way at odds, you can begin to understand the rhetorical task before you—both in developing terms appropriate to your emerging claims and in finding ways to make your claims intelligible and persuasive to peers committed to other beliefs and rhetorics. Newton, as we have seen, had to struggle with the Baconian empiricism and the Cartesian skepticism he perceived around him in order to find ways first to present his findings and then to assert arguments of great certainty and generality.

- CONSIDER THE STRUCTURE OF THE LITERATURE, THE STRUCTURE OF THE COMMUNITY, AND YOUR PLACE IN BOTH

At any particular moment the literature of a field is structured around issues and themes historically evolving and of current moment. The prior literature establishes a conversation that has established accepted understandings, visions of the world, topics of concern and open questions. As you step in to add your utterance, it necessarily must address the rhetorical situation established by that literature, for certainly it will be received and measured against that communal construction. Even a newly emerging field with a small and loosely structured literature draws on the literary capital of other specialties out of which it emerged; however, the protean possibilities of a newly emerging field offer opportunities for direction-setting innovations. In more established fields, more must be uprooted to significantly alter the rhetorical dynamics.

The explicit recognition of the importance of prior statements has been realized through the techniques of overt intertextuality developed over the last few centuries (such as references and citations, article in-

Writing Well, Scientifically and Rhetorically

troductions, reviews of literature, eponymity, and shared theory). Thus in addition to evaluating the state of the discussion to evaluate the rhetorical moment, you must represent that state of the discussion so as to locate and justify your contribution. Swales's schematic analysis of the four moves of a typical article introduction (establishing the field, summarizing previous research, preparing for present research, and introducing present research) is precisely an elaboration of the standard current strategics of this generic task.

Explicit intertextuality also helps mobilize a range of literature to support and extend the new claim. The more firmly you can tie the claim to the accepted intertextual web, the more persuasive the claim appears. The more centrally the claim can be placed at a crucial juncture in the web, the greater significance it will have. Finally, explicit intertextuality offers opportunities for rewriting history from your vantage point. The opportunities for persuasive restructuring of the literature depend both on how tightly and convincingly the literature currently seems structured to members of the community and on the powerfulness of the new perspective from which you wish to re-view the prior conversation.

The need to assert your work against an explicitly recognized literature heightens the need to know how and why you are reading that literature. Reading the literature against a developing schematic view of what problems the discipline has addressed, what the discipline has learned, where it is going, who the major actors are, and how all these things contribute to your own project, helps you interpret the literature actively in support of your developing project. The highly developed and self-conscious reading behaviors of the physicists interviewed for chapter 8 indicate the importance these individuals had placed on becoming skilled, active readers of their discipline.

The rhetorical moment one speaks to is shaped not only by a history of paper, but by living persons whom you wish to move in some manner by your written comments. These individuals share, to differing extents, communal assumptions and projects as well as a familiarity with the disciplinary literature. However, these individuals are also driven by their own active projects and view the communal legacy through their own interests and schema.

To some extent, you can know parts of your audience as individuals, through face-to-face interaction and familiarity with their writing; however, except in the most contained and tightly structured fields, you can come to know only a few individuals well, a wider group superficially, and the greater number of colleagues not at all. Through coming to know how statuses, roles, and relations tend to be structured in a field you can, nevertheless, gain a fairly good idea of your audience, and

Five: Scientific Writing as a Social Practice

even more of yourself in relation to that audience. Familiarity with the social structure of a community surrounds you with statuses, roles, norms, rights, obligations, appropriate attitudes, and acceptable actions. You learn what you must do and how you must act to participate in the activity of the community, what the acceptable degrees and ranges of variation are, and what sanctions are likely for violation.

In most cases, accepting your place within the social structure grants you sufficient voice to assert your projects, particularly if the projects are conceived and carried out well within the standards of the community; occasionally, however, establishing new social relations can have revolutionary impact on the community. Compton's arguments were credible to his peers because he acted as a physicist in his time and place should; he did so precisely because he himself was a committed physicist according to the standards of the time. On the other hand, Newton, who adopted the guise of a Baconian in his "New Theory" article, got into more fundamental persuasive difficulties, because he wished to carry a different kind of argument than was currently allowed Baconians. He had to rewrite the social structure and social relations, with himself at the top of a compelling hierarchy, in order to persuade the community of his experience and beliefs. The consequences of the restructuring extended far beyond the acceptance of his claims about colors.

- CONSIDER YOUR IMMEDIATE RHETORICAL SITUATION AND RHETORICAL TASK

Within all the fundamental frameworks of disciplinary and personal assumptions and goals, of structured literature and structured community, the rhetorical moment presents itself and you must define an immediate rhetorical task. Large issues coalesce into a specific question, large research goals take shape in a specific project, a local environment of immediately relevant claims and counterclaims emerges from the literature, and you find yourself positioned in a certain relationship with your colleagues. The more clearly you understand this emergent rhetorical situation, the more precisely and effectively you can choose what you do next. Assessing the situation helps you judge what kind of statement is called for, if any. The situation may seem to call for an immediate written response, it may call for further experiments to address unresolved questions and criticisms and to result in a compelling published answer, or it may call for fundamental investigations out of which whole new kinds of statements will grow.

Within the conversation of communal science, all choices have rhetorical import, for they help shape the next statement to be made. Compton's sequence of investigations and papers reveals consistent rhetorical choices as to how more satisfactory and persuasive claims might be developed and pressed. Newton's hand-to-hand combat over his optical claims reveals continual rhetorical choice making. No less do all the cases discussed here show the impact of rhetorical attempts to address the rhetorical moment, although the agonistic struggle may not be nearly as dramatic. With greater or lesser clarity, each writer has set out to make some argumentative gain within the field at a particular moment in the communal discussion.

- Consider Your Investigative and Symbolic Tools

The tools available to pursue goals of asserting claims within science are dialectically related empirical experience and symbols. The textual analyses here have revealed some of the resources available within scientific use of language and the kinds of impacts and actions realized by these resources. The genre of experimental article has found ways of bringing to bear on any particular argument the literature of the field, the currently accepted theory, deductive reasoning, representations of method, and representation of empirical experience. At particular moments, other forces are also brought to bear.

No less are the panoply of investigative tools rhetorically significant, for the scientific argument hangs on the quality and character of the evidence. Experimental and observational techniques are precisely ways of transforming nature into symbolic representations, which then have meaning for claims and arguments asserted on the symbolic plane. Choice of the investigative tool determines the kind of evidence available to generate new claims and to bolster old claims. A new method of investigation can bring a powerful resource to an argument by generating data of more exact relevance to the issues in question, by exposing new issues, and by creating a new kind of symbolic grounds on which to carry out the argument.

Thus a key issue in developing rhetorically effective science is considering how you may make nature your ally. On the most simple level, of course, this means advancing claims consonant with the available evidence (symbolic representations of empirical experience). More deeply, however, this means several other kinds of strategic choices. You can choose to pursue investigations that are likely to result in strong and striking evidence for the emergent claims. You can choose investigations where you suspect the emerging evidence is likely to expose new issues or reopen old ones. Or you can choose to employ new or different

Five: Scientific Writing as a Social Practice

investigative tools that you suspect will make the familiar look different, that will make the hidden visible in powerful ways. By shaping your research program you can use empirical experience as a heuristic to generate new statements about nature. Thus which research program to pursue and which means to use to pursue that program are important rhetorical choices affecting the kinds of claims and arguments that will emerge at the end. Shrewd guesses as to what kind of researches will produce empirical leverage against symbolic issues can generate much ultimate rhetorical power. Zuckerman's study of Nobel Prize winners reveals how much conscious thought these eminent scientists put into choosing what to investigate and how to investigate it so as to produce those powerful statements that win prizes *(Scientific Elite)*.

- CONSIDER THE PROCESSES OF KNOWLEDGE PRODUCTION

Although it is the final, publically stated claim that has rhetorical power, one cannot simply think only in terms of the final shape claims will take. Early choices of questions to consider, claims to pursue, literature to read, colleagues to discuss ideas with, investigative techniques to employ, analyses to carry out, and so on will all affect what kind of product will emerge at the end. These choices will generate thinking, data, formulations, and arguments which may well find expression in the final article. Moreover, the experimental article requires a certain amount of explicit representation of selected parts of the process that goes into its creation, such as the after-the-fact reconstruction of the intellectual genealogy in the review of the literature, the focused procedural account of methods, and the selective narrative of results. Finally, the representation of the final paper implies a web of activities and relations engaged in by the author as part of the construction of the argument, implicit activities that may be summed up by saying that the author has in the preparation of the article acted as a scientist (with whatever local meaning that takes on within the relevant specialty).

Because the final text is so dependent on the process by which it is produced, it is important to consider how you should go about producing the text so as to wind up with the kind of statement you hope for, without leaving yourself open to charges of fraud for representing a process that did not occur or improper conduct for not living up to implied behavior. In fact, process is so important to the production of persuasive scientific arguments that the final representation or writing-up seems a limited activity, with all the major parameters of the text determined by prior decisions. Well-considered procedure is not only good science, it results in good rhetoric.

Writing Well, Scientifically and Rhetorically

Just as a consideration of the process of text production helps gain control of the final text, an anticipation of a text's reception helps gain control of the meaning likely to be attributed to a text. Given the positions held by colleagues, the kinds of arguments previously used, and the dynamics of competing research programs, you can often gain some sense of what kind of impact a reported experiment or a newly framed argument is likely to have. Anticipation of the impact can help you shape the presentation to forestall unwanted responses and heighten the desired ones. You can cut opposition off at the pass, press your advantages, draw in desired audiences, and provoke desired follow-up work.

- ACCEPT THE DIALECTICS OF EMERGENT KNOWLEDGE

Despite the attempt to understand and control all the dynamics of written communication, we are always reaching into the unknown. The outcomes of investigations, writing processes, and social interaction can never be anticipated with clarity and certainty. Having made our best guesses as to how to proceed, we must then be ready to notice what develops and revise our plans accordingly.

As events unfold we discover that our nascent formulations match and mismatch in curious ways with the data we pursue in order to explore those investigations. The dialectical struggle to find ways of generating data significant for our formulations and to then reconcile that data with those formulations can lead to manifold discoveries of new kinds of data, new kinds of claims, new issues to investigate and new methods of investigation. We do not know what we will find, and what we will be led to say by what we find. Although we need issues, assumptions, methods, hypotheses to drive our discovery process, we must be ready to accept the worlds revealed to us in our attempt to come to terms with what we discover. Otherwise, we may throw away our most promising stories.

Similarly, as we start to draw all the elements of our investigation together in the single location of a text, we are forced to reexamine how the parts fit together. Again this is a moment that calls for an openness of imagination, as we create a coherent account of the literature, issues, theoretical positions, investigative goals, empirical events, and conclusions we draw. Creating a single text provides us with a retrospective vision that can tighten threads of connections, reveal new issues and anomalies, excite new insights, and define new projects. As well, the formal requirements of completing a text puts us on the spot, forcing us to fill in the blanks as best we are able, to dig deeply into our thought

Five: Scientific Writing as a Social Practice

and experience to fulfill the outlines of the argument demanded by the rhetorical situation and our rhetorical tools. If we do not have the means to live up to the rhetorical demands, we are forced back into both the library and the laboratory.

And finally, having sent our text out into the world, we need to be open to what experience and thought others bring to the published formulations. We need to understand what kind of social reality the text becomes, so as to pursue the conversation of knowledge to the best advantage. Sometimes this may mean buttressing arguments, closing loopholes, and clearing up misunderstandings. These acts in themselves may lead to new discoveries or more powerful formulations. But often responses can teach us new contexts which generate new meaning for the work. Interaction with new realms of ideas, problems, and data can transform the claims. And the evolution of continuing work will assign a social meaning and pragmatic role for our formulations; our understanding of and reactions to that social meaning will influence our future investigations and formulations. To keep the conversation going, we must constantly reread the dynamics and meaning of the conversation and our place in it. An inability to recognize the continuing evolution of the communal projects will leave us singing the same old song, a song that may lose its meaning when sung out of season.

As we create the formulations that gain communal meaning as knowledge, we are bringing worlds into being. By identifying, selecting, recording, and making claims about empirical experience, we are bringing our experience of the world of objects into the human-created world of symbolic actions. Although our procedures of generating and using statements through our empirical experience bring the world of objects in relation to the world of symbols, the world of symbols does not exist and our knowledge does not exist until we make them within the social world, as protean and transient as that is. We cannot fully know what we bring into being until it has taken its place in our world, but then, since the world immediately starts changing around it, what we have made changes. To gain what limited mastery we can over this changing social world of symbols, we should follow Odysseus, who must catch Proteus in his own lair on the edge of the ever-changing sea.

In short, writing well in science means to apply to one's own situation and tasks the same rhetorical understanding applied in this book to a wide range of texts and writers. Playing chess well involves an analytical knowledge of the most interesting and informative of prior games, and then applying that knowledge to the position in front of you. By recognizing the power of different moves in different contexts, you can then mobilize that power. To see the practices and institutions of scientific writing as protean and evolving is not to discredit them as transitory, but to grant them the proper respect for the great power they realize.

The Limits of Rhetorical Self-Awareness and the Teaching of Writing

To hold every statement up for rhetorical examination is, of course, an unrealistic demand. Both art and science are long, and life is short. We must make choices as to where we devote our energies. It seems enough to ask a physicist to learn physics and the symbol system of mathematics. Should we then also demand competence in the other symbol system of words? And how much competence? Certainly not a Ph.D. in rhetoric. On the other hand, more than a junior high school course in grammar and spelling seems required.

Just as scientists in different specialties and of different personal bents master mathematics in different areas and to different depths, depending on applied need and theoretical grasp, so too will rhetorical needs and command vary. In fields with restricted, slowly evolving, and apparently adequate rhetorical practices, a thorough practical command within the regularized domain may need to be supplemented only by an analysis of the implications and a cursory knowledge of basic rhetorical concepts. Then, if the rhetorical problems heat up, the individual scientist can at least recognize the problem and know where to begin looking for answers. Interdisciplinary fields that draw on several bodies of knowledge may require greater virtuosity and understanding of the technologies of literature discussion, synthesis, and citation; as well, the ability to analyze the communicative dynamics of different fields may aid both interpretation of the varied literatures and the formulation of arguments for different venues. Fields with rapidly evolving theory require other skills, such as complex argumentative structuring and organizational flexibility. Fields that depend on descriptive taxonomies or historical reconstructions may call for large depictional and narrative repertoires, while other fields need tricks of aggregation. Choices are necessary as to which parts of rhetoric are likely to have the biggest payoffs in each case.

However, here is where the big difference currently exists between rhetoric and mathematics. Needs for mathematics are well recognized and often well-defined. Scientists are likely to know when they have need of additional mathematical tools, what those tools are, and where to go to find out about them. They know which books they must study, which courses to regret having by-passed, which colleagues in the

Five: Scientific Writing as a Social Practice

mathematics department to talk to. Moreover, their colleagues in the math department are used to applying their abstract knowledge to problems in the natural and quantitative social sciences, so that a broad and useful common ground for discussion exists. Even when the mathematics is new or exotic or when the application is unusual, so that the right tool does not immediately come to hand, at least the scientist and mathematician know they can and should be talking with each other.

Scientists, however, are unlikely to recognize difficulties in framing successful investigations and claims as rhetorical, unlikely even to be aware of rhetoric as a relevant field. Even if they are aware that their claim making can be fruitfully conceived in rhetorical terms, they may have little idea of what the relevant branches of rhetoric are, what books to read, or whom to talk to. Finally, even if they find a willing rhetorician to talk to, very few of those rhetoricians have had any experience in talking to scientists and applying rhetorical knowledge to problems of scientific communication.

Rhetoric has only recently begun to take up the challenge of scientific use of language. While classical rhetoric does have a well-defined body of knowledge of several discrete parts and well-known procedures of application, appropriate to different kinds of situations, that rhetorical technology applies only to politics, the courts, and similar contexts. No such technology exists for knowledge-generating disciplines, or more particularly the sciences. Few rhetoricians have attempted serious studies of scientific use of language. While a few interesting propositions have been put forward, substantiated claims based on examination of actual language practices in science have been rare.

We need thoroughgoing and wide-ranging research into the historical and current rhetoric within the sciences and other knowledge-generating communities to gain a grasp of the range of practices, the thematic interactional concerns, the local emergence of typified forms and actions, and the implications for socially produced knowledge. We need far more than one writer's idiosyncratic glimpses into only a few scattered rhetorical locations, such as offered here. Only with a communally shared, reliable set of formulations will we be able to develop intelligent curricula to meet the local rhetorical needs of students entering into specific knowledge-generating communities, to frame efficient analytical procedures to allow writers to analyze their rhetorical situations and rhetorical options, and to present to other disciplines a knowledge and technology that will be of obvious use and power. Only then may other disciplines recognize the deeply rhetorical character of their enterprises, realize that the discipline of rhetoric can offer them important tools for their symbol-creating tasks, and wish to talk with us. Then the fun will begin.

REFERENCES

(References to primary articles examined in chapters 2, 3, 6, 7, 9, and 10 are provided in end-of-chapter reference lists and appendices or are mentioned in passing within the text.)

Abt, Helmut A. "Some Trends in American Astronomical Publications." *Publications of the Astronomical Society of the Pacific* 93 (1981): 269-73.
Agassi, Joseph. *Faraday as a Natural Philosopher.* Chicago: University of Chicago Press, 1971.
American Heritage Dictionary of the English Language. Boston: Houghton Mifflin, 1976.
Aristotle. *Posterior Analytics.* Tr. Hugh Tredennick. Cambridge: Harvard University Press, 1960.
Aristotle. *Rhetoric.* Tr. Lane Cooper. Englewood Cliffs: Prentice-Hall, 1932.
Austin, J. L. *How to Do Things with Words.* Cambridge: Harvard University Press, 1962.
Bacon, Francis. *The Advancement of Learning.* London, 1603.
Bacon, Francis. *Magna Instauratio.* London, 1620.
Bachelard, G. *Le Materialisme Rationnel.* Paris: PUF, 1953.
Baddam. *Memoirs of the Royal Society,* vol. 1. London, 1738.
Baldauf, R. B., and B. H. Jernudd. "Language of Publications as a Variable in Scientific Communication." *Australian Review of Applied Linguistics* 6 (1983): 97-108.
Barnes, Barry. *Scientific Knowledge and Sociological Theory.* London: Routledge and Kegan Paul, 1974.
Barnes, Barry, and Steven Shapin. *Natural Order: Historical Studies of Scientific Culture.* Beverly Hills: Sage, 1979.
Barnes, Sherman B. "The Editing of Early Learned Journals." *Osiris* I (1936): 155-72.
Bazerman, Charles. "How Natural Philosophers Can Cooperate." In *Text and Profession,* ed. Bazerman and Paradis, Madison: University of Wisconsin Press, forthcoming.
Bazerman, Charles. *The Informed Writer.* 3d edition. Boston: Houghton Mifflin, 1989.
Bazerman, Charles. "Scientific Writing as a Social Act." In *New Essays in Technical Writing and Communication,* ed. Anderson, Brockmann, and Miller. Farmingdale: Baywood, 1983: 156-84.
Bazerman, Charles. "Studies of Scientific Writing: E Pluribus Unum?" *4S Review* 3, 2 (1985): 13-20.
Beach, Richard, and Lillian Bridwell. *New Directions in Composition Research.* New York: Guilford, 1984.

References

Becher, Tony. "Disciplinary Discourse." *Studies in Higher Education* 12 (1987): 261–74.
Bechler, Zev. " 'A Less Agreeable Matter': The Disagreeable Case of Newton and Achromatic Refraction." *British Journal for the History of Science* 8 (1975): 101–26.
Bechler, Zev. "Newton's Search for a Mechanistic Model of Colour Dispersion: A Suggested Interpretation." *Archive for History of Exact Sciences* 11 (1973): 1–37.
Behn, Aphra. *The Emperor in the Moon. Works*, vol. 3. London: Heinnemann, 1915.
Bellone, Enrico. *A World on Paper.* Cambridge: MIT, 1980.
Ben-David, Joseph. *The Scientist's Role in Society.* Chicago: University of Chicago Press, 1974.
Biddle, Bruce, and E. Thomas. *Role Theory: Concepts and Research.* New York: John Wiley & Sons, 1966.
Birch, Thomas. *The History of the Royal Society.* London, 1746.
Bitzer, Lloyd F. "The Rhetorical Situation." *Philosophy and Rhetoric* 1 (1968): 1–14.
Bizzell, Patricia. "Cognition, Convention, and Certainty: What We Need to Know About Writing." *Pre/Text* 3 (1982): 213–43.
Bleich, David. *Subjective Criticism.* Baltimore: Johns Hopkins University Press, 1978.
Bloor, David. *Knowledge and Social Imagery.* London: Routledge and Kegan Paul, 1976.
Brannigan, Augustine. *The Social Basis of Scientific Discovery.* Cambridge: Cambridge University Press, 1981.
Brown, Richard. *A Poetic for Sociology.* Cambridge: Cambridge University Press, 1977.
Bruce, Bertram. *A Social Interaction Model of Reading.* Research Report 218. Urbana, Ill.: Center for the Study of Reading, 1981.
Bruner, Jerome. "The Ontogenesis of Speech Acts." *Journal of Child Language* 2: 1–20.
Butler, Samuel. "Elephant in the Moon" and "On the Royal Society." *Genuine Remains in Verse and Prose.* London, 1759.
Callon, Michel, John Law, and Arie Rip, eds. *Mapping the Dynamics of Science and Technology.* London: Macmillan, 1986.
Cattell, J. M., and J. Cattell, eds. *American Men of Science.* 4th ed. New York: Science Press, 1927.
Chomsky, Noam. *Language and Mind.* New York: Harcourt, Brace, Jovanovich, 1968.
Chubin, Daryl, and S. Moitra. "Content Analysis of References." *Social Studies of Science* 5 (1975): 423–41.
Clifford, J. "On Ethnographic Authority," *Representations* 1 (1982): 118–46.
Cohen. I. B. *The Newtonian Revolution.* Cambridge: Cambridge University Press, 1980.
Cohen, I. B., ed. *Isaac Newton's Papers and Letters on Natural Philosophy.* Cambridge: Harvard University Press, 1958.

References

Cohen, Ralph. "History and Genre." *New Literary History* 17, 2 (1986): 203-19.
Cohen, Robert, and Thomas Schnelle. *Cognition and Fact: Materials on Ludwik Fleck*. Dordrecht: D. Reidel, 1986.
Cole, Jonathan, and Steven Cole. *Social Stratification in Science*. Chicago: University of Chicago Press, 1973.
Collins, Harry. "The Sociology of Scientific Knowledge: Studies of Contemporary Science." *Annual Review of Sociology* (1983): 265-85.
Collins, Harry. *Changing Order: Replication and Induction in Scientific Practice*. Beverly Hills: Sage, 1985.
Collins, Harry, and Trevor Pinch. *Frames of Meaning: The Social Construction of Extraordinary Science*. London: Routledge and Kegan Paul, 1982.
Collins, Randall. "On the Microfoundations of Macrosociology." *American Journal of Sociology* 86 (1981): 984-1014.
Consigny, Scott. "Rhetoric and Its Situations." *Philosophy and Rhetoric* 7 (1974): 175-86.
Cooper, Marilyn. "The Ecology of Writing." *College English* 48 (1986): 364-75.
Coser, Rose Laub. "Role Distance, Sociological Ambivalence, and Transitional Status Systems." *American Journal of Sociology* 72 (1966): 173-87.
Cozzens, Susan. "Comparing the Sciences: Citation Context Analysis of Papers from Neuropharmacology and the Sociology of Science." *Social Studies of Science* 15 (1985): 127-53.
Cozzens, Susan. "The Life History of a Knowledge Claim: The Opiate Receptor Case." Paper delivered at the meeting of the Society for the Social Studies of Science, Atlanta, November 1981.
Cozzens, Susan. "Taking the Measure of Science: A Review of Citation Theories." *International Society for the Sociology of Knowledge Newsletter*, March 1981.
Crane, Diana. *Invisible Colleges*. Chicago: University of Chicago Press, 1972.
Crick, F. H. C., and J. D. Watson. "The Complementary Structure of Deoxyribonucleic Acid." *Proceedings of the Royal Society*. A223 (1954): 80-96.
Crosland, Maurice. "Explicit Qualifications as a Criterion for Membership in the Royal Society: A Historical Review." *Notes and Records of the Royal Society* 37 (1982): 167-87.
Crystal, David, and Derek Davy. *Investigating English Style*. London: Longmans, 1969.
Day, Robert. *How to Write a Scientific Paper*. Philadelphia: ISI, 1983.
Dear, Peter. "Totius in Verba: Rhetoric and Authority in the Early Royal Society." *Isis* 76 (1985): 145-61.
Derrida, Jacques. *Of Grammatology*, trans. Gayatri Chakravorty Spivak. Baltimore: Johns Hopkins University Press, 1977.
Dictionary of Scientific Biography. New York: Scribners, 1970-80.
DiMaggio, Paul. "Classification in Art." *American Sociological Review* 52 (1987): 440-55.
Dolezal, Frederic Thomas. *The Lexicographical and Lexicological Procedures and Methods of John Wilkins*. Ph.D. thesis, University of Illinois at Urbana-Champaign, 1983.

References

Dubrow, Heather. *Genre.* London: Methuen, 1982.
Eastwood, B. "Descartes on Refraction: Scientific versus Rhetorical Method." *Isis* 75 (1984): 481–502.
Eco, Umberto. *The Role of the Reader.* Bloomington: Indiana University Press, 1979.
Ede, Lisa. "Audience: An Introduction to Research." *College Composition and Communication* 35 (1984): 140–54.
Edge, David. "Is There Too Much Sociology of Science?" *Isis* 74 (1983): 250–56.
Eisenstein, Elizabeth. *The Printing Press as an Agent of Change.* 2 vols. Cambridge: Cambridge University Press, 1979.
Ennis, M. "The Design and Presentation of Informational Material." *Journal of Research Communication Studies* 2 (1980): 67–82.
Fabian, J. *Time and the Other: How Anthropology Makes Its Object.* New York: Columbia University Press, 1983.
Fahnestock, Jeanne. "Accommodating Science: The Rhetorical Life of Scientific Facts." *Written Communication* 3 (1986): 275–96.
Faigley, Lester. "Competing Theories of Process: A Critique and a Proposal." *College English* 48 (1986): 275–96.
Faigley, Lester, Roger Cherry, David Jolliffe, and Anna Skinner. *Assessing Writers' Knowledge and Processes of Composing.* Norwood, N.J.: Ablex, 1985.
Fear, David. *Technical Communication.* Glenview, Ill.: Scott Foresman, 1977.
Finocchario, M. A. *Galileo and the Art of Reasoning: Rhetorical Foundations of Logic and Scientific Method.* Boston Studies in the Philosophy of Science 61. Dordrecht: D. Reidel, 1980.
Firth, J. R. *Papers in Linguistics, 1934–1951.* London: Oxford University Press, 1957.
Fleck, Ludwik. *Genesis and Development of a Scientific Fact.* Chicago: University of Chicago Press, 1979.
Flower, Linda, and Richard Hayes. "The Pregnant Pause: An Inquiry into the Nature of Planning." *Research in the Teaching of English* 15 (1983): 229–43.
Foucault, Michel. *The Order of Things: An Archeology of the Human Sciences,* trans. Alan Sheridan. New York: Pantheon, 1970.
Fowler, Alastair. *Kinds of Literature: An Introduction to the Theory of Genres and Modes.* Cambridge: Harvard University Press, 1982.
Fox, Theodore. *Crisis in Communication.* London: Athlone Press, 1965.
Frank, Joseph. *Beginnings of the English Newspaper, 1620–1660.* Cambridge: Harvard University Press, 1961.
Garfinkel, Harold. *Studies in Ethnomethodology.* Englewood Cliffs, N.J.: Prentice Hall, 1967.
Garfinkel, Harold, Michael Lynch, and Eric Livingston. "The Work of a Discovering Science Construed with Materials from the Optically Discovered Pulsar." *Philosophy of the Social Sciences* 11 (1981): 131–58.
Garvey, William D. *Communication: The Essence of Science.* Oxford: Pergamon Press, 1979.
Geertz, Clifford. *The Interpretation of Cultures.* New York: Basic Books, 1973.
Gieryn, Tom. "Boundary-Work and the Demarcation of Science from Non-Sci-

ence: Strains and Interests in Professional Ideologies of Scientists." *American Sociological Review* 48 (1983): 781–95.
Gilbert, G. Nigel. "Referencing as Persuasion." *Social Studies of Science* 7 (1977): 113–22.
Gilbert, G. Nigel. "The Transformation of Research Findings into Scientific Knowledge." *Social Studies of Science* 6 (1976): 281–306.
Gilbert, G. Nigel, and Michael Mulkay. *Opening Pandora's Box*. Cambridge: Cambridge University Press, 1984.
Goffman, Erving. *Encounters*. Indianapolis: Bobbs-Merrill, 1961.
Goody, Jack. *The Domestication of the Savage Mind*. Cambridge: Cambridge University Press, 1977.
Goody, Jack, and Ian Watt. *Literacy in Traditional Societies*. Cambridge: Cambridge University Press, 1968.
Gopnik, Myrna. *Linguistic Structures in Scientific Texts*. The Hague: Mouton, 1972.
Graff, Harvey, ed. *Literacy and Social Development in the West*. Cambridge: Cambridge University Press, 1981.
Greenstein, Fred I., and Nelson Polsby, eds. *Handbook of Political Science*. Vol. 1, *Political Science: Scope and Theory* and vol. 7, *Strategies of Inquiry*. Reading, Mass.: Addison-Wesley.
Guerlac, Henry. *Newton on the Continent*. Ithaca: Cornell University Press, 1981.
Gumperz, John. *Discourse Strategies*. Cambridge: Cambridge University Press, 1982.
Gusfield, Joseph, "The Literary Rhetoric of Science," *American Sociological Review* 41 (1976): 16–34.
Hacking, Ian. *The Emergence of Probability*. Cambridge: Cambridge University Press, 1975.
Hacking, Ian. *Representing and Intervening*. Cambridge: Cambridge University Press, 1983.
Hahn, Roger. *The Anatomy of a Scientific Institution: The Paris Academy of Sciences, 1666–1803*. Berkeley: University of California Press, 1971.
Hall, Marie Boas. "Henry Oldenburg and the Art of Scientific Communication." *British Journal for the History of Science* 2 (1965): 277–90.
Hall, Marie Boas. "Oldenburg, the *Philosophical Transactions*, and Technology." *The Uses of Science in the Age of Newton*, ed. John G. Burke. Berkeley: University of California Press, 1983.
Halliday, Michael. *An Introduction to Functional Grammar*. London: Edward Arnold, 1985.
Halliday, Michael. "Language as Code and Language as Behavior." *Semiotics of Culture and Language*, vol. 1. Ed. R. Fawcett et al. London: Frances Pinter, 1984.
Halliday, Michael. *Language as Social Semiotic*. London: Edward Arnold, 1978.
Halliday, Michael. *Learning to Mean: Explorations in the Development of Child Language*. London: Edward Arnold, 1975.
Halloran, S. Michael. "The Birth of Molecular Biology: An Essay in the Rhetorical Criticism of Scientific Discourse." *Rhetoric Review* 3 (1984): 70–83.

References

Handel, Warren. "Normative Expectations and the Emergence of Meaning as Solutions to Problems: Convergence of Structuralist and Interactionist Views." *American Journal of Sociology* 84 (1979): 855–81.

Hanson, N. R. *Patterns of Discovery.* Cambridge: Cambridge University Press, 1958.

Hartman, Geoffrey H. *Wordsworth's Poetry, 1787–1814.* New Haven: Yale University Press, 1964.

Havelock, Eric. *The Greek Concept of Justice.* Cambridge: Harvard University Press, 1976.

Havelock, Eric. *Origins of Western Literacy.* Toronto: Ontario Institute for Studies in Education, 1976.

Havelock, Eric. *Preface to Plato.* Cambridge: Harvard University Press, 1967.

Hayes, Louis D., and Ronald D. Hedlund, eds. *The Conduct of Political Inquiry.* Englewood Cliffs, N.J.: Prentice-Hall, 1970.

Hjelmslev, Louis. *Prolegomena to a Theory of Language.* Madison: University of Wisconsin Press, 1961.

Herrington, Anne. "Writing in Academic Settings." *Research in the Teaching of English* 19 (1985): 331–61.

Hogben, Lancelot. *The Vocabulary of Science.* London: Heinemann, 1969.

Holland, Norman. *The Dynamics of Literary Response.* New York: Oxford University Press, 1968.

Holmes, Frederic. "Scientific Writing and Scientific Discovery." *Isis* 78 (1987): 220–35.

Holub, R. C. *Reception Theory.* London: Methuen, 1984.

Horner, Winifred B., ed. *The Present State of Scholarship in Historical and Contemporary Rhetoric.* Columbia: University of Missouri Press, 1983.

Houghton, Bernard. *Scientific Periodicals.* London: Clive Bingley, 1975.

Houp, Kenneth, and Thomas Pearsall. *Reporting Technical Information,* 4th ed. Encino, Calif.: Glencoe, 1980.

Huddleston, R. D. *The Sentence in Written English.* Cambridge: Cambridge University Press, 1971.

Hudson, R. A. *Sociolinguistics.* Cambridge: Cambridge University Press, 1980.

Hunter, Michael. "Early Problems in Professionalizing Scientific Research." *Notes and Records of the Royal Society* 36 (1981): 189–209.

Hunter, Michael. *Science and Society in Restoration England.* Cambridge: Cambridge University Press, 1981.

Hunter, Michael, and Paul B. Wood. "Towards Solomon's House: Rival Strategies for Reforming the Early Royal Society." *History of Science* 24 (1986): 49–108.

Huygens, Christiaan. *Treatise on Light.* Chicago: University of Chicago Press, 1945.

Institute for Scientific Information. *A Citation Index for Physics: 1920–1929.* Philadelphia: ISI, 1980.

Isaak, Alan C. *Scope and Methods of Political Science.* Homewood, Ill.: Dorsey, 1969.

References

Iser, Wolfgang. *The Act of Reading: A Theory of Aesthetic Response.* Baltimore: Johns Hopkins University Press, 1978.

Jamieson, Kathleen M. Hall. "Generic Constraints and the Rhetorical Situation." *Philosophy and Rhetoric* 7 (1974): 162–70.

Johnston, P. *Implications of Basic Reading Research for the Assessment of Reading Comprehension.* Research Report 206. Urbana, Ill.: Center for the Study of Reading, 1981.

Judson, Horace Freeland. *The Eighth Day of Creation.* New York: Simon and Schuster, 1979.

Kennedy, George A. *The Art of Persuasion in Greece.* Princeton: Princeton University Press, 1963.

Kennedy, George A. *The Art of Rhetoric in the Roman World: 300 B.C.–A.D. 300.* Princeton: Princeton University Press, 1972.

Kennedy, George A. *Classical Rhetoric and Its Christian and Secular Tradition from Ancient to Modern Times.* Chapel Hill: University of North Carolina Press, 1980.

Kevles, Daniel. *The Physicists: The History of a Scientific Community in Modern America.* New York: Knopf, 1977.

Kinneavy, James. *A Theory of Discourse.* Englewood Cliffs, N.J.: Prentice Hall, 1971.

Kittredge, Richard, and John Lehrberger, eds. *Sublanguage: Studies of Language in Restricted Semantic Domains.* Berlin: De Gruyter, 1982.

Knorr-Cetina, Karin. *The Manufacture of Knowledge: An Essay on the Constructivist and Contextual Nature of Science.* Oxford: Pergamon, 1981.

Knorr-Cetina, Karin. "Producing and Reproducing Knowledge: Descriptive or Constructive." *Social Science Information* 16 (1977): 669–96.

Knorr-Cetina, Karin. "Tinkering Toward Success: Prelude to a Theory of Scientific Practice." *Theory and Society* 8 (1979): 347–76.

Knorr, Karin D., and Dietrich W. Knorr. *From Scenes to Scripts: On the Relationship Between Laboratory Research and Published Paper in Science.* Vienna: Institute for Advanced Studies Research Memorandum 132 (1978).

Knorr-Cetina, Karin, and A. V. Cicourel, eds. *Advances in Social Theory and Methodology: Toward an Integration of Micro- and Macro-Sociologies.* Boston: Routledge and Kegan Paul, 1981.

Kozulin, Alex. *Psychology in Utopia: Toward a Social History of Soviet Psychology.* Cambridge: MIT Press, 1984.

Kronick, David A. *A History of Scientific and Technical Periodicals: The Origin and Development of the Scientific and Technical Press, 1665–1790.* 2nd ed. Metuchen, N.J.: Scarecrow Press, 1976.

Kristeva, Julia. *Desire in Language: A Semiotic Approach to Literature and Art.* New York: Columbia University Press, 1980.

Kuhn, Thomas S. *The Structure of Scientific Revolutions.* Chicago: University of Chicago Press, 1962.

La Capra, Dominick. *History and Criticism,* Ithaca: Cornell University Press, 1985.

References

Lakatos, Imre. *The Methodology of Scientific Research Programs.* Cambridge: Cambridge University Press, 1978.
Latour, Bruno. "Essai de Science-Fabrication." *Etudes Françaises* 19 (1984): 114-33.
Latour, Bruno. *Science in Action.* Cambridge: Harvard University Press, 1987.
Latour, Bruno, and P. Fabbri. "La Rhetorique de la Science." *Actes de la Recherche* 13 (1981): 81-95.
Latour, Bruno, and Steve Woolgar. *Laboratory Life: The Social Construction of Scientific Facts.* Beverly Hills: Sage, 1979.
Law, John, and R. J. Williams. "Putting Facts Together: A Study of Scientific Persuasion." *Social Studies of Science* 12 (1982): 535-58.
Lee Kok Cheung, *Syntax of Scientific English.* Singapore: Singapore University Press, 1978.
Leech, Geoffrey N. *Principles of Pragmatics.* London: Longmans, 1983.
LeFevre, Karen Burke. *Invention as a Social Act.* Carbondale: Southern Illinois University Press, 1987.
LePage, Robert, and Andree Tabouret-Keller. *Acts of Identity.* Cambridge: Cambridge University Press, 1985.
Lohne, J. A. "Experimentum Crucis." *Notes and Records of the Royal Society of London* 23 (1968): 169-99.
Lohne, J. A. "The Increasing Corruption of Newton's Diagrams." *History of Science* 6 (1967): 69-89.
Lynch, Michael. *Art and Artifact in Laboratory Science: A Study of Shop Work and Shop Talk in a Research Laboratory.* London: Routledge and Kegan Paul, 1985.
Lyons, John. *Semantics.* 2 vols. Cambridge: Cambridge University Press, 1977.
McCloskey, Donald N. *The Rhetoric of Economics.* Madison: University of Wisconsin Press, 1986.
McConkie, G. W., K. Rayner, and S. J. Wilson. "Experimental Manipulation of Reading Strategies." *Journal of Educational Psychology* 65 (1973): 1-8.
Maimon, Elaine P., et al. *Writing in the Arts and Sciences,* 2nd ed. Boston: Little Brown, 1985.
Malinowski, Bronislaw. "The Problem of Meaning in Primitive Languages." Supplement to *The Meaning of Meaning.* C. K. Ogden and I. A. Richards. London: Kegan Paul, 1923.
Manier, Edward. "Darwin's Language and Logic." *Studies in History and Philosophy of Science* 11 (1980): 305-323.
Marcus, G., and D. Cushman. "Ethnographies as Texts." *Annual Review of Anthropology* 11 (1982): 25-69.
Marwell, G., and J. Hage. "The Organization of Role Relationships: A Systematic Description." *American Sociological Review* 35 (1970): 884-900.
Mead, George Herbert. *Mind, Self, and Society.* Chicago: University of Chicago Press, 1934.
Meadows, A. J. *Communication in Science.* London: Butterworths, 1974.
Meadows, A. J., ed. *Development of Science Publishing in Europe.* Amsterdam: Elsevier Science Publishers, 1980.

References

Medawar, P. B. "Is the Scientific Paper Fraudulent?" *Saturday Review* (1 August 1964): 42–43.
Menzel, H. "Planned and Unplanned Scientific Communication." In *Sociology of Science*, ed. Barber and Hirsch. New York: Free Press, 1962.
Merton, Robert K. "The Matthew Effect in Science." *Science* 159 (1968): 56–63.
Merton, Robert K. *On the Shoulders of Giants*. New York: Free Press, 1965.
Merton, Robert K. "The Role-Set: Problems in Sociological Theory." *British Journal of Sociology* 8 (1957), 106–20.
Merton, Robert K. *Social Theory and Social Structure*. New York: Free Press, 1968.
Merton, Robert K. *Sociological Ambivalence and Other Essays*. New York: Free Press, 1976.
Merton, Robert K. *The Sociology of Science*, ed. Norman Storer. Chicago: University of Chicago Press, 1973.
Merton, Robert K., and A. Kitt Rossi. "Contributions to the Theory of Reference Group Behavior." In *Continuities in Social Research*, ed. Merton and Lazaresfeld, 40–105. New York: Free Press, 1950.
Merton, Robert K., and Harriet Zuckerman. "Patterns of Evaluation in Science: Institutionalization, Structure and Functions of the Referee System." *Minerva* 9 (1971): 66–100.
Messeri, Peter. "Obliteration by Incorporation." Paper delivered at the meeting of the American Sociological Association, San Francisco, September 1978.
Meyer, Bonnie Jean. *The Organization of Prose and Its Effect on Memory*. New York: American Elsevier, 1975.
Miller, Carolyn R. "Genre as Social Action." *Quarterly Journal of Speech* 70 (1984): 151–67.
Mills, Gordon, and John Walter. *Technical Writing*, 3d ed. New York: Holt, Rinehart and Winston, 1970.
Mittroff, Ian, and Darryl Chubin. "Peer Review at the NSF." *Social Studies of Science* 9 (1979): 199–232.
Moravcsik, Michael, and P. Murugesan. "Some Results on the Function and Quality of Citations." *Social Studies of Science* 5 (1975): 86–92.
Mulkay, Michael. *Science and the Sociology of Knowledge*. London: Allen & Unwin, 1979.
Mulkay, Michael. "The Scientist Talks Back: A One-Act Play." *Social Studies of Science* 14 (1984): 265–84.
Mulkay, Michael. *The Word and the World: Explorations in the Form of Sociological Analysis*. London: George Allen & Unwin, 1985.
Murphy, James J., ed. *A Synoptic History of Classical Rhetoric*. Davis, Calif.: Hermagoras Press, 1983.
Myers, Greg. "The Social Construction of Two Biologists' Proposals." *Written Communication* 2 (1985): 219–45.
Myers, Greg. "Texts as Knowledge Claims: The Social Construction of Two Biology Articles." *Social Studies of Science* 15 (1985): 595–630.
Nabokov, Vladimir. *Speak Memory*. New York: Putnam. 1966.

References

Nelson, C. E., and D. K. Pollock, eds. *Communication Among Scientists and Engineers*. Lexington, Mass.: D. C. Heath, 1970.

Nelson, John S., Allan Megill, and Donald N. McCloskey, *The Rhetoric of the Human Sciences*. Madison: University of Wisconsin Press, 1987.

Newton, Isaac. *Certain Philosophical Questions: Newton's Trinity Notebook*, ed. J. E. McGuire and M. Tamny. Cambridge: Cambridge University Press, 1983.

Newton, Isaac. *Correspondence*, ed. H. W. Turnbull, J. F. Scott, A. R. Hall, and L. Tilling. 7 vols. Cambridge: Cambridge University Press, 1959–77.

Newton, Isaac. *Mathematical Principles of Natural Philosophy*. Trans. and ed., Andrew Motte; rev. Florian Cajori. Berkeley: University of California Press, 1934.

Newton, Isaac. *Optical Papers*. Vol 1, *The Optical Lectures*, ed. Alan Shapiro. Cambridge: Cambridge University Press, 1984

Newton, Isaac. *Opticks*. London, 1704, 1717, 1721, 1730; New York: Dover, 1979.

Newton, Isaac. Add 3958. Cambridge University Library.

Newton, Isaac. Add 3970. Cambridge University Library.

Nicholson, Marjorie Hope. *Newton Demands the Muse*. Princeton: Princeton University Press, 1946.

North, Stephen M. "Writing in a Philosophy Class: Three Case Studies." *Research in the Teaching of English* 20 (1986): 225–62.

Nystrand, Martin. "Rhetoric's 'Audience' and Linguistics' 'Speech Community.'" In *What Writers Know*, ed., Mosenthal et al., 220–35. New York: Longman, 1983.

Ochs, Kathleen H. "The Failed Revolution in Applied Science: Studies of Industry by Members of the Royal Society of London, 1660–1688." University of Toronto, 1981.

Odell, Lee, and Dixie Goswami, eds. *Writing in Non-Academic Settings*. New York: Guildford, 1985.

Oehler, Kay, and Nicholas Mullins. "Mechanisms of Reflexivity in Science." Unpublished manuscript, 1986.

Olby, Robert, *The Road to the Double Helix*. London: Yoxen, 1976.

Oldenburg, Henry, *Correspondence*. 9 vols. Ed. M. B. Hall and R. Hall. Madison: University of Wisconsin Press, 1965–73. Vols. 10–11: Chicago: Mansell, 1975 and 1977. Vols. 12–13: London: Taylor and Francis, 1986.

Orr, Leonard."Intertextuality and the Cultural Text in Recent Semiotics." *College English* 48 (1986): 811–23.

Overington, Michael. "The Scientific Community as Audience: Toward a Rhetorical Analysis of Science." *Philosophy and Rhetoric* 10 (1977): 143–62.

Oxford English Dictionary. Oxford: Oxford University Press, 1971.

Paradis, James. "Montaigne, Boyle, and the Essay of Experience." In *One Culture*, ed. George Levine, 59–91. Madison: University of Wisconsin Press, 1987.

Pauling, Linus. *The Nature of the Chemical Bond*. Ithaca: Cornell University Press, 1939; 1940; 1960.

Perelman, Les. "The Context of Classroom Writing." *College English* 48 (1986): 471–79.

References

Physics Survey Committee, National Research Council. *Physics in Perspective IIB. The Interfaces.* Washington, D. C.: National Academy of Sciences, 1973.

Pickering, Andrew. *Constructing Quarks.* Chicago: University of Chicago Press, 1984.

Plato. *Gorgias.* Trans. W. C. Helmbold. Indianapolis: Bobbs-Merrill, 1952.

Plato. *Phaedrus.* Trans. W. C. Helmbold and W. G. Rabinowitz. Indianapolis: Bobbs-Merrill, 1956.

Plato. *The Republic.* Trans. Allan Bloom. New York: Basic Books, 1968.

Polyanyi, Michael. *Personal Knowledge.* Chicago: University of Chicago Press, 1962.

Popper, Karl. *Objective Knowledge: An Evolutionary Approach.* Rev. ed. Oxford: Clarendon Press, 1979.

Price, Derek de Solla. *Little Science, Big Science.* New York: Columbia University Press, 1963.

Price, Derek de Solla, and D. Beaver. "Collaboration in an Invisible College." *American Psychologist* 21 (1966): 1011–18.

Priestley, Joseph. *History and Present State of Electricity.* London, 1775.

Reynolds, R. E., and R. C. Anderson. *Influence of Questions on the Allocation of Attention during Reading.* Research Report 183. Urbana, Ill.: Center for the Study of Reading, 1980.

Reynolds, R. E., M. A. Taylor, M. S. Steffensen, L. L. Shirey, and R. C. Anderson. *Cultural Schemata and Reading Comprehension.* Research Report 201. Urbana, Ill.: Center for the Study of Reading, 1981.

Rosaldo, Renato. "Where Objectivity Lies: The Rhetoric of Anthropology." In Nelson et al., 87–110.

Rubin, Donald, and Bennett Rafoth. "Social Cognitive Ability as a Predictor of the Quality of Expository and Persuasive Writing Among College Freshmen." *Research in the Teaching of English* 20 (1986): 9–21.

Rudwick, Martin. *The Great Devonian Controversy.* Chicago: University of Chicago Press, 1985.

Rumelhart, D. E., and A. Orotony. "The Representation of Knowledge in Memory." *Schooling and the Acquisition of Knowledge.* Ed. R. Anderson, R. Spiro, and W. Montague. Hillsdale, N.J.: Erlbaum, 1977.

Sabra, A. I. *Theories of Light from Descartes to Newton.* Cambridge: Cambridge University Press, 1981.

Sapir, Edward. *Language.* New York: Harcourt Brace, 1921.

Saussere, F. de. *Course in General Linguistics.* Tr. Roy Harris. London: Duckworth, 1983.

Savory, T. H. *The Language of Science.* London: Andre Deutsch, 1967.

Sayre, Anne. *Rosalind Franklin and DNA.* New York: Norton, 1975.

Scaliger, Julius Caesar. *Poetices Libri Septem.* Heidelberg, 1581.

Schaffer, Simon. "Glass Works." Unpublished manuscript, 1986.

Schutz, Alfred, and Thomas Luckmann. *The Structures of the Life-world.* Evanston, Ill.: Northwestern University Press, 1973.

Scribner, Sylvia, and Michael Cole. *The Psychological Consequences of Literacy.* Cambridge: Harvard University Press, 1981.

References

Searle, John. *Speech Acts*. Cambridge: Cambridge University Press, 1969.
Searle, John, Ferenc Kiefer, and Manfred Bierwisc, eds. *Speech Act Theory and Pragmatics*. Dordrecht: D. Reidel, 1980.
Shadwell, Thomas. *The Virtuouso*. London, 1676.
Shapin, Steven. "Pump and Circumstance: Robert Boyle's Literary Technology." *Social Studies of Science* 14 (1984): 481-520.
Shapin, Steven. "History of Science and Its Sociological Reconstructions." *History of Science* 20 (1982): 157-211.
Shapin, Steven, and Simon Schaffer. *The Leviathan and the Air-Pump: Hobbes, Boyle, and the Experimental Life*. Princeton: Princeton University Press, 1985.
Shapiro, Alan. "The Evolving Structure of Newton's Theory of White Light and Colors." *Isis* 71 (1989): 197-210.
Shapiro, Alan. "Experiment and Mathematics in Newton's Theory of Color." *Physics Today* 37 (1984): 34-42.
Shapiro, Alan. "Newton's 'Achromatic' Dispersion Law: Theoretical Background and Experimental Evidence." *Archive for the History of the Exact Sciences* 21 (1979): 91-128.
Shapiro, Barbara. *Probability and Certainty in Seventeenth-Century England*. Princeton: Princeton University Press, 1983.
Skinner, B. F. *Verbal Behavior*. London: Methuen, 1957.
Small, Henry G. "Cited Documents as Concept Symbols." *Social Studies of Science* 8 (1978): 327-40.
Small, Henry G. "A Co-Citation Model of a Scientific Specialty: A Longitudinal Study of Collagen Research." Social Studies of Science 7 (1977): 139-66.
Smith, Laurence D. "Psychology and Philosophy: Toward a Realignment, 1905-1935." *Journal of the History of the Behavioral Sciences* 17 (1981): 28-37.
Sommerfeld, Arnold. *Atembau und Spektralinien*. Braunschweig: Friederich Vieweg, 1919; 1921; 1922; 1924; 1929; 1931; 1944.
Spiro, R. J. *Schema Theory and Reading Comprehension: New Directions*. Research Report 191. Urbana, Ill.: Center for the Study of Reading, 1980.
Sprat, Thomas. *History of the Royal Society*. London, 1667.
Steffenson, M. S. *Register, Cohesion, and Cross-Cultural Reading Comprehension*. Research Report 220. Urbana, Ill.: Center for the Study of Reading, 1981.
Stuewer, Roger. *The Compton Effect*. New York: Science History, 1975.
Stimson, Dorothy. *Scientists and Amateurs*. London: Sigma, 1949.
Stinchcombe, Arthur. "Merton's Theory of Social Structure." In *The Idea of Social Structure*, ed. L. Coser. New York: Harcourt Brace Jovanovich, 1975: 11-34.
Stryker, Sheldon. "Status Inconsistency and Role Conflict." *Annual Review of Sociology* 1978: 57-90.
Suleiman, S., and I. Crossman, eds. *The Reader in the Text*. Princeton: Princeton University Press, 1980.
Sullivan, Harry Stack. *The Interpersonal Theory of Psychology*. New York: Norton, 1953.
Sutherland, James. *The Restoration Newspaper and Its Development*. Cambridge: Cambridge University Press, 1986.

Swales, John. *Aspects of Article Introductions*. Birmingham: Aston University ESP Research Report, 1983.
Swales, John. "English as the International Language of Research." *RELC Journal* 16 (1986): 1–7.
Swift, Jonathan. *Gulliver's Travels*. London, 1726.
Tate, Gary, ed. *Teaching Composition: Twelve Bibliographic Essays*. Fort Worth: Texas Christian University Press, 1987.
Tompkins, J., ed. *Reader Response Criticism*. Baltimore: Johns Hopkins University Press, 1980.
Toulmin, Stephen. *Human Understanding*. Princeton: Princeton University Press, 1972.
Toulmin, Stephen, and David Leary. "The Cult of Empiricism in Psychology." In *A Century of Psychology as Science: Retrospections and Assessments*. New York: McGraw Hill, in press.
Turner, R. "The Navy Disbursing Officer as Bureaucrat." *American Sociological Review* 12 (1947): 342–48.
Turner, R. "Role-taking, Role-Standpoint, and Reference Group Behavior." *American Journal of Sociology* 61 (1956): 316–28.
Tyler, Stephen. *The Unspeakable: Discourse, Dialogue, and Rhetoric in the Postmodern World*. Madison: University of Wisconsin Press, 1987.
Vatz, Richard E. "The Myth of the Rhetorical Situation." *Philosophy and Rhetoric* 6 (1973): 154–61.
Vygotsky, Lev. *Mind in Society: The Development of Higher Psychological Processes*. Cambridge: Harvard University Press, 1978.
Vygotsky, Lev. *Thought and Language*. Cambridge: MIT Press, 1962.
Watson, James. *The Double Helix*. New York: Atheneum, 1968.
Watson, J. D., and F. H. C. Crick. "Genetical Implications of the Structure of Deoxyribonucleic Acid." *Nature* 171 (1953): 942–67.
Watson, J. D., and F. H. C. Crick. "The Structure of DNA." *Cold Spring Harbor Symposia on Quantitative Biology* 18 (1953): 123–31.
Webster's New Collegiate Dictionary. Springfield, Mass.: G. & C. Merriam, 1953.
Weber, Max. *The Theory of Social and Economic Organization*. London: Oxford University Press, 1947.
Weimer, Walter B. "Science as a Rhetorical Transaction: Toward a Nonjustificational Conception of Rhetoric." *Philosophy and Rhetoric* 10 (1977): 1–19.
Wertsch, James. *Vygotsky and the Social Formation of Mind*. Cambridge: Harvard University Press, 1985
Wertsch, James, ed. *The Concept of Activity in Soviet Psychology*. Armonk, N. Y.: M. E. Sharpe, 1981.
Wertsch, James, ed. *Culture, Communication, and Cognition: Vygotskian Perspectives*. New York: Cambridge University Press, 1985.
Westfall, Richard. "The Development of Newton's Theory of Color." *Isis* 53 (1962): 339–58.
Westfall, Richard. "Isaac Newton's Coloured Circles twixt two Contiguous Glasses." *Archive for the History of the Exact Sciences* 2 (1965): 181–96.

References

Westfall, Richard. *Never at Rest*. Cambridge: Cambridge University Press, 1980.
Westfall, Richard. "Uneasily Fitful Reflections on Fits of Easy Transmission." *Texas Quarterly* 10 (1967): 86–102.
White, James Boyd. *Heracles' Bow: Essays on the Rhetoric and Poetics of the Law*. Madison: University of Wisconsin Press, 1986.
White, Hayden. *Tropics of Discourse*. Baltimore: Johns Hopkins University Press, 1978.
White, Leslie. *The Science of Culture*. New York: Farrar, Strauss and Cudahy, 1949.
Whorf, Benjamin Lee. *Language, Thought, and Reality*. Cambridge: MIT, 1956.
Wilkins, John. *An Essay Towards a Real Character and a Philosophical Language*. London, 1668.
Wittgenstein, Ludwig. *Philosophical Investigations*. New York: Macmillan, 1953.
Woolgar, Steve. "Discovery: Logic and Sequence in a Scientific Text." In *The Social Processes of Scientific Discovery*. Dordrecht: D. Reidel, 1981.
Woolgar, Steve. "Irony in the Social Study of Science." *Science Observed*. Ed. Karin D. Knorr-Cetina and Michael Mulkay. London: Sage, 1983: 239–66.
Woolgar, Steve. "Writing an Intellectual History of Scientific Development: The Use of Discovery Accounts." *Social Studies of Science* 6 (1976): 395–422.
Yearley, Stephen. "Textual Persuasion: The Role of Scientific Accounting in the Construction of Scientific Arguments." *Philosophy of the Social Sciences* 11 (1981): 409–35.
Ziman, John. *Public Knowledge*. Cambridge: Cambridge University Press, 1968.
Zuckerman, Harriet. "Cognitive and Social Processes in Scientific Discovery." Unpublished manuscript.
Zuckerman, Harriet. *Scientific Elite*. New York: Free Press, 1977.

INDEX

Abt, Helmut, 162
Abel, Othenio, 81
Abstract: Compton's writing of, 220–22; use of in reading, 241; in psychology, 262–63, 272–74
Academia del Cimento, 130
Academie des Sciences, 130, 139
Accounts and accountability, 60–62, 145, 184, 190, 191, 198, 253, 305–13, 328–30
Achard, Franz Karl, 81
Acid, changing definition of, 29
Acknowledgments within articles, 33, 175, 176
Active constraints, 61, 222, 312–13. *See also* Passive constraints
Adanson, Michel, 81
Aepinus, Franz, 81
Agassi, Joseph, 4*n*3
Agonistic structure of scientific communication, 82, 130, 147, 155–56, 191, 200, 316, 319, 321, 323
Aitken, Robert, 81
Alzate y. Ramírez, José Antonio, 81
American Journal of Psychology, 263–64, 265, 268, 273
American Physical Society, 162
American physics, compared to European, 157, 162
American Political Science Review, 280–88
American Psychological Association Publication Manual, 17, 253–63, 275, 316; prior versions, 261–63, 271; prescriptiveness, 259–63, 271
Ampere, Andre, 81
Anomalous findings, 141
Anschuetz, Richard, 81
Anthropology, 21, 279–80
Antoniadi, Eugen, 81

Apparatus, passive constraints on, 311
Appleton, Edward, 81
Arbitrariness of sign, 296
Aristotle, 188, 256
Audience: role of in texts, 24–25, 46, 325, 326; in Watson and Crick, 28, 32–33; in Merton, 34, 36; in Hartman, 39, 41, 44; in Newton, 86, 87, 99, 124–25; differentiation of in early journals, 131–33, 135–36; in Compton 214–19; in psychology, 264–65, 268, 270, 274–75
Austin, J. L., 299
Author: role of, 131–35. *See also* Role conflicts
Authorial presence (persona), 24–26, 46–47, 140–41, 325–26, 328; in Watson and Crick, 28, 30, 31; in Merton, 34, 36, 39; in Hartman, 39–40, 44–45; in Newton, 123, 124–25; in *Philosophical Transactions*, 133–36; in Compton, 214; in psychology, 268, 270, 272–73; in political science, 287

Bacon, Francis, 15–16, 91–92, 138, 140, 293
Baconianism, 91–92, 293, 326
Baldauf, R. B., 18*n*1
Baldwin, J. M., 263
Balmer, Johann Jakob, 161
Barkla, C. G., 194, 196
Barnes, Barry, 129*n*1, 294
Barnes, Sherman, 138*n*8
Barrow, Isaac, 85, 86, 88
Bechler, Zev, 95*n*15
Behaviorism in psychology, 263, 268, 323
Behaviorist rhetoric, 259, 263, 269–75
Behn, Aphra, 134

Index

Bellone, Enrico, 293
Ben-David, Joseph, 149n13
Bible, 20, 21
Biddle, Bruce, 136
Bitzer, Lloyd, 8n10
Bizzell, Patricia, 322n2
Black boxes. *See* Reading
Bleich, David, 236
Bloor, David, 294
Bohr, Nils, 158, 179, 182, 206-7
Books as scientific communication medium, 80-82, 158
Boring, Edwin, 273
Boyle, Robert, 69, 76, 80, 81, 85, 88, 134
Bragg, Lawrence, 194, 196-97, 198
Brannigan, Augustine, 35n15
Bromley Report, 157n6, 162, 281
Brouncker, Lord, 105n19
Brown, Richard, 11n12
Bruce, Bertram, 236
Bruner, Jerome, 304
Butler, Samuel, 134

Callon, Michel, 190n1
Cambridge University: Newton at, 85, 86
Cattell, J. M., 263
Cavallo, Tiberius, 74
Chomsky, Noam, 297, 299, 300
Chubin, Darryl, 164n8
Citations and References, 25, 319; in *Physical Review*, 164-67; style in psychology, 262-63, 274; in *American Political Science Review*, 282-83; citation studies, 235-36
Clifford, J., 280
Closed text, 83, 124
Codification, 25, 31, 46, 126, 274, 316, 319
Cognitivist influence on psychological style, 275
Cohen, I. B., 82n3, 125, 127n25
Cohen, Ralph, 7n8
Cole, Jonathan, 141, 146
Collins, Harry, 4n2, 23, 129n1, 141, 156n3, 174, 294
Collins, Randall, 129n1
Composition, 3n1
Compton, Arthur Holly, 17, 91, 191-99, 203, 282, 284, 314, 316, 326, 327; "Quantum Theory," 192, 197-99, 203-4; "Measurement of Beta Rays," 192-93, 201, 205-24; "Secondary Radiations," 199; follow-up publications, 203-7; research program, 194-96; theory change, 199, 203-4; use of cloud chamber experiments, 204-7; notebooks, 207-22; corrections to data, 208-11, 219; writing-up of article, 209-10; revisions of text, 210-22
Compton effect: defined, 191-92
Conclusions section, 175, 243, 267
Conklin, E., 267
Conklin, V., 267
Consigny, Scott, 8n10
Constraints: empirical, 182, 192, 194-95, 200, 201, 205-6, 208; from research program, 194-99. *See also* Active and passive constraints; Empirical experience
Construction: of theory, 178-79, 180, 181; of linguistic representation of objects, 215; of textual object, 219-22, 224; of reading, 235-37; of literature through reading, 250-53; of scientific language, 313
Constructivism, 129, 148-49, 294-95
Contributor, *See* Author
Cooper, Charles, 322n2
Cornell University, 162
Correspondence theory of reference, 295, 298
Correspondent, role of, 131-32
Coser, Rose, 148n11
Cozzens, Susan, 25n6, 157n4,5, 164n8, 203, 235, 283
Crane, Diana, 22, 145
Crick, Frances. *See* Watson, James
Critic, role of, 135-36
Criticism, 138-39, 147

Index

Crosland, Maurice, 136
Crucial experiment. *See* Newton, Isaac
Crystal, David, 155
Cultural forms, history of, 314–16

Dallenbach, K. M., 267
Darwin, Charles, 6
Dear, Peter, 5n4, 123, 140
deBroglie, Louis, 158
Debye, Peter, 197–98
Decorum, 320
Deixis, 299
Derrida, Jacques, 10, 60n2
Desaguliers, Jean T., 67, 70, 72, 74, 76, 80
DesCartes, Rene, 85, 88
DiMaggio, Paul, 7n8
Discovery account, 78, 90–95
Discussion sections, 175
Documentation. *See* Citations
Dolezal, Frederic, 16
Dolland, John, 95n15
Dooley, Lucille, 268
Dubrow, Heather, 7n8

Eastwood, B., 5n4
Eco, Umberto, 236
Economics, rhetoric of, 279
Editor: role and role conflicts of, 132–33, 136–37; impact on article length, 163. *See also* Oldenburg, Henry
Editorial board, 137
Einstein, Albert, 124, 158, 194
Eisenstein, Elizabeth, 5n4, 128
Electrodynamic theory, Compton's commitment to, 194–95, 197–99
Empirical experience: as authority, 147; as constraint and resource, 188, 312–13, 327; embedded in language use, 188, 190–91, 309–11, 321
Empiricism, 149, 295, 323; cult of, 268
Ennis, Michael, 202
Epistemic level, Compton's control of, 214–15

Epistemology, 26–27, 174, 177–81, 323–24
Equations in text, 173
Ethnographies, rhetorical complexity of, 279–80
Ethnomethodology, 21, 299
Ethos, 140
Everyday exprience, in political science, 285–86
Experimental subjects in psychology, 260, 266, 267, 270, 271. *See also* Introspection
Experiments: reports of 22, 59, 202; changing definition of, 65–68; controls, 71; series of, 76–77, 274, 282; demonstration, 98
Extending, 211–12

Fabian, J., 280
Fact, Fleck's definition of, 61
Facticity, persuasion of, 140
Fahnestock, Jeanne, 322n3
Faigley, Lester, 322n2
Fine-tuning, 212–13
Finocchario, M. A., 4n3
First person: in Watson and Crick, 30, 31; in Newton, 91, 96; in Compton, 216–17; in psychology, 265, 269; in political science, 287
Firth, J. R., 297, 300
Fleck, Ludwik, 4n3, 61, 108, 200, 222, 296, 312–13
Footnotes. *See* Citations
Foucault, Michel, 10, 60n2
Fowler, Alistair, 7n8
Fowlkes, Diane, 281, 283–88
Fox, Theodore, 138n8
Fraunhofer lines, 160
Freud, Sigmund, 270
Friedrich, Max, 265

Garfinkel, Harold, 293
Garvey, William, 59n1
Gascoines, John, 117–18
Gatekeeping, 136–38, 141, 144n10
Geertz, Clifford, 280

Index

Genre: theory and definition of, 6–8, 62–63, 319–20; experimental report as a, 6–8, 59–60, 62–63, 77–78, 260, 315–17
Geometrical demonstrations. *See* Mathematics
George, S. S., 267
Gieryn, Tom, 294
Gilbert, Nigel, 25*n*6, 156*n*3, *156n3*, 202
Goffman, Erving, 148*n*11
Goody, Jack, 5*n*4, 128
Gopnik, Myrna, 155, 298
Graff, Harvey, 128
Graphic features, charts, and illustrations: in Watson and Crick, 32; in *Philosophical Transactions*, 71–72; in Newton, 85, 94*n*14, 122; in *Physical Review*, 172–74, 177; in Compton, 198, 208
Greenstein, Fred, 281
Group formation and integration, 138–41
Guerlac, Henry, 118*n*23
Gumperz, John, 298
Gusfield, Joseph, 25*n*8, 156*n*3

Hack, 12
Hacking, Ian, 4*n*3, 7*n*9, 123, 140
Hahn, Roger, 139
Hall, Captain, 76
Hall, G. Stanley, 263, 264, 265
Hall, Marie Boas, 66*n*3, 130, 147
Hall, Rupert, 130*n*3
Halliday, Michael, 300–301
Halloran, S. Michael, 322*n*3
Handel, Warren, 149
Hanson, N. R., 144, 294
Hartman, Geoffrey: discussion of, 27, 39–48, 61; text excerpt, 54–55
Hauksbee, Francis, 67, 72, 80
Havelock, Eric, 5*n*4, 128
Hayes, Louis D., 281
Hemingway, Ernest, 36
Henry, William, 68, 71, 77
Henshaw, Thomas, 69
Herrington, Anne, 322*n*2

Herschel, William, 67*n*4, 71, 72, 73, 76, 79
Hewson, William, 77
Hill, John, 137
Historical analysis of language use, 313–15
History, 4, 10
Hjelmslev, Louis, 297
Hogben, Lancelot, 155
Holland, Norman, 236
Holmes, Frederic, 5*n*4
Holub, R. C., 236
Hooke, Robert, 100–101, 104–9, 111, 123, 134
Horner, Winifred B., 332*n*1
Huddleston, R. D., 155, 298
Hudson, R. A., 298
Hunter, Michael, 133, 135, 136
Huygens, Christiaan, 66, 107, 108, 110–16, 123
Hypothesis: testing, 7*n*9, 66–68, 281; placement in article in psychology, 273–74
Hyslop, James, 265

Incremental encyclopedism in experimental psychology, 261, 273–75
Institutionalization of genres and style, 316, 319
Integrative apparatus, 126–27, 157. *See also* Citations; Intertextuality; Theory
Interdisciplinarity of this study, 3–6
Intersubjectivity in language use 107*n*21, 251–52, 258, 303–11, 314–16
Intertextuality, 236, 316, 324–25. *See also* Citations; Theory
Introspection in early experimental psychology, 265, 267, 270
Isaak, Alan C., 281
Iser, Wolfgang, 236

Jameison, Kathleen, 8*n*10
Jastrow, Joseph, 266
Jennings, Edward T., 281, 283–88
Johnston, P., 236

Index

Johnstone, James, 73
Journal des Scavans, 129n2
Journal of Experimental Psychology, 267, 269
Judson, Horace F., 28n11

Kayser, H. J. G., 161
Kennedy, George, 322n1
Kevles, Daniel, 157
Key, V. O., 283, 285
Kinneavy, James, 24
Kirchoff, Gustav Robert, 161
Kittredge, Richard, 155, 297
Knorr (Knorr-Cetina), Karin, 23, 25n8, 129n1, 156n3, 200, 209, 294, 298
Knowledge, textual form and, 18-24, 47
Kozulin, Alex, 302
Kristeva, Julia, 236
Kronick, David A., 59n1, 63, 80, 81
Kuhn, Thomas, 4n3, 82n3, 295, 308

Lakatos, Imre, 4n3, 284, 308
Language: philosophy and theory of, 5, 19, 289-315; spoken vs. written, 21-23; underdetermination and overdetermination of, 28, 36-37, 45; and activity, 295-96, 303-15. *See also* Scientific language
Langue-parole distinction, 296-97
Latour, Bruno, 22, 23, 25n8, 28n10, 130n4, 145, 156n3, 190n1, 200, 207, 287, 294, 298
Law, John, 156n3
Lee, Kok Cheung, 155, 298
Leech, Geoffrey, 299
Leeuwenhoek, Anton, 80
LeFevre, Karen, 322n2
Legal language, 61
Leibniz, Gottfried W., 67, 76
Length of articles: in *Physical Review*, 162-64, 183; in experimental psychology, 273, 274; in *American Political Science Review*, 281-82
LePage, Robert, 298

Lewis, William, 73
Line, Francis, 69-70, 72-73
Linguistics, 10, 11, 21, 187, 295; code orientation of, 296-97; contrary developments, 297-300; sociolinguistics, 298; pragmatics, 299; developmental, 299, 300, 304-5; text grammars, 299-300
Literary criticism, method of, 159
Literary studies, 3n1, 11, 61, 189, 236
Literary theory, 3n1, 10, 20, 60, 62, 187
Literature (of a field): use of in writing, 24-25, 324-26, 328; In Watson and Crick, 27, 30-31; in Merton, 34, 35-36, 38; four levels of in Hartman, 39, 42-44; in *Philosophical Transactions*, 78-79; Newton's lack of, 126-27; Compton's merging of self into, 216; construction of in reading, 250-53; in psychology, 265, 267, 270, 273; reconstruction of in political science, 282-84
Logic, 188-89
Lohne, J. A. 94n14
Lucas, Anthony, 118-19
Lynch, Michael, 23, 304n1
Lyons, John, 299

McCloskey, Donald, 11n12
McConkie, G. W., 236
McGuire, J. E., 83, 85, 93
Malinowski, Bronislaw, 297, 299, 300
Manier, Edward, 4n3
Marcus, G., 280
Marwell, G., 134-36
Material activity and language, 305-15
Mathematics: in scientific writing, 6; as model of discourse for Newton, 87, 98-99, 113-16; selective reading of, 246-47; in economics, 279; in political science, 280, 288; compared to rhetoric as a need of scientists, 331-32
Mead, George Herbert, 5n5
Meadows, A. J., 59n1, 67, 139

Index

Medawar, P. B., 200, 209
Menzel, H., 145
Mersenne, Marin, 130
Merton, Robert King: writing analyzed, 27, 34–39, 46–48; text excerpt, 51–53; theory and findings, 4n2, 25n7, 129, 134–37, 139, 142, 146, 148, 157n6
Messeri, Peter, 203, 283
Metaphor, 36–37
Method: of studies in this book, 24–27, 63–65, 83–84, 128–29, 159–62, 167n10, 192–93, 237–38, 263, 281; in *Philosophical Transactions*, 68–72, 284, 285; sections, 175, 243, 328; in psychology, 260, 267, 272, 285; in political science, 284–85; in Compton, 285
Meyer, Bonnie Jean, 155
Microstructure, relation to macrostructure, 129, 148–49
Miller, Carolyn, 7n8
Mind, 266n1
Mittroff, Ian, 145
Modelling approach to physics, 181, 183
Moravcsik, Michael, 164n8
Moray, Robert, 100, 101–2
Mortimer, Cromwell, 137
Moving minds, 188–91
Mozart, Wolfgang Amadeus, 65
Mulkay, Michael, 11n12, 129n1, 141, 294
Munsterberg, Hugo, 264
Murphy, James J., 322n1
Myers, Greg, 144n10, 145, 146, 184

Names in reading search, 239–40
Negotiation of claims, 144–46, 184, 309–11
Nelson, John, 11n12
Newton, Isaac: crucial experiment, 7n9, 91, 92, 94, 103, 104, 109, 118; as rhetorical innovator and influence, 15, 16, 82–83, 92, 125–26, 191, 314, 317, 320, 324; answer to Line, 69–70, 72–73, 117; his "New Theory," 82, 89, 90–99, 126, 133; refusal to publish, 82, 119; his *Opticks*, 83, 84, 104, 108, 119–26, 157; his notes and notebooks, 83, 85, 88, 93, 94, 97; his "Of Colours," 83, 85–86, 93, 94; his *Optical Lectures*, 83, 86–87, 89, 93, 94, 98–99, 119; relation to audience, 86–87, 99, 102, 110, 116; mathematical argument, 87, 98-99, 113–16, 119–23, 308; reflecting telescope, 88, 95–96; and Royal Society, 88, 100, 111n22, 134; Correspondence with Oldenburg, 88–90, 97–98, 100, 110, 111n22, 113; certainty of facticity, 89–90, 97–98, 99, 123; use of Baconian ethos, 91–92, 93, 95, 123, 326; conflicts among presentations of findings, 92–95; purchase of prisms, 93; doctrine, 96–99; proscription on hypothesizing, 97–98, 101n18; his "Queries," 100, 101, 110; answer to Moray, 100, 101–2; answer to Pardies, 100, 102–4; answer to Hooke, 100–101, 104–9; controversy over "New Theory," 100–119, 327; empiricist reduction of questions, 101n18, 103, 104, 105–7, 110, 114; and compound colors, 101n18, 108-9, 111–16; analogies, use of and argument against, 106, 112; concept of abstraction, 106–7; answer to Huygens, 107, 108, 110–16; compulsion through system, 114–17, 121–27; answer to Gascoines, 117–18; labelling disagreement as error, 117–19; answer to Lucas, 118–19; with Aubrey, 119; his *Principia*, 119, 125–26, 317; public persona, 125, 134; acceptance of criticism and cooperation, 189n9
Nicholson, Marjorie Hope, 124
Nonfiction, 60–123
Norms of science, 148
North, Stephen, 322n2

Index

Nystrand, Martin, 322n2

Object of study, 24–25, 47–48; in Watson and Crick, 27–34; in Merton, 34–39; in Hartman, 39–40, 42
Ochs, Kathleen, 66n3, 147
Ocular proof, 73–75
Odell, Lee, 322n2
Odysseus, 330
Oehler, Kay, 294
Olby, Robert, 28n11
Oldenburg, Henry: as editor of *Philosophical Transactions*, 69, 75, 129–34; and Newton, 84, 88–90, 97–98, 100, 110, 113; mentioned, 15, 314, 320
Organization: in *Philosophical Transactions*, 75–77; in Newton's *Opticks*, 119–23; in *Physical Review*, 174–81; in *APA Manual*, 260–63; in psychological journals, 264, 266, 268, 269; in *American Political Science Review*, 280, 287, 288
Orr, Leonard, 236
Overington, Michael, 322n3

Page, Benjamin, 281, 283–88
Papin, Denis, 66
Paradis, James, 63, 82n1, 123
Pardies, Ignace, 100, 102–4
Passive constraints, 61, 108, 200, 205–7, 223, 311–13
Pauling, Linus, 31, 81
Perelman, Les, 322n2
Peterson, Margaret Jean, 274
Petty, William, 76
Philosophical Transactions of the Royal Society of London, 16, 63–80, 130–38, 140, 282, 284, 320; Newton's perception of, 82, 84, 87–90; early readership, 133, 134; mentioned, 80. See also Oldenburg, Henry
Philosophische Studien, 263, 264–65, 266
Philosophy, 4, 10, 62, 187
Photographic plates, used by Compton, 314

Physical Review, 16, 80, 157–83, 204, 282, 283
Physical Review Letters, 158
Pickering, Andrew, 294
Plato, 21, 189, 264, 293
Political science, 17, 189; rhetoric of, 280–88; scope and methods books, 281; compared to political philosophy, 287
Polyanyi, Michael, 307n2
Popper, Karl, 4n3, 7n9, 25n5, 78
Popular journals, 135
Postmodernism, 12, 13
Postponing, 210–11
Practice, language as, 13, 301–4
Precision, in Compton's revisions, 212–14, 223
Prescriptiveness in writing of psychology, 259–63, 271
Price, Derek de Solla, 145, 158, 283
Priestley, Joseph, 68, 80
Problem: definition in articles, 177, 179, 181, 197–98, 243, 244, 284; area, 194–96
Pruette, Lorine, 268
Psychological Review, 263, 264, 266, 269
Psychology, experimental: diffusion point for social scientific style, 259; official style, 259–61, 271, 285, 320; early journals of, 263; early philosophic character of, 263–68; split from philosophy, 266–69; rise of behaviorism, 268–71; mentioned, 5, 10, 17, 189. See also *American Psychological Association Publication Manual*; Experimental subjects
Public identity of scientists, 134–36, 138

Quantum theory: in spectroscopy articles, 157, 161, 171, 182–83; and Compton effect, 191–92, 194, 197, 199, 203–4, 216; mentioned, 154n1

Reader, role of, 131–33

Index

Reading, 235–53, 317, 325; as contextualized constructive activity, 235; purpose in, 237–38, 242–50; schema in, 237, 239–50; relation to research program, 236, 238; search for articles to read, 238–41; comprehension difficulties, 244–45; black boxes in, 244–47; evaluation of texts; 246–49; psychological and sociological variables of, 250–52
Reception of texts, 203–4, 329, 330
Referee, institution of, 137
Reference, problem of, 187–88, 224–25, 294, 299
References. *See* Citations
Reflexivity. *See* Rhetorical self-consciousness
Reliably reconstitutable phenomena, 173, 307–11, 314–15
Religious discourse, 61, 253, 312
Replication, difficulties of, 27n9, 69–70, 74, 141, 174, 310
Results section, 72–73, 175, 243, 272, 328
Review of literature essay, function for Compton, 199
Revision, *See* Compton, Arthur H.
Reynolds, R. E., 236
Rhetoric, 3n1, 6n7, 62; rhetorical theory, 15; rhetorical character of knowledge writing, 23–24; rhetorical situations and problems, 8, 206–7, 271, 316, 326–27; rhetorical universe, stabilization of, 217, 275; rhetorical self-consciousness, 317, 320–21, 323–32; rhetorical tradition, 321–22
Rich, G. J., 267
Rizzetti, John, 70
Role: conflict, 134–36, 142–43, 147; conflict mediation, 143–48; unification in science, 144, 146–49; distance, 148n11
Roller, Duane H. D., 119
Rosaldo, Renato, 280
Royal Society of London: and Newton, 88–89, 111n22, 134; early membership, 133; mentioned, 130, 131, 133. *See also* Oldenburg, Henry; *Philosophical Transactions*
Rubin, Donald, 322n2
Rudwick, Martin, 5n4, 156n4
Rumelhart, D. E., 236
Runge, Carl, 161
Rydberg, Johannes Robert, 161

Sabra, A. I., 91
Sacred texts, 61, 253
Sapir, Edward, 297
Saussure, F. de, 296–97, 300
Savory, T. H., 155
Sayre, Anne, 33n14
Scaffolding, 306
Scaliger, Julius Caesar, 7n8
Schaffer, Simon, 118n23
Schema. *See* Reading
Schools, formation of in social sciences, 39
Schrodinger, Erwin, 158
Scientific language: invisibility of rhetorical character, 6, 14, 257, 320, 322; accomplishment of, 6, 13–15, 257; studies of, 155–56, 189–90, 315; and activity of science, 155; as construction, 155, 298; migration to social sciences, 257–58, 278–79; as complex persuasive system, 258; simplified models of in social sciences, 259–61, 279–81; privileging in common sense view, 292–93; reasons to distrust, 293–95; lack of adequate model of, 298, 300–302; proposed model, 302–7. *See also* Genre
Scribner, Sylvia, 5n4, 128, 313
Searle, John, 299
Section headings: in *Physical Review*, 174–75; in Compton, 220; in psychology, 260–62, 268, 269, 272; in political science, 287–88
Selectivity of data by Compton, 201, 208

Index

Semiotics, 294, 297
Sentence length in *Physical Review*, 167–69
Shadwell, Thomas, 134
Shapin, Steven, 5n4, 75, 140, 144
Shapiro, Alan, 84, 89, 95n15, 108
Shapiro, Barbara, 123, 140
Shared understanding in communication, 303–11
Simon, Alfred W., 192
Situation, structured, 193–94
Skinner, B. F., 268, 293
Sloan, Hans, 136–37
Small, Henry, 21–22, 25n6, 128, 164n8, 203, 303–5
Smith, Lawrence D., 268n2
Social structure, relation to language and cognition, 128–29, 182, 301, 319–21, 324–26
Social system of science, 33, 148–49
Socialization into language use, 304–7
Sociology, 4, 10, 37, 38, 187, 189
Sommerfeld, Arnold, 81
Sophists, 20, 293
Spectroscopy, 160–61
Speech acts, 299
Spiro, R. J., 236
Sprat, Thomas, 15, 134, 293
Statistical indicators problem, 159, 285–86
Statistics in psychology, 272
Steffenson, M. S., 236
Stimson, Dorothy, 136, 137
Stinchcombe, Arthur, 129, 134, 136
Stryker, Sheldon, 134, 136
Stuewer, Roger, 192, 196, 199, 203
Suleiman, S., 236
Sullivan, Harry Stack, 5n5
Summary. *See* Abstract
Swales, John, 18n1, 155n2, 325
Swift, Jonathan, 149–50, 293
Synchronic-diachronic distinction, 296

Tate, Gary, 4n2
Texts: importance of, arguments against, 19–24; activity of, 21–22; relation to social structure, 22; transience, 23; as objects, 219–22. *See also* Genre; Scientific language; Writing
Theory, scientific, as an intertextual integrating device, 157, 179–81, 182–84
Theory section, 175
Thomson, J. J., 194, 196
Thought collective, 61, 312–13
Titles, in reading, 239–42
Tompkins, Jane, 236
Toulmin, Stephen, 4n3, 268, 308
Turner, R., 134, 136
Tyler, Stephen, 280

Usefulness, as a criterion, 252, 312

Vatz, Richard, 8n10
Virtual experience, representation as, 74–75, 140
Vocabulary, technical, 28–29, 37, 43–44, 169–71, 274
Vygotsky, Lev, 5n5, 295–96, 302–7, 312–14, 317

Wallis, John, 88
Watson, James D.: discussion of article, 27–34, 46–48, 154n1, 282; text of article, 49–50
Watson, John B., 268–70, 271
Watson, Richard, 70
Weber, Max, 137n8, 184n15
Weimer, Walter B., 322n3
Wertsch, James, 302, 303, 313
Westfall, Richard, 82n1, 83, 93, 105n20, 108
White, Hayden, 10n11
White, James Boyd, 11n12
White, Leslie, 153
Whorf, Benjamin Lee, 297
Wilkins, John, 15, 16, 293
Willoughby, Francis, 69
Wilson's cloud chamber experiments, use by Compton, 204–8
Witness of experiments, 73–75, 140

Index

Wittgenstein, Ludwig, 10, 299
Woolgar, Steve, 156n3, 294
Word choice. *See* Vocabulary, technical
Wordsworth, William, 27
Writing: study of, 9–13; practice of, 10, 155, 322–23; unreflective, 15, 320; bad, 244; teaching of 320, 332; advice for, 323–31
Writing-up, 201–3, 209–10

Wundt, Wilhelm, 263, 265, 266, 267, 314

Yearley, Stephen, 156n3, 294

Zeeman, Pieter, 160
Ziman, John, 139
Zone of proximal development, 306
Zuckerman, Harriet, 25n7, 32n13, 328